T0339601

Nanocrystalline Materials

Nanocrystalline Materials
Their Synthesis—Structure-Property Relationships and Applications

Second Edition

Sie-Chin Tjong

Department of Physics & Materials Science,
City University of Hong Kong,
Hong Kong

AMSTERDAM • BOSTON • HEIDELBERG • LONDON • NEW YORK • OXFORD
PARIS • SAN DIEGO • SAN FRANCISCO • SINGAPORE • SYDNEY • TOKYO
ELSEVIER

Elsevier
32 Jamestown Road, London NW1
7BY, UK 225 Wyman Street, Waltham, MA
02451, USA

First Edition 2006
Second Edition 2014

British Library Cataloguing-in-Publication Data
A catalogue record for this book is available from the British Library

Library of Congress Cataloging-in-Publication Data
A catalog record for this book is available from the Library of Congress

ISBN: 978-0-08-101509-4

For information on all Elsevier publications
visit our website at store.elsevier.com

This book has been manufactured using Print On Demand technology. Each copy is produced to order and is limited to black ink. The online version of this book will show color figures where appropriate.

Contents

Preface

Nanocrystalline materials with different shapes, morphologies, and structures possess excellent chemical, physical, or mechanical properties compared to their microcrystalline counterparts. Thus the synthesis, development, and material characterization of novel nanocrystalline materials have attracted the research interest of chemists, physicists, and materials engineers. This book reviews the latest synthesis, development, and characterization of nanocrystalline materials, as well as the potential use of such novel materials for bioimaging, drug delivery, electronic device, solar cell, and structural engineering applications. The chapters were written by leading scientists in their respective subject fields. The purposes are to provide the readers, including all scientists, engineers, and graduate students working in academia and industry, new information and knowledge on nanocrystalline materials to enhance their understanding and innovation.

The first edition of this book was published in 2006. The synthesis and characterization of nanomaterials have developed very rapidly in the past 7 years. Therefore, I have removed a few chapters of the first edition which have become a little dated. These include: Chapter 1 (Solution route to semiconducting nanomaterials), Chapter 2 (Synthesis architecture of inorganic nanomaterials), Chapter 4 (Fabrication and structural characterization of ultrathin nanoscale wires and particles), Chapter 5 (Synthesis and characterization of various one-dimensional semiconducting nanostructures), and Chapter 7 (Synthesis of hyberbranced conjugative polymers and their applications as photoresists and precursors for magnetic ceramics).

The contents of this second edition have been carefully revised and updated to expand coverage of several new key topics relating to the energy, biomedical science, and electronic device. Chapter 1 presents the synthesis of multiwalled carbon nanotubes using bamboo charcoal as one of the precursor materials. Bamboo charcoal is an eco-friendly biocarbon obtained by pyrolyzing bamboos at high temperatures. Bamboo charcoal contains a considerable amount of mineral elements, acting as effective catalysts for growing carbon nanotubes at high temperatures. This eliminates the use of transition metal nanoparticles commonly used for the synthesis of nanotubes. Chapter 2 addresses the structure and property of one-dimensional (1D) ZnO nanowire array for use as the photoanode of dye-sensitized solar cells. ZnO possesses high electron mobility, low crystallization temperature, and anisotropic growth behavior rendering it a very promising candidate for the flexible photoanode. Chapter 3 reviews the fundamentals and the most recent developments in the preparation methods, basic property characterizations as well as the applications of 1D nanostructured materials in the energy transformation into electricity by solar cells, piezoelectric generators, thermal conversion devices and energy storage in fuel cells, lithium ion batteries, and supercapacitors. Chapter 4 concerns the synthesis, property, and application of lanthanide-doped fluoride nanocrystals. Such doped nanoparticles can be excited with near-infrared (NIR) radiation, thereby effectively converting the NIR light to ultraviolet and visible regions, a process known as upconversion. Upconversion generally causes the emission of higher energy photons through the sequential absorption of lower energy photons. Lanthanide-doped upconversion nanoparticles with appropriate surface modification find a wide range of biomedical applications, including biodetection, cancer therapy, and bioimaging. Chapter 5 is the revised text of the first edition, concerning molecular self-assembled monolayers on hydrogen-terminated silicon. The self-assembly approach facilitates the fabrication of ultrathin films using organic molecules as the building blocks, thereby permitting the design of new functional nanodevices at the supramolecular level. Chapter 6 introduces the synthesis and application of polymer nanocomposite as a dielectric layer for flexible organic electronics. Novel dielectric material based on aluminum titanate nanoparticles functionalized with n-octadecylphosphonic acid and the poly(4-vinylphenol) matrix exhibits superior electronic performance in both the n-type and p-type organic thin film transistors. From the flexibility test, such devices are mechanically stable and environmentally robust. Chapter 7 discusses the use of specific types of nanomaterials, such as semiconductor quantum dots, magnetic nanoparticles, layered double hydroxides, organic nanoparticles, and metal nanostructures, as the vehicles for drug delivery systems. Chapter 8 presents the fabrication, structure, and mechanical properties of aluminum and magnesium-based composites reinforced with ceramic nanoparticles. The mechanical performance of metal matrix nanocomposites depends greatly on the attainment of homogeneous dispersion of ceramic nanoparticles in the metal matrix. Particular attention is paid to the processing techniques for achieving uniform dispersion of nanoparticles in the composites and their structure—property relationships. Chapter 9 gives an introduction to the use of ceramic, organic, metal, and carbon nanoparticles to enhance dielectric permittivity of polymers. This chapter describes the fundamental aspects, processing techniques, and the underlying mechanisms for attaining high permittivity in polymer nanocomposites. Chapter 10 reports the synthesis and mechanical properties of graphene—polymer nanocomposites. Two-dimensional graphene exhibits exceptionally high elastic modulus and mechanical strength. It is an ideal nanomaterial to reinforce polymers for fabricating structural composites at very low filler contents. The chapter presents the strategies used to improve the mechanical performance of the polymer nanocomposites. Finally, Chapter 11 discusses the electrical conducting behavior of polymers filled with carbonaceous nanomaterials, including carbon nanofibers, carbon nanotubes, and graphene nanoplatelets. Under the application of an electric field, the percolating nanocomposites exhibit nonlinear electrical conductivity and Zener tunneling effect.

Sie Chin Tjong, CEng CSci FIMMM
City University of Hong Kong, Hong Kong and
King Abdulaziz University, Jeddah, Saudi Arabia

List of contributors

Xianfeng Chen   Center of Super-Diamond and Advanced Films (COSDAF) and Department of Physics and Materials Science, City University of Hong Kong, Kowloon Tong, Hong Kong SAR, PR China

Zhi-Min Dang   Laboratory of Dielectric Polymer Material and Device, Department of Polymer Science and Engineering, School of Chemistry and Biological Engineering, University of Science & Technology Beijing, Beijing, China

Ning Han   Department of Physics and Materials Science, and Centre for Functional Photonics (CFP), City University of Hong Kong, Tat Chee Avenue, Kowloon, Hong Kong SAR, PR China

Su-Ting Han   Department of Physics and Materials Science and Center of Super-Diamond and Advanced Films (COSDAF), City University of Hong Kong, Hong Kong SAR, PR China

Linxiang He   Department of Physics and Materials Science, City University of Hong Kong, Hong Kong, PR China

Johnny C. Ho   Department of Physics and Materials Science, and Centre for Functional Photonics (CFP), City University of Hong Kong, Tat Chee Avenue, Kowloon, Hong Kong SAR, PR China

Juncai Jia   Department of Chemistry, The Chinese University of Hong Kong, Shatin, Hong Kong, China

Chen-Hao Ku   Department of Chemical Engineering, National Cheng Kung University, Tainan, Taiwan

Wen-Pin Liao   Department of Chemical Engineering, National Cheng Kung University, Tainan, Taiwan

V.A.L. Roy   Department of Physics and Materials Science and Center of Super-Diamond and Advanced Films (COSDAF), City University of Hong Kong, Hong Kong SAR, PR China

Hiroyuki Sugimura   Department of Materials Science and Engineering, Kyoto University, Sakyo, Kyoto, Japan

Sie Chin Tjong   Department of Physics and Materials Science, City University of Hong Kong, Tat Chee Avenue, Kowloon, Hong Kong, PR China; Department of Physics, Faculty of Science, King Abdulaziz University, Jeddah, Saudi Arabia

Feng Wang   Department of Physics and Materials Science, City University of Hong Kong, Hong Kong SAR, PR China

Hongli Wen   Department of Physics and Materials Science, City University of Hong Kong, Hong Kong SAR, PR China

Chun-Te Wu   Department of Chemical Engineering, National Cheng Kung University, Tainan, Taiwan

Jih-Jen Wu   Department of Chemical Engineering, National Cheng Kung University, Tainan, Taiwan

Li Yan   Center of Super-Diamond and Advanced Films (COSDAF) and Department of Physics and Materials Science, City University of Hong Kong, Kowloon Tong, Hong Kong SAR, PR China

Ye Zhou   Department of Physics and Materials Science and Center of Super-Diamond and Advanced Films (COSDAF), City University of Hong Kong, Hong Kong SAR, PR China

Jiangtao Zhu   School for Engineering of Matter, Transport and Energy, Arizona State University, Tempe, AZ

1 Preparation, Structure, and Application of Carbon Nanotubes/ Bamboo Charcoal Composite

Jiangtao Zhu[1], Juncai Jia[2] and Sie Chin Tjong[3]

[1]School for Engineering of Matter, Transport and Energy, Arizona State University, Tempe, AZ, [2]Department of Chemistry, The Chinese University of Hong Kong, Shatin, Hong Kong, China, [3]Department of Physics and Materials Science, City University of Hong Kong, Tat Chee Avenue, Kowloon, Hong Kong, China

1.1 Introduction

Naturally grown plants, such as wood and bamboo, have unique and sophisticated structures after evolution by mother nature for ages. Functional materials can be fabricated by mimicking the hierarchically build structural morphologies of renewable bioresource materials. Bamboo plants are indigenous to East Asia but are now planted in worldwide subtropical regions. The total area of bamboo forest is about 22 million ha (1 ha is 10,000 m^2), accounting for about 1.0% of the total global area of forest [1]. Although the total forest areas in many countries decrease drastically in recent years, but the bamboo vegetation increases at a rate of 3% annually. There are about 1200 types of bamboo globally. As bamboo does not need fertilizers, or pesticides, bamboo products are considered as inexpensive and eco-friendly materials. Once the bamboo is carbonized or pyrolyzed, bamboo charcoal with profuse porosity is produced. The bamboo charcoal has attracted intense attention in the past few years due to their attractive properties, including absorption, catalyst support, medical electrode, and agricultural function [2–16].

Carbon nanotubes (CNTs) exhibit excellent mechanical, thermal, and electrical properties having many functional applications [17–19]. CNTs are synthesized in large quantities by means of chemical vapor deposition (CVD) of hydrocarbon gases using transition metal or rare earth metal catalysts [18]. However, those heavy metals are toxic and pose serious problems to human health and the environment. In addition, loose CNTs are unsuitable for some applications, such as gas phase catalysis and liquid phase absorption [20]. In this regard, the successful synthesis of CNTs from bamboo charcoals is considered of technological importance. This is because bamboo charcoals containing many minerals can serve as effective

Nanocrystalline Materials. DOI: http://dx.doi.org/10.1016/B978-0-12-407796-6.00001-4

catalysts for the nucleation and growth of nanotubes. Accordingly, CNTs can be grown on a charcoal supporting material with eco-friendly characteristic.

In this chapter, the morphology, structure, and composition of the bamboo charcoal are first addressed. The processes for growing CNTs on the bamboo charcoal surfaces via CVD and arc-discharge techniques are reviewed. The role of extra catalyst added to the bamboo charcoal on the growth of CNTs at low temperatures is also discussed. Furthermore, bamboo charcoal is an excellent biomass adsorbent that has strong adsorption ability for organic pollutants and heavy metal ions. The applications of CNT/bamboo charcoal for water purification and hydrogen storage are also presented.

1.2   Bamboo Charcoal

1.2.1   Morphology, Phase, and Compositions of Bamboo Charcoal

The bamboo charcoal is usually obtained by pyrolysis bamboo in an inert atmosphere at temperatures above 700°C. The process is also termed as the carbonization in the preparation of activated carbon. During carbonization, biological bamboo releases contained water and organic gases followed by the reconstruction of carbon structure. Figure 1.1 shows typical thermogravimetry analysis (TGA) of several bamboos in a nitrogen atmosphere from room temperature to 900°C. The first mass loss at temperatures below 100°C is related to water vaporization. At 80−250°C, the mass loss of the bamboo remains stable. Then the mass drops drastically at 250−400°C due to the fast thermal decomposition of cellulose, lignin, and semicellulose. Large CO_2 and CO are released at this stage. Above 400−900°C, the mass decreases slowly. This final stage is mainly associated with the carbonization during which the aromatic rings rearrange into charcoal [21]. Four types of bamboos as shown in Figure 1.1 are used for the TGA, i.e., *Phyllostachys edulis*, *Dendrocalamus*, *Dendrocalamus brandisii*, and *Bambusa inteimedia*. Clearly, the most popular *P. edulis* has a slight different thermal property compared with other types of bamboos. Even *P. edulis* from different locations in China displays different behaviors. The *B. inteimedia* has a highest residue after the TGA process.

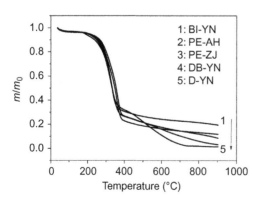

Figure 1.1 TGA of five types of bamboo: (1) BI-YN, *Bambusa*; (2) PE-AH, *Phyllostachys*; (3) PE-ZJ, *P. edulis*; (4) DB-YN, *D. brandisii*; and (5) D-YN, *Dendrocalamus*. YN, AH, and ZJ refer to Yunnan, Anhui, and Zhejiang provinces (China) where the bamboos obtained.

Generally, there is a significant volume shrinkage during the pyrolysis, however, the tubular structure of bamboo is still retained. Scanning electron microscopy (SEM) cross-sectional and lateral views show the porous feature of the bamboo charcoal carbonized at 1000°C [22] (Figure 1.2). The bamboo charcoal still retains porous nature of original bamboo. Mineral aggregates are found on the walls of vessel lumen. Jiang et al. [23] studied the shrinkage of the bamboo at different carbonization temperatures. The scaffolding of the bamboo charcoal remained the same as original bamboo, however, the microstructure of the bamboo was slightly changed during the shrinkage. The shrinkage ratios of the bamboo were about 21, 38, and 40% at 500, 750, and 1000°C, respectively. Jiang et al. also found that the wall of the parenchyma, the basic unit inside the bamboo, became rough. The outer surface of the bamboo became smooth and the whole bamboo charcoal became hard. Very recently, Zhu et al. studied the morphologies of bamboo charcoal treated at temperatures up to 1500°C. The morphologies of bamboo charcoal carbonined at 1000−1500°C retain porous feature of fresh bamboo [24]. Moreover, the bamboo charcoal exhibits a wide range of pore distribution from <1 nm to 1 μm based on the mercury porosimetry measurements (Figure 1.3). The charcoal has several dominant pore sizes (at 30, 200, 2000, and 20,000 nm), confirming hierarchical pore feature of the bamboo charcoal. The total volume of the mercury absorbed by the bamboo charcoal is about 1.7 mL/g.

The bamboo charcoal usually contains mainly carbon with a small amount of impurities. The X-ray diffraction (XRD) patterns of several bamboo charcoals prepared at

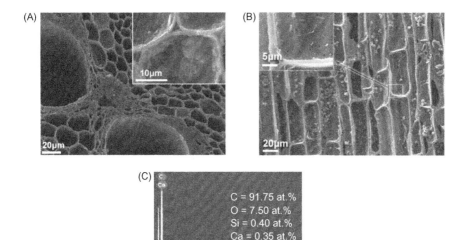

(A)

10μm

20μm

(B)

5μm

20μm

(C)

C = 91.75 at.%
O = 7.50 at.%
Si = 0.40 at.%
Ca = 0.35 at.%

Full Scale 1518 cts Cursor 0.000 keV

Figure 1.2 SEM images showing the (A) cross-sectional and (B) lateral views of raw bamboo biotemplate after pyrolysis and (C) corresponding EDS spectrum.
Source: Reprinted from Ref. [22] with permission of Wiley.

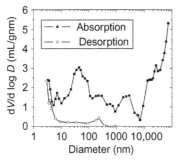

Figure 1.3 The pore structure of bamboo charcoal (*P. edulis*) revealed by absorption/desorption of mercury porosimetry.

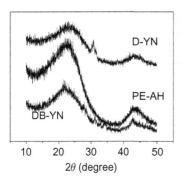

Figure 1.4 XRD patterns of D-YN, *Dendrocalamus*; PE-AH, *P. edulis*; and DB-YN, *D. brandisiss*. YN and AH refer to Province of Yunnan and Anhui (China), respectively, where the bamboos obtained.

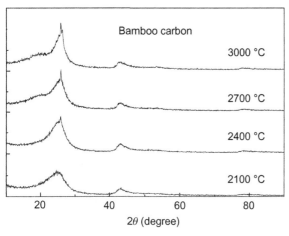

Figure 1.5 The XRD profiles of bamboo charcoal after heat treatment at different temperatures.
Source: Reprinted from Ref. [25] with permission of Springer.

800°C are shown in Figure 1.4. There are only two broad peaks located at 22 and 43°, corresponding to C_{0002} and C_{0004} reflections of carbon, respectively. Those broad peaks confirm they are mainly amorphous carbon. Chen et al. [25] (Figure 1.5) studied graphitization behavior of bamboo charcoal after carbonization up to 3000°C using XRD. The d_{0002} spacing of the bamboo charcoal decreased while the apparent graphite crystallite size L_c (0002) increased with the increase of graphitization temperatures. However, even after heat treatment at 3000°C, the d_{0002} and L_c (0002) values were only about

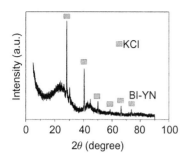

Figure 1.6 XRD patterns of BI-YN, *B. inteimedia*. YN is Yunnan province where the bamboo comes from.

0.341 and 9 nm, respectively. The bamboo charcoal is typical nongraphitizable carbon or hard carbon.

The types and amount of minerals or impurities of the bamboo charcoal depend on geological vegetation source and heat treatment history. The amount of the impurities is usually <10 wt%. X-ray energy dispersive spectroscopy (EDS) is widely recognized to be a fast and convenient route to detect the impurities. Quantitative analysis of the impurities in charcoal with higher accuracy can be determined from atomic emission spectroscopy [24]. As shown in Figure 1.2, Ca, Si, and O are found in the aggregate adhered on the pore surface of the bamboo charcoal. Jiang et al. [23] demonstrated that the bamboos (*P. edulis*) from Fujian province (China) contain many mineral elements, such as Si, N, P, K, Mg, Al in addition to C, O, H. Zhu et al. [24] identified the elements found in several types of bamboos. Bamboo (*P. edulis*) from Zhejiang province contains K, Cl, Fe, etc. Bamboo (*P. edulis*) from Anhui province contains K, Cl, Fe, Si, P, etc. Bamboo (*Dendrocalamus*) from Yunnan province contains K, Ca, Ti, Fe, Cr, Si, S, etc. Bamboo (*D. brandisii*) from Yunnan province contains Si, Ca, K, P, Mg, Fe, etc. Bamboo (*B. inteimedia*) from Yunan province contains K, Ti, Fe, Cr, Cl, Si, S, etc. The amount of KCl in bamboo (*B. inteimedia*) is so large that even it can be detected in its XRD pattern (Figure 1.6). The crystalline size is about 50 nm. The KCl should come from the soil for growing bamboos. The high amount of KCl could be one reason that its residues in TGA are higher than other types of bamboo charcoals (Figure 1.1).

1.2.2 Phases of the Minerals Inside Bamboo Charcoal

It is necessary to understand the distributions of minerals inside bamboo charcoals for designing novel bioresource materials for practical applications. Jiang et al. [23] reported that the minerals are preferrably accumulated at the outer surface of the bamboo. Zhu et al. [24] studied the evolution of minerals with carbonization temperatures. The uneven distribution of the minerals is observed for the bamboo charcoal treated at 1000°C as evidenced by EDS. Some minerals, such as K, P, Mg, S, and Ca, tend to form agglomerates. Figure 1.7 shows SEM images of bamboo charcoals pyrolyzed at high temperatures of 1200−1500°C. From the surface of charcoal pyrolyzed at 1200°C (Figure 1.7A), Mg_2SiO_4 faceted crystals of submicrometer sizes can be readily seen in the micrograph. Only spheres are found on the charcoal treated at 1300°C (Figure 1.7B) and 1400°C (Figure 1.7C). The EDS

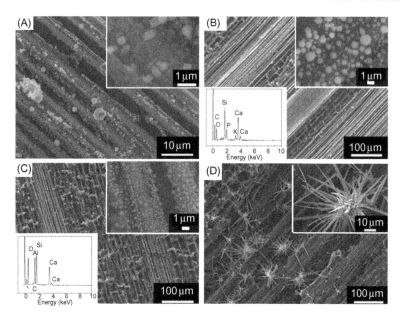

Figure 1.7 SEM images of bamboo specimens pyrolyzed at (A) 1200, (B) 1300, (C) 1400, and (D) 1500°C. Magnified images are shown in insets (top right). The bottom left insets in (B) and (C) are EDS spectra.
Source: Reprinted from Ref. [24] with permission of Elsevier.

results (insets) reveal that the spheres consist mainly of Ca, Si, and O. Trace amounts of K and P are also found inside the spheres of the charcoal treated at 1300°C. Al is detected for samples treated at 1400°C. By increasing temperature to 1500°C, SiC whiskers with lengths of about tens of micrometers are formed on the charcoal. It is interesting to note that these structures do not contain Fe.

The phases of minerals of the bamboo charcoal treated at 1000−1500°C were identified by XRD (Figure 1.8). The samples were first calcined in air at 550°C to partially remove carbon inside the charcoal. Only calcined residuals were analyzed. The residuals in the charcoal pyrolyzed at 1000°C match well with crystalline K_2SO_4 (PDF 05-0613). For the residuals of charcoals pyrolyzed at 1200°C (Figure 1.8B), three phases based on $MgSiO_3$ (PDF 88-1926), $CaSO_4$ (PDF 37-1496), and K_2SO_4 can be identified. MgO (PDF 78-0430) with low peak intensity is also detected. $MgSiO_3$ results from the decomposition of Mg_2SiO_4 during calcination at 550°C. By increasing temperature to 1300°C, only K_2SO_4 and $CaSO_4$ are detected in the residuals (Figure 1.8C). At 1400°C (Figure 1.8D), other crystalline phases such as SiC (PDF 89-1977), SiO_2 (PDF 88-2487), and CaO (PDF 77-2376) are found, while peaks due to K_2SO_4 and $CaSO_4$ become very weak. The CaO phase generates from the decomposition of $CaSO_4$. Crystalline SiO_2 is resulted from the melting and solidification of amorphous SiO_2 inside the charcoal. Crystalline SiC yields from carbothermal reduction of amorphous silica or crystalline SiO_2 phase during high-temperature pyrolysis. The above results demonstrate that the mineral phase formation in bamboo charcoals depends greatly on the heat treatment temperatures.

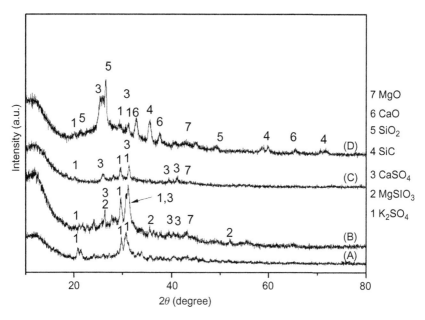

Figure 1.8 XRD spectra of the residuals of bamboo charcoals pyrolyzed at (A) 1000, (B) 1200, (C) 1300, and (D) 1400°C. (1) K_2SO_4, (2) $MgSiO_3$, (3) $CaSO_4$, (4) SiC, (5) SiO_2, (6) CaO, and (7) MgO.
Source: Reprinted from Ref. [24] with permission of Elsevier.

1.3   Functional Materials Derived from Charcoal

There is very scarce information in the literature reporting the structure of minerals inside the bamboo charcoal treated above 1500°C. Saito and Arima [26] reported the formation of cone-shaped graphitic whiskers formed inside the wood charcoal cell heat treated above 2000°C (Figure 1.9). High-resolution transmission election microscopy (HRTEM) was used to analyze the structure of synthesized graphitic whiskers. The whiskers are composed of conical stacked hexagonal carbon layers with a cone apex angle of 136°. This angle probably assisted successive helical growth of the whiskers. The growth mechanism of these whiskers was proposed. Pyrolyzed gas from the walls of the wood cells was initially released and localized mainly at the wood cell cavities. Once the gas concentration reached a supersaturation level, the whisker grew then followed the vapor–solid mode.

As discussed above, the main ingredient of the bamboo charcoal is carbon. This kind of biocarbon with special structure attracts considerable attention for preparing biomorphic materials. For example, Cheung et al. [22] converted bamboo into biomorphic composites having high purity cristobalite SiO_2 and β-SiC nanowires by sintering bamboo charcoal infiltrated with tetraethyl orthosilicate (TEOS) at 1200 and 1400°C, respectively. The fabrication process was very simple without adding

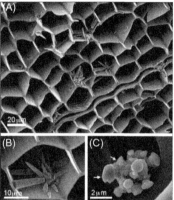

Figure 1.9 (A) SEM photographs of *Cryptomeria japonica* heat treated at 2500°C without SiC addition. (B) Magnified image of carbon deposits grown inside a wood cell. (C) Magnified image of nongrowth, nucleated substrate-like carbon. The arrows denote the cone-shaped head of nongrowth carbon clearly seen at an apex angle of 136°.
Source: Reprinted from Ref. [26] with permission of Elsevier.

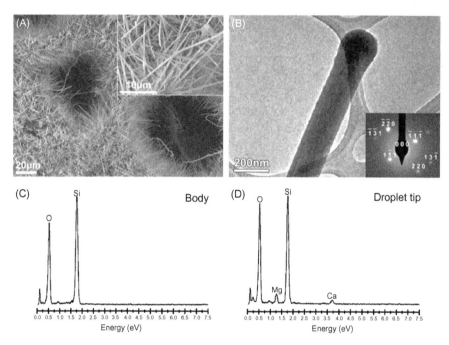

Figure 1.10 (A) SEM image of the bamboo charcoal infiltrated with TEOS sintered at 1200°C for 10 h. (B) TEM image of the top portion of a nanowire extracted from the sample shown in (A). (C and D) The TEM-EDS spectra taken at the body and droplet tip of silica nanowire shown in (B), respectively.
Source: Reprinted from Ref. [22] with permission of Wiley.

any catalysts. This is because the minerals inside bamboo charcoal can act as the catalysts. In the case of SiO_2, after sintering at 1200°C for 10 h, a thin white layer was found on the surface of the biotemplate. Figure 1.10 shows the formation of dense SiO_2 nanowires on the walls of tubular structure. EDS spectrum indicates

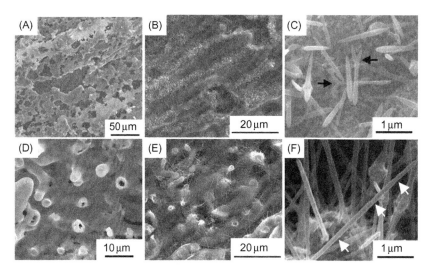

Figure 1.11 SEM images of a TEOS infiltrated bamboo leaf at 1200°C for 1 h: (A) top and (D) bottom surfaces. SEM images of a TEOS infiltrated leaf at 1300°C for 3 h: (B) the view of top surface, (C) nanowires collected from top surface, (E) the view of bottom surface, and (F) the nanowires collected from bottom surface.
Source: Reprinted from Ref. [28] with permission of Elsevier.

that the body of the nanowires contains Si and O, while its tip has <10 at.% impurities, including Mg and Ca. SiC nanowires can also be fabricated on the surface of the bamboo charcoal by a similar process at 1400°C. Zhu et al. [27] converted entire bamboo charcoal monolith into biomorphic SiC recently. Due to the complicated structure of various textures inside the bamboo, biomorphic SiC displayed different microstructures at different locations of the textures of original bamboo. Near the outer surface and large vessels of the bamboo charcoal, β-SiC nanowires were observed. In the inner part of the bamboo charcoal, such as parenchyma, sieve tube, and fibers, SiC particle aggregates that provide a high surface area for a potential support were observed.

Our recent study also reported that 6H-SiC nanowires can be synthesized from the charcoal derived from bamboo leaves [28]. Apparently, different morphologies and structures of the leaf surfaces can affect the yield and the structure of the SiC nanowires drastically. The yield of SiC grown on the top surface of the bamboo leaf is higher and with a branched structure. The SiC on the bottom surface displays a bamboo-like structure (Figure 1.11).

1.4 Synthesis of CNTs/Bamboo Charcoal

CNTs exhibit excellent mechanical, thermal, and electrical properties. In addition, CNTs also exhibit high absorbent properties. CNTs can effectively remove different metal ions, such as cadmium, copper, lead, zinc, and nickel. However, a support is needed to immobilize CNTs due to their tiny size dimensions. Bamboo charcoal is a good support due to its low-cost, novel hierarchical structure, and high absorption ability. Due to the porous nature of bamboo charcoal, it is quite straightforward to grow CNTs on the bamboo charcoal by *in situ* synthesis method.

1.4.1   Low-Temperature CVD Method Assisted by Extra Catalyst

To grow CNTs on bamboo charcoal at low temperatures, it is necessary to load catalyst onto the charcoal initially, followed by the CVD with carbon precursor. Zhang et al. [29] impregnated bamboo charcoal initially with a ferrocene solution, then admitted volatile xylene/ferrocene through H_2/Ar carrier gas at 820°C for 20–200 min. The presence of the ferrocene was very critical for the growth of CNTs. CNTs would not form on the bamboo charcoal without using ferrocene. The prepared CNTs with a multiwall structure had curved and entangled features. The specific surface area of the bamboo charcoal dropped from 7.0 to 3.2 m^2/g after CVD growth of CNTs for 20 min. Zhang et al. also studied the effect of carbonization temperature on the growth of CNTs. Low carbonization temperature was unfavorable for the growth of CNTs.

We studied the CNTs growth on the bamboo charcoal infiltrated with $Fe(NO_3)_3$ or $Ni(NO_3)_2$ by ethanol CVD method. As shown in Figure 1.12, with the catalyst precursor $Fe(NO_3)_3$, CNTs tended to grow on the cells in both the cross-sectional and lateral directions of the bamboo charcoal after CVD process at 750°C for 30 min. The CNTs exhibited a bamboo-like structure as confirmed by the HRTEM image. Few small crystals at the end of the CNTs were observed. The EDS of the

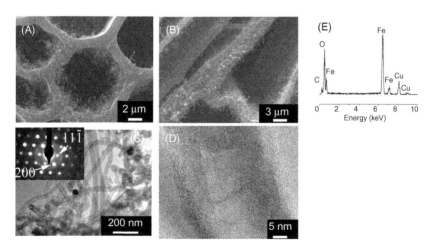

Figure 1.12 SEM images of CNTs grown on (A) cross-sectional and (B) lateral directions of bamboo charcoal by $Fe(NO_3)_3$ as catalyst precursor. (C) TEM image of CNTs with SAED from the circled crystal. (D) HRTEM of part of the CNTs showing its bamboo-like structure. (E) EDS from the circled part of (C).

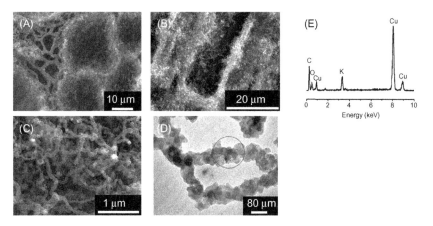

Figure 1.13 SEM images of CNTs grown on (A) cross-sectional and (B) lateral directions of the bamboo charcoal by $Ni(NO_3)_2$ as catalyst precursor. (C) SEM image of CNTs at higher magnification. (D) TEM of CNTs showing particles on their surface. (E) EDS from the circled part of (D).

crystals revealed the presence of iron and oxygen. Their selected area electron diffraction (SAED) pattern confirmed the Fe_3O_4 phase structure. When $Ni(NO_3)_2$ (Figure 1.13) was used as the catalyst precursor, CNTs of higher yield were formed on the surface of the bamboo charcoal. The CNT features were similar to those produced by Zhang et al. [29] with curved and entangled noodle appearances. The main difference was that K_2O particles from the minerals of bamboo charcoal tended to cover the surfaces of CNTs.

Recently, Chen et al. [30] reported the synthesis of carbon nanofibers (CNFs) on activated carbon produced from agricultural wastes using CVD method without adding any catalysts. They claimed that the iron inside the ash can function as the catalyst. The iron is a biologically essential element and ubiquitous in plants. In this regard, the step to upload iron catalyst onto activated carbon for the growth of CNFs is eliminated. Thus, the overall process is simplified. In the process, three activated carbons prepared from palm kernel shell (AC-P), coconut (AC-C), and wheat straw (AC-W) were first calcinated to reduce the amount of carbon but enhance the amount of impurities. Then ethylene and hydrogen were admitted during the CVD process. The iron was reduced and CNFs were grown on the surface of the three ACs (Figure 1.14). TEM image of the CNFs showed the presence of round shape of iron at their ends, revealing that iron was melted during the fiber growth (Figure 1.15). The growth mechanism of CNFs was proposed according to the tip-growing model. HRTEM examinations showed that the surface of the CNFs is rather rough. The carbon layers were not well ordered but nearly perpendicular to the axis of CNFs. Chen et al. also reported that the yield of CNFs on the ACs was not proportional to the iron content, suggesting that iron was buried in the bulk of ACs, hence not all iron took part in the CNF growth.

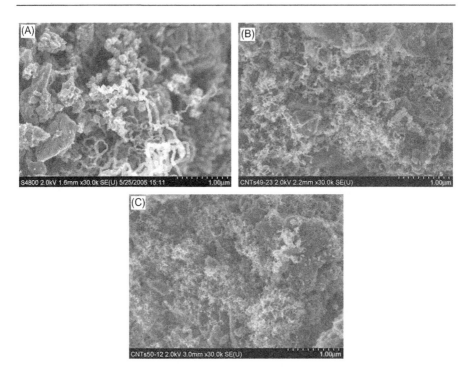

Figure 1.14 SEM images of CNFs grown on the surface of (A) AC-P, (B) AC-C, and (C) AC-W.
Source: Reprinted from Ref. [30] with permission of Elsevier.

1.4.2 Higher Temperature CVD Method Assisted by Silicate Inside Bamboo Charcoal

As discussed above, Chen et al. [30] indicated that iron element of activated carbon derived from biomass can assist the growth of CNFs at 700°C. Since iron also exists inside the bamboo, it is beneficial to use it to synthesize CNFs. To the best of our knowledge, there is no report for such synthesis of CNFs in the literature. Zhang et al. [29] mentioned that CNTs cannot grow on the bamboo charcoal without adding any iron catalyst. On the contrary, Zhu et al. [24] demonstrated that the minerals inside the bamboo charcoal can aid the growth of CNTs at temperatures of 1200–1400°C. Without adding metal catalyst, the bamboo charcoal itself was proved to be a good support and catalyst for the growth of CNTs using ethanol CVD method. CNTs were found on the surface of the bamboo charcoal between 1200 and 1400°C (Figure 1.16). Most CNTs prepared at 1300°C had round calcium silicate droplet, indicating that calcium silicate was melted during the high-temperature synthesis process. The CNTs prepared at 1400°C did not show a clear silicate droplet

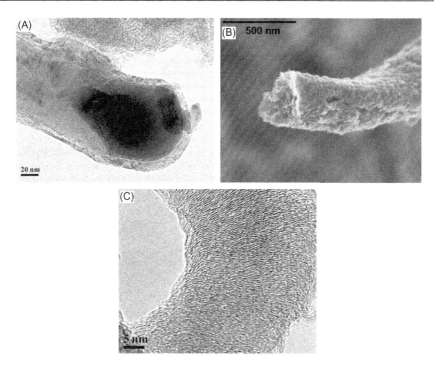

Figure 1.15 (A) TEM image of a CNF with an iron particle on its tip, (B) SEM image of a carbon nanofiber, and (C) high-resolution image of a carbon nanofiber.
Source: Reprinted from Ref. [30] with permission of Elsevier.

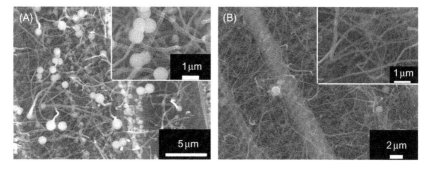

Figure 1.16 SEM images of CNTs formed on the surface of bamboo charcoal pyrolyzed at (A) 1300 and (B) 1400°C. Insets are their respective magnified images.

tip, but most CNTs were filled with silicate. TEM images were taken from the pre-pared CNTs, confirming hollow structure of the CNTs (Figure 1.17). The growth of CNTs was proposed to follow the vapor−liquid−solid model.

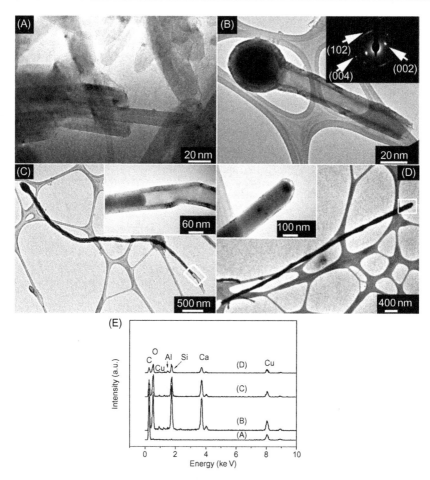

Figure 1.17 TEM images of CNTs collected from charcoal pyrolyzed at (A) 1200, (B and C) 1300, and (D) 1400°C. Inset in (B) is SAED pattern of the wall of the CNT. Insets in (C) and (D) are magnified images of rectangles. (E) EDS spectra of (A): tube in (a), (B): droplet in (b), (C): droplet in (c), and (D): filling inside the tube in (d).
Source: Reprinted from Ref. [24] with permission of Elsevier.

To exclude the role of the iron of bamboo charcoal for nucleating CNTs, Zhu et al. [31] used pure graphite as the substrate and pre-prepared calcium silicate film as a catalyst. CNT droplets filled with calcium silicate were found on the surface of the graphite, similar to the results on the bamboo charcoal. A typical CNT with a droplet end is shown in Figure 1.18. The bright and dark images confirmed that the outer shell is carbon. Electron energy-loss spectra of carbon around the droplet end and carbon in the tube body showed the existence of carbon of different bondings (Figure 1.19) [32,33]. An intense and sharp ($1s-\pi^*$) transition at 285 eV of

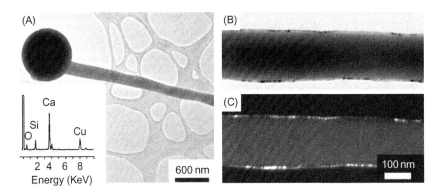

Figure 1.18 (A) TEM image of the CNTs filled with silicate, (B) bright and (C) dark field images of part of a CNT. EDS of the filling is shown as the inset.
Source: Reprinted from Ref. [31] with permission of Elsevier.

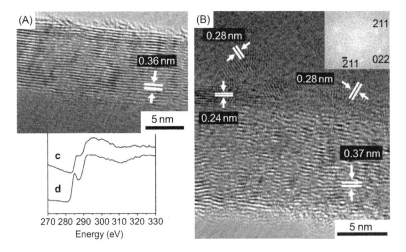

Figure 1.19 HRTEM images of CNTs (A and B). Fast Fourier transform of lattice fringe is given as the inset in (B). EELS collected from (C) the carbon around the catalyst droplet and (D) the CNT wall along the tubes.
Source: Reprinted from Ref. [31] with permission of Elsevier.

the tube wall indicated good graphene nanotexture of the CNTs. However, carbon formed around the droplet was quite amorphous. This suggested that the graphitic tube was converted from amorphous carbon around the droplet. The solid state transformation of carbon may also involve during the growth of CNTs. Since no iron is needed during the CVD process, calcium silicate apparently serves as the catalyst for the growth of CNTs. From the HRTEM observation of the wall of

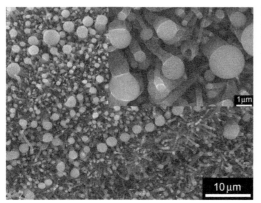

Figure 1.20 SEM image of carbon tubes grown on the surface of the bamboo charcoal by CVD method using TEOS as the source.

CNTs, the carbon layer is well crystalline, unlike the rough surface and poor crystalline CNFs prepared on the surface of activated carbon [30].

For the growth mechanism of CNTs by silicate, Zhu et al. [24] proposed that the carbon solubility in the silicate is critical for their growth. As the carbon solubility in silicate increases with the acidity, TEOS vapor acting as both carbon and silica sources is employed for increasing the acidity of silicate. During the CVD process, TEOS decomposes into silica, thereby adding more silicate into the bamboo charcoal [22]. As the same time, diethylether of TEOS also supplies carbon. Hollow CNTs with droplet ends tend to grow on the surface of the bamboo charcoal with well alignment after admitting TEOS vapor (Figure 1.20). The diameter of the carbon tubes can reach micrometer level. Small CNTs with diameter <100 nm are also found. The carbon tubes are fully filled with silicate if the amount of TEOS supplied is high enough.

1.4.3 Arc Discharge and Current Heating of Charcoal

Arc discharge is believed to be a simple and effective technique to produce high quality CNTs [18]. It is well known that crystalline CNTs can be produced by arc-discharge method using graphite as carbon source material [34,35]. In recent years, some researchers attempted to synthesize CNTs by the arc-discharge process using charcoal. For example, Fan et al. [36] successfully produced single wall carbon nanotubes (SWNTs) by direct current arc-discharge method using charcoal as the carbon source and FeS (20 wt%) as the catalyst. The diameters of synthesized SWNT bundles ranged from 10 to 40 nm (Figure 1.21). Raman spectrum showed that the ratio of integral peak intensity of G to D k was approximately 5.1, revealing that the nanotubes had a small quantity of disordered carbon or structural defect [37]. From the radial breathing mode (RBM) of SWNT, the bands at 123, 152, 167, and 187 cm^{-1} corresponded to the tube diameters of 1.2−2 nm.

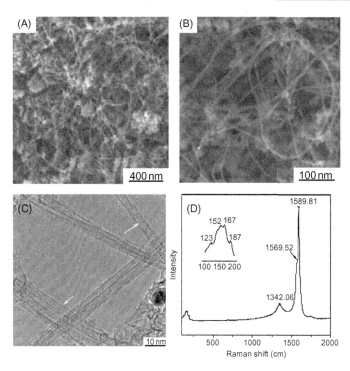

Figure 1.21 Typical SEM images of the as-prepared SWNTs obtained using FeS as catalyst by arc-discharge method: (A) low magnification image; (B) high magnification image; (C) HRTEM image; and (D) Raman spectra of the SWNTs.
Source: Reprinted from Ref. [36] with permission of Springer.

Singjai et al. [38] developed a simple and inexpensive method by heating charcoal with electrical current (CHC) to fabricate nanofibers consisting of CNTs and silicon carbide. The CHC synthesis utilizes arc-discharge facility consisting of two copper-bar electrodes, a DC power supply, and a charcoal rod. The charcoal rod was deliberately, completely, or partly broken across its middle section in order to cause poor electrical conductivity at the crack, thus inducing large current heating. After CHC, a white film was formed on the surface of charcoal rod at the cathode and at the induced crack of the rod. As shown in Figure 1.22, the fibers take the form of nanobeads. The diameter of the fibers is about 50 nm and the sizes of the beads vary from several hundred nanometers down to about 100 nm. The length of the fibers can be tens of micrometers and thus the aspect ratio is up to 1000. Singjai et al. also compared the fibers prepared by the arc discharge. The arc-discharge products have a larger diameter than the CHC fibers but with lower yield. Singjai et al. claimed that SiC was present in CHC-beaded nanofibers on the basis

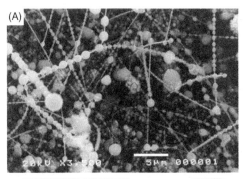

Figure 1.22 (A) SEM micrograph of nanofibers produced by the current heating of charcoal techniques in air. (B) Zoom-in image of their beads. *Source*: Reprinted from Ref. [38] with permission of Elsevier.

of XRD results. The silicon came from the charcoal. Those beaded nanofibers are excellent candidates for the reinforcements in the composte materials.

1.5 Applications of CNT/Bamboo Charcoal

CNT/bamboo charcoal composite holds multifunctional properties from both CNT and bamboo charcoal. Figure 1.23 shows the pore size distribution and nitrogen adsorption–desorption isotherms for the bamboo charcoal pyrolyzed at 1300°C with and without CNTs [24]. The average pore diameter determined from the isotherms using Barret–Joyner–Halenda (BJH) approach decreases from 7.0 to 4.5 nm after uploading of CNTs. In contrast, the adsorption volume of N_2 increases from 0.052 to 0.351 cm³/g (inset of Figure 1.23). Moreover, the Brunauer–Emmett–Teller (BET)-specific surface area increases from 98 to 655 m²/g.

However, Chen et al. [30] found that the BET-specific surface area, total pore and micropore volume of activated carbon, decreased drastically by growing CNFs on activated carbon. The BET value dropped from 1200 to ~100 m²/g as the CNFs filled in the pores of ACs. The size of micropores that contributed to specific surface area of activated carbon is <2 nm in diameter. But the diameters of synthesized CNFs are within 30–300 nm. There is little chance that CNFs can fill into micropores of ACs. Further study is needed to elucidate this problem.

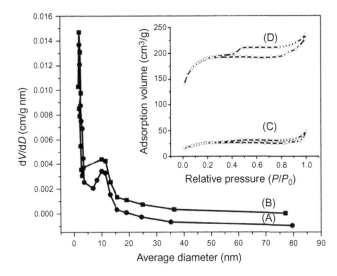

Figure 1.23 Pore size distribution of the charcoal pyrolyzed at 1300°C (A) with and (B) without the growth of CNTs. Inset: N_2 adsorption−desorption isotherms at 77 K (C) before and (D) after the growth of CNTs.
Source: Reprinted from Ref. [24] with permission of Elsevier.

1.5.1 Water Purification

The presence of excessive amounts of metal ions (copper, lead) in drinking water may accumulate in the liver of humans, thereby causing gastrointestinal problems. Activated carbon with high adsorption capability is an ideal material for removing heavy metal pollutants. Bamboo, because of its abundance and the availability, has attracted considerable attention as a carbonaceous absorbent precursor for water purification or treatment of industrial and municipal effluent. CNT/bamboo charcoal with excellent absorption properties is increasingly used in water treatment and water purification.

Zhu et al. [39] studied the absorption kinetics of bamboo-based activated carbon for phenol inside water. The absorption of phenol on the bamboo-based activated carbon generally followed the pseudo-second-order equation in which intraparticle diffusion is the rate-controlling step. The small size of activated carbon would increase the absorption rates. Zhang et al. [29] synthesized CNTs on the bamboo charcoal surface and then studied their absorption behavior for water purification. They indicated that the absorption capacity toward copper ions can be significantly improved by growing CNTs. From their results (Figure 1.24), the adsorption capacity of the CNTs/bamboo charcoal to copper ion is much higher than that of commercial activated bamboo charcoal or bamboo charcoal. An adsorption capacity of 16.34 mg/g is twice of the pristine bamboo charcoal after introducing CNTs.

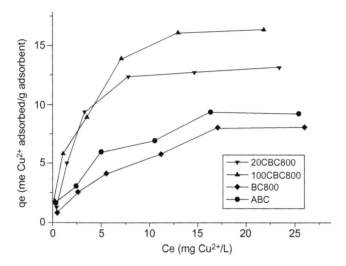

Figure 1.24 Absorption isotherm of Cu^{2+} onto four carbon samples: ABC, commercial activated bamboo charcoal; BC800, bamboo charcoal after carbonization at 800°C; 20CBC800, CNTs grown on BC800 for 20 min; 100CBC800, CNTs grown on BC800 for 100 min.
Source: Reprinted from Ref. [29] with permission of The American Chemical Society.

Huang et al. [40] studied the absorption properties of the CNT/bamboo charcoal for Pb^{2+}. As illustrated in Figure 1.25, for pristine bamboo charcoal, the amount of Pb^{2+} absorbed increased with increasing equilibrium concentration up to 24 mg/L. In contrast, the composites of the CNT/bamboo charcoal showed much larger adsorption capacity for Pb^{2+}. The amount of Pb^{2+} absorbed reached ~ 44 mg/g for the CNT/bamboo charcoal after CVD processing for 40 min. The absorption capacity of the CNT/bamboo charcoal for lead ions is even higher than that of activated carbon and multiwall carbon nanotubes (MWCNTs). The enhanced absorption obviously benefited from the CNT additions and the increase of the acidic functional groups on the surface of the composites. The hexagonal arrays of carbon atoms in the graphite sheet of the CNT surface favor strong interactions with other molecules or atoms. Thus, CNTs grown on the bamboo charcoal surface enhance its absorption ability.

1.5.2 Hydrogen Storage

Global energy demand and weather warming have led scientists to search for alternative fuel instead of petroleum. Hydrogen is the most abundant element in the earth, has great potential as an energy source. Unlike petroleum, it can be easily generated from renewable energy sources. Thus, the safety storage of hydrogen for

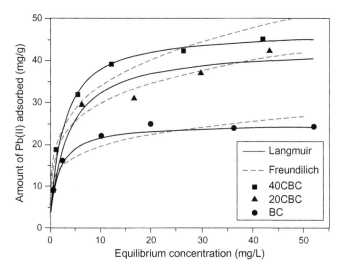

Figure 1.25 Absorption isotherms of BC and CNT/BC composites, 20CBC (20 min CVD) and 40CBC (40 min CVD), toward Pb^{2+} in aqueous solution. Curves shown in solid line are based on Langmuir equation and those in dotted line based on Freundlich equation. *Source*: Reprinted from Ref. [40] with permission of Elsevier.

practical application is quite challenging. Carbon is widely known to be a promising material for hydrogen storage, especially in the form of CNTs. Porous bamboo charcoal in combination with CNTs is an ideal candidate for hydrogen storage. Miao et al. [41] studied hydrogen storage properties of SWCNTs, MWCNTs, and CNTs/bamboo charcoal (Table 1.1). The hydrogen storage of SWCNTs is the highest among those materials while the MWCNTs have the lowest value. For carbonized bamboo charcoal, its hydrogen storage is enhanced by the acid and alkali erosion, owing to the elimination of amorphous carbon and fullerence inside the materials. However, Miao also found that the hydrogen storage of bamboo charcoal is similar with or without loading CNTs.

With the aid of metal, porous materials, such as activated carbon and charcoal, are prominent hydrogen storage medium via the spillover effect. As shown in Figure 1.26, for the spillover effect, the hydrogen molecules are firstly dissociated by the doped metal clusters on the carbon surface. The resulting hydrogen atoms then diffuse away from the metal clusters and bind to the surface of the porous materials. Tsao et al. [42] studied the hydrogen spillover effect using inelastic neutron scattering. They reported that the hydrogen storage increases from 0.018% mass fraction for undoped activated carbon to 1.0% mass fraction for Pt-doped activated carbon at 66.7 kPa (500 torr). The enhancement factor is about 50. As bamboo charcoal is also a porous carbon and can be converted to activated carbon easily, it is also a promising candidate for the hydrogen storage.

Table 1.1 The Comparison of Hydrogen Storage in SWCNTs, MWCNTs, and Taiwan Bamboo Charcoal

Materials		Hydrogen Storage
Carbon nanotubes	SWCNTs	0.668 wt%
	MWCNTs (grown by arc discharge)	0.219 wt%
	MWCNTs (grown by arc discharge and purified)	0.588 wt%
Taiwan bamboo charcoal	Virgin	0.431 wt%
	Pretreatment by 70% nitric acid after 24 h	0.522 wt%
	Pretreatment by 6 M KOH after 24 h	0.569 wt%
	Bamboo charcoal with CNTs grown by arc discharge	0.462 wt%
	Bamboo charcoal with CNTs grown by MPECVD	0.520 wt%

Source: Reproduced courtesy of The Electromagnetics Academy [41].

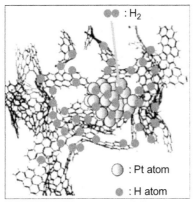

: H$_2$

: Pt atom

: H atom

Figure 1.26 Scheme to show the hydrogen spillover on Pt cluster uploaded on activated carbon.
Source: Reprinted from Ref. [42] with permission of The American Chemical Society.

1.6 Conclusions

Novel, inexpensive, multifunctional composites based on CNT/bamboo charcoal are introduced in this chapter. Their fabrication method, structure, and application are addressed. Bamboo charcoal is an eco-friendly biocarbon obtained by pyrolyzing bamboos at high temperatures. It exhibits porous structure and contains amorphous carbon and a small amount of minerals. The types of mineral elements depend mainly on the geological regions of bamboo vegetation. The minerals inside the bamboo charcoal can transform into various phases and structures under different calcination temperatures. For example, SiC nanowhiskers are formed on the bamboo charcoal surface by heating at 1500°C. To grow CNTs on bamboo charcoal at low temperatures (700–800°C), ethanol CVD assisted by extra metal catalysts (e.g., iron, nickel) is employed. At 1200–1500°C, the minerals inside charcoal are beneficial for the nucleation and growth of CNTs. The CNT/bamboo charcoal exhibits excellent absorption properties for removing metal ion pollutants. Finally, bamboo charcoals with or without CNTs are ideal materials for hydrogen storage.

References

[1] Zhou GM, Meng CF, Jiang PK, Xu QF. Review of carbon fixation in bamboo forests in China. Bot Rev 2011;77(3):262−70.

[2] Liao P, Yuan S, Xie W, Zhang W, Tong M, Wang K. Adsorption of nitrogen-heterocyclic compounds on bamboo charcoal: kinetics, thermodynamics, and microwave regeneration. J Colloid Interface Sci 2013;390(1):189−95.

[3] Liao P, Yuan S, Zhang W, Tong M, Wang K. Mechanistic aspects of nitrogen-heterocyclic compound adsorption on bamboo charcoal. J Colloid Interface Sci 2012;382(1):74−81.

[4] Zhang H, Zhu G, Jia X, Ding Y, Zhang M, Gao Q, et al. Removal of microcystin-LR from drinking water using a bamboo-based charcoal adsorbent modified with chitosan. J Environ Sci (China) 2011;23(12):1983−8.

[5] Wang M, Huang ZH, Liu G, Kang F. Adsorption of dimethyl sulfide from aqueous solution by a cost-effective bamboo charcoal. J Hazard Mater 2011;190 (1−3):1009−15.

[6] Wang FY, Wang H, Ma JW. Adsorption of cadmium (II) ions from aqueous solution by a new low-cost adsorbent—bamboo charcoal. J Hazard Mater 2010;177 (1−3):300−6.

[7] Mizuta K, Matsumoto T, Hatate Y, Nishihara K, Nakanishi T. Removal of nitrate-nitrogen from drinking water using bamboo powder charcoal. Bioresour Technol 2004;95(3):255−7.

[8] Chuang CS, Wang MK, Ko CH, Ou CC, Wu CH. Removal of benzene and toluene by carbonized bamboo materials modified with TiO₂. Bioresour Technol 2008;99 (5):954−8.

[9] Kosuwon W, Laupattarakasem W, Saengnipanthkul S, Mahaisavariya B, Therapongpakdee S. Charcoal bamboo as a bone substitute: an animal study. J Med Assoc Thai 1994;77(9):496−500.

[10] Venkatesan N, Yoshimitsu J, Ito Y, Shibata N, Takada K. Liquid filled nanoparticles as a drug delivery tool for protein therapeutics. Biomaterials 2005;26(34):7154−63.

[11] Hsieh MF, Wen HW, Shyu CL, Chen SH, Li WT, Wang WC, et al. Synthesis, in vitro macrophage response and detoxification of bamboo charcoal beads for purifying blood. J Biomed Mater Res A 2010;94(4):1133−40.

[12] Lin CC, Ni MH, Chang YC, Yeh HL, Lin FH. A cell sorter with modified bamboo charcoal for the efficient selection of specific antibody-producing hybridomas. Biomaterials 2010;31(32):8445−53.

[13] Yang S, Jia B, Liu H. Effects of the Pt loading side and cathode biofilm on the performance of a membrane-less and single-chamber microbial fuel cell. Bioresour Technol 2009;100(3):1197−202.

[14] Hua L, Chen Y, Wu W. Impacts upon soil quality and plant growth of bamboo charcoal addition to composted sludge. Environ Technol 2012;33(1−3):61−8.

[15] Xu T, Lou L, Luo L, Cao R, Duan D, Chen Y. Effect of bamboo biochar on pentachlorophenol leachability and bioavailability in agricultural soil. Sci Total Environ 2012;414:727−31.

[16] Gao H, Zhang Z, Wan X. Influences of charcoal and bamboo charcoal amendment on soil-fluoride fractions and bioaccumulation of fluoride in tea plants. Environ Geochem Health 2012;34(5):551−62.

[17] Thostenson ET, Ren ZF, Chou TW. Advances in the science and technology of carbon nanotubes and their composites: a review. Compos Sci Technol 2001;61 (13):1899−912.

[18] Tjong SC, Chen H. Nanocrystalline materials and coatings. Mater Sci Eng R 2004;45 (1−2):1−88.

[19] Zhu JT, Tjong SC, Li XQ. Spark plasma sintered hydroxyapatite/multiwalled carbon nanotube composites with preferred crystal orientation. Adv Eng Mater 2010;12 (11):1161−5.

[20] Lam CW, James JT, McCluskey R, Arepalli S, Hunter RL. A review of carbon nanotube toxicity and assessment of potential occupational and environmental health risks. Crit Rev Toxicol 2006;36(3):189−217.

[21] Zuo SL, Gao SY, Yuan XG, Xu BS. Carbonization mechanism of bamboo (Phyllostachys) by means of Fourier transform infrared and elemental analysis. J Forestry Res 2003;14(1):75−9.

[22] Cheung TLY, Ng DHL. Conversion of bamboo to biomorphic composites containing silica and silicon carbide nanowires. J Am Ceram Soc 2007;90(2):559−64.

[23] Jiang ZH, Zhang DS, Fei BH, Yue YD, Chen XH. Effects of carbonization temperature on the microstructure and electrical conductivity of bamboo charcoal. New Carbon Mater 2004;19(4):249−53.

[24] Zhu JT, Jia JC, Kwong FL, Ng DHL, Tjong SC. Synthesis of multiwalled carbon nanotubes from bamboo charcoal and the roles of minerals on their growth. Biomass Bioenerg 2012;36:12−9.

[25] Cheng HM, Endo H, Okabe T, Saito K, Zheng GB. Graphitization behavior of wood ceramics and bamboo ceramics as determined by X-ray diffraction. J Porous Mater 1999;6(3):233−7.

[26] Saito Y, Arima T. Features of vapor-grown cone-shaped graphitic whiskers deposited in the cavities of wood cells. Carbon 2007;45(2):248−55.

[27] Zhu JT, Kwong FL, Ng DHL. Synthesis of biomorphic SiC ceramic from bamboo charcoal. J Nanosci Nanotechnol 2009;9(2):1564−647.

[28] Zhu J, Jia J, Kwong FL, Ng DHL. Synthesis of 6 H-SiC nanowires on bamboo leaves by carbothermal method. Diamond Relat Mater 2013;33:5−11.

[29] Zhang JN, Huang ZH, Lv R, Yang QH, Kang FY. Effect of growing CNTs onto bamboo charcoals on adsorption of copper ions in aqueous solution. Langmuir 2009;25 (1):269−74.

[30] Chen XW, Timpe O, Hamid SBA, Schlogl R, Su DS. Direct synthesis of carbon nanofibers on modified biomass-derived activated carbon. Carbon 2009;47(1):340−3.

[31] Zhu JT, Jia JC, Kwong FL, Ng DHL, Crozier PA. Metal-free synthesis of carbon nanotubes filled with calcium silicate. Carbon 2012;50(7):2666−9.

[32] Egerton RF. Electron energy-loss spectroscopy in the TEM. Rep Prog Phys 2009;72 (1): 016502 (25pp.).

[33] Egerton RF. Electron energy-loss spectroscopy in the electron microscope. 3rd ed. New York: Springer; 2011.

[34] Ajayan PM, Redlich P, Ruhle M. Balance of graphite deposition and multishell carbon nanotube growth in the carbon arc discharge. J Mater Res 1997;12(1):244−52.
[35] Gamaly EG, Ebbesen TW. Mechanism of carbon nanotube formation in the arc-discharge. Phys Rev B 1995;52(3):2083−9.
[36] Fan WW, Zhao J, Lv YK, Bao WR, Liu XG. Synthesis of SWNTs from charcoal by arc-discharging. J Wuhan Univ Technol 2010;25(2):194−6.
[37] Zhu JT, Kwong FL, Lei M, Ng DHL. Synthesis of carbon coated silica nanowires. Mater Chem Phys 2010;124(1):88−91.
[38] Singjai P, Wongjamras A, Yu LD, Tunkasiri T. Production and characterization of beaded nanofibers from current heating of charcoal. Chem Phys Lett 2002;366 (1−2):51−5.
[39] Zhu JT, Huang ZH, Kang FY, Fu JH, Yue YD. Adsorption kinetics of activated bamboo charcoal for phenol. New Carbon Mater 2008;23(4):326−30.
[40] Huang ZH, Zhang FZ, Wang MX, Lv RT, Kang FY. Growth of carbon nanotubes on low-cost bamboo charcoal for Pb(II) removal from aqueous solution. Chem Eng J 2012;184:193−7.
[41] Miao HY, Chang LW, Lue JT, editors. PIERS Proceedings, Hangzhou, China, March 24−28. Cambridge, MA: The Electromagnetics Academy; 2008.
[42] Tsao CS, Liu Y, Chuang HY, Tseng HH, Chen TY, Chen CH, et al. Hydrogen spill-over effect of Pt-doped activated carbon studied by inelastic neutron scattering. J Phys Chem Lett 2011;2(18):2322−5.

2 Hierarchically Nanostructured One-Dimensional Metal Oxide Arrays for Solar Cells

Jih-Jen Wu, Chen-Hao Ku, Chun-Te Wu and Wen-Pin Liao

Department of Chemical Engineering, National Cheng Kung University, Tainan, Taiwan

2.1 Introduction

Metal oxide nanostructured electrodes have attracted considerable attention for their potential applications in many energy conversion and storage devices, such as Li-ion batteries [1−3], proton exchange membrane (PEM) fuel cells [4], solar cell [5−7], electrochromic devices [8,9], and photoelectrochemical cell [10−12]. To achieve superior performances of the devices, the electrodes for use in these applications all need to exhibit good electron transporting properties and a high surface area even though the working principles of the devices are different from each other.

Nanotechnology and solar energy technology are the topics that attract lots of attentions over recent years. The dye-sensitized solar cell (DSSC) which features a relatively high efficiency and a low cost is a successful product combining these two research fields. DSSCs still attract great interest now after a breakthrough achieved by O'Regan and Gratzel in 1991 [13]. As shown in Figure 2.1A, typical DSSC is composed of a dye-sensitized and high-surface-area TiO_2 nanoparticle (NP) film on a transparent conducting oxide (TCO) and a platinized counter electrode sandwiched together with an I^-/I_3^- redox electrolyte solution. Solar radiation is absorbed by a monolayer of dye on the surface of the TiO_2 anode. As illustrated in Figure 2.1B, electrons are transferred from the highest occupied molecular orbital (HOMO) to the lowest unoccupied molecular orbital (LUMO) of the dye to form an exciton. The exciton dissociation can only occur at the dye/TiO_2 interface and the electrons are then injected into the TiO_2 conduction band which is just below the LUMO band of the dye. Once the TiO_2 matrix receives electrons from the dye, the electrons diffuse through the interconnecting network of the porous TiO_2 anode and are collected at the TCO-coated glass substrate. The oxidized dye

Nanocrystalline Materials. DOI: http://dx.doi.org/10.1016/B978-0-12-407796-6.00002-6

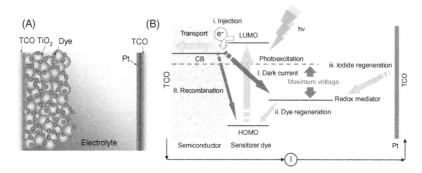

Figure 2.1 Schematic (A) and working principle (B) of DSSC.

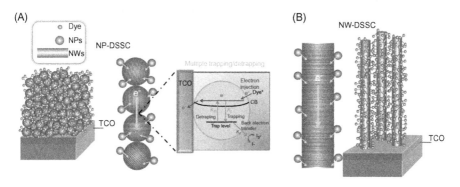

Figure 2.2 Schematics of electron transport paths in (A) NP DSSC and (B) NW DSSC.

is regenerated by electron transfer from the I^- ions, resulting in the formation of I_3^-. The I_3^- ions diffuse through the electrolyte and are regenerated by electron transfer from the Pt counter electrode.

TiO_2 NP film is typically employed to be the photoanode of DSSC and the solar energy conversion efficiency (η) higher than 12% has been achieved using such anodes [14]. TiO_2 NP film provides large enough surface area for dye adsorption, however, it exhibits relatively slow electron transport rate attributed to the multiple trapping/detrapping events occurring within grain boundaries [15–17], as illustrated in Figure 2.2A. Moreover, electrons in the trap states will suffer from significant recombination to electrolyte if electron detrapping rate is slow. The charge transport/recombination behaviors may limit the device efficiency. Superior device efficiency is thus anticipated by replacing the NP film with a vertical single-crystalline nanowire (NW) array, as shown in Figure 2.2B, for enhancing the electron transport rate. Owing to the unsuccessful development of vertical single-crystalline TiO_2 NW array on the TCO substrate in the earlier age, the ZnO NW array DSSC is the first demonstrated NW array DSSC [18]. Law et al. [18] found the electron diffusion coefficient of ZnO NWs is several hundred times higher than those of ZnO or TiO$_2$ NP film. However, the performance of the ZnO NW DSSCs ($\eta \sim 1.5\%$) is still limited by insufficient surface for dye adsorption. Afterwards, the performances of the rutile TiO$_2$ NW DSSCs were examined when the formation of well-aligned and single-crystalline rutile TiO$_2$ NW arrays has been successfully demonstrated on fluorine-doped tin oxide (FTO) substrates [19−21]. The rutile TiO$_2$ NW DSSCs also perform rather well, however, the electron transport rate in the TiO$_2$ NW DSSC is considerably inferior to that in ZnO NW cell [21,22].

The incident photon-to-current efficiency (IPCE) of a solar cell is the product of the light harvesting efficiency, electron injection efficiency, and electron collection efficiency. The electron collection efficiency (η_c) is determined by the performance of the anode of DSSC. Assuming that injected electrons are the sum of electrons collected on TCO and electrons lost by recombination and the rate constants for these processes are independent of each other, η_c is given by the expression [23]:

$$\eta_c = \frac{J_{sc}}{J_{inj}} = \frac{J_{sc}}{J_{sc} + J_r} = 1 - \frac{\tau_{sc}}{\tau_{oc}} \tag{2.1}$$

where J_{sc}, J_{inj}, and J_r are the short-circuit current density, the electron injection current density from dye to anode, and the recombination current density, respectively; τ_{sc} is the electron transit time and τ_{oc} is the electron recombination time. Moreover, the useful thickness of the photoanode shall be enlarged when the effective electron diffusion length (L_n) is enhanced [24]. The surface area for dye adsorption and thus light harvesting efficiency as well as the overall efficiency of the DSSC would be improved. The effective electron diffusion length is function of effective diffusion coefficient (D_{eff}) and electron recombination time as shown in Eq. (2.2). On the other hand, the effective diffusion coefficient is proportional to the product of the square of the photoanode thickness (L) and the inverse electron transport time [25,26]. Therefore the effective electron diffusion length is proportional to square root of the ratio of the electron recombination time to transit time:

$$L_n = (D_{eff}\tau_{oc})^{1/2} = c_1 L \left(\frac{\tau_{oc}}{\tau_{sc}} \right)^{1/2} \tag{2.2}$$

where c_1 is a constant. Electrochemical impedance spectroscopy (EIS) and optical modulation spectroscopy are typically employed to investigate the dynamics of charge transport and recombination in the DSSCs.

Superior electron transport properties and therefore charge collection efficiencies have been demonstrated in the ZnO-based DSSCs with various anode structures, such as single-crystalline NW [18], porous crystalline film [26], NW/NP composite film [27,28], and nanodendrite (ND)/NP composite film [29]. In this chapter, a review is given on the formations of hierarchical nanostructures on the ZnO NW arrays and their applications to DSSCs. The charge transport/

recombination properties in the hierarchical nanostructured ZnO DSSCs investigated by electrochemical impedance and optical modulation spectroscopies will be discussed as well.

2.2 Measurements of Charge Transport and Recombination in the DSSCs

2.2.1 Electrochemical Impedance Spectroscopy

EIS is a powerful technique for the characterization of electrochemical systems. The fundamental approach of the impedance analysis is to apply a small amplitude sinusoidal excitation signal to the system under investigation and measure the response. With a single experimental procedure conducting in a sufficiently broad range of frequencies, the influence of the governing physical and chemical phenomena is possible to be isolated and distinguished at a given applied potential using the EIS method. Typical Nyquist plot of the impedance data of DSSCs obtained at the open-circuit condition under illumination is shown in Figure 2.3A. There are three semicircles appearing in the spectra. Figure 2.3B illustrates an equivalent circuit representing the DSSCs, which is based on the diffusion−recombination model proposed by Bisquert [30]. R_s is a lumped series resistance for the transport resistance of TCO and all resistances out of the cell, W_1 is the impedance of diffusion of redox species in the electrolyte, and R_{Pt} and C_{Pt} are the charge-transfer resistance and the interfacial capacitance at the Pt counter electrode/electrolyte interface, respectively. Moreover, with a photoanode thickness of L, the electron transport resistance in photoanode ($R_w = r_w L$), charge-transfer resistance related to recombination of an electron ($R_k = r_k/L$), and the chemical capacitance ($C_\mu = c_\mu L$) can be extracted from the impedance analyses.

The three semicircles from high to low frequency in the typical DSSC Nyquist plot are corresponding to the charge transfers at the Pt/electrolyte and the photoanode/electrolyte interfaces, as well as electrolyte diffusion impedance in the cell

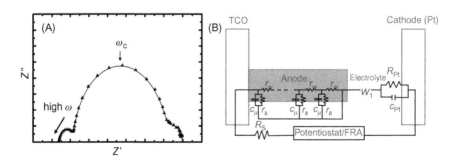

Figure 2.3 (A) Typical Nyquist plot of the impedance data of DSSCs. (B) An equivalent circuit representing the DSSCs.

[31]. The characteristic frequency (ω_c) of the central semicircle, as shown in Figure 2.3A is therefore to be employed to calculate the first-order reaction rate constant for electrons being lost ($k_{eff} = \omega_c/2\pi$) and lifetime of an electron in photoanode ($\tau = 2\pi/\omega_c$).

The continuity equation for the electron density (n) in the conduction band of the photoanode can be established based on the diffusion−recombination model [30] and the following relations are therefore obtained [30−32]:

$$R_w = \frac{k_B T}{q^2 An} \frac{L}{D_{eff}} = Con \frac{L}{D_{eff}} \tag{2.3}$$

$$R_k = Con \frac{1}{Lk_{eff}} \tag{2.4}$$

where k_B, T, q, and A are the Boltzman constant, absolute temperature, charge of a proton, and the electrode area, respectively. Combining Eqs. (2.3) and (2.4), the relation of R_w, R_k, and effective electron diffusion coefficient (D_{eff}) is obtained as

$$D_{eff} = \frac{R_k}{R_w} \frac{L^2}{\tau} \tag{2.5}$$

The procedure of estimation of the electron transport parameters in the DSSC photoanode from the Nyquist plot is summarized as follows [31]:

1. k_{eff} is estimated from ω_c of the central semicircle.
2. R_k is estimated from the diameter of the central semicircle.
3. L is the measured anode thickness.
4. Con and n are estimated from Eqs. (2.3) and (2.4).
5. R_w value is determined by fitting using equivalent circuit shown in Figure 2.3B.
6. D_{eff} is estimated from Eq. (2.5).

2.2.2 Optical Modulation Spectroscopy—Intensity-Modulated Photocurrent Spectroscopy and Intensity-Modulated Photovoltage Spectroscopy

In addition to EIS, optical modulation spectroscopies, i.e., intensity-modulated photocurrent spectroscopy (IMPS) and intensity-modulated photovoltage spectroscopy (IMVS), have been employed to measure the dynamics of electron transport and recombination in DSSCs. The techniques are based on the application of a small-signal harmonic perturbation of the irradiation intensity. The phase of the photocurrent/photovoltage response is shifted with respect to that of the impinging irradiation. The optoelectrical response of the system composed of the phase shift and the amplitude of the photocurrent/photovoltage, which is a function of the applied modulation frequency, can be measured over a broad frequency range. The charge transport dynamics in the solar cells can be determined using IMPS

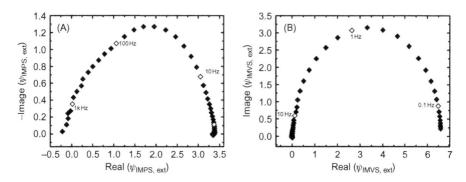

Figure 2.4 Typical (A) IMPS and (B) IMVS responses of DSSCs.

technique at short-circuit conditions. On the other hand, IMVS measurements provide information regarding the electron lifetime and charge recombination dynamics at open-circuit conditions.

Typical IMPS and IMVS responses of DSSC are shown in Figure 2.4. The electron transit time and lifetime in the anodes can be extracted from the IMPS and IMVS results, respectively. As shown in Figure 2.4A, the IMPS plot goes toward the negative real/negative imaginary quadrant at high frequency. The measured IMPS response at high frequency is affected by RC attenuation (R: cell series resistance and C: photoelectrode capacitance) [33] that limits the dynamical response of the photocell. The effect is suggested to be corrected by an attenuation function of $A(\omega)$ [33,34]:

$$A(\omega) = \frac{1 - i\omega RC}{1 + \omega^2 R^2 C^2} \tag{2.6}$$

The measured IMPS and IMVS responses take the forms [23]:

$$\psi_{\text{IMPS,ext}}(\omega) = \Delta J_{\text{sc}} \cdot A(\omega)$$

$$= \frac{q\phi I_0 \alpha}{\gamma^2 - \alpha^2} \left[\frac{2\alpha \exp(-\alpha L) - (\alpha - \gamma)\exp(\gamma L) - (\alpha + \gamma)\exp(-\gamma L)}{\exp(\gamma L) + \exp(-\gamma L)} \right]$$

$$\left(\frac{1 - i\omega RC}{1 + \omega^2 R^2 C^2} \right) \tag{2.7}$$

$$\psi_{\text{IMVS,ext}}(\omega) = \Delta V_{\text{oc}} = \frac{\phi I_0 \alpha}{k_{\text{r}} - m D_{\text{cb}} \alpha^2 + i\omega}$$

$$\left[1 + \frac{\alpha}{\gamma} \frac{2\exp(-\alpha L) - \exp(\gamma L) - \exp(-\gamma L)}{\exp(\gamma L) + \exp(-\gamma L)} \right] \tag{2.8}$$

where ΔJ_{sc} is the photocurrent conversion efficiency, ΔV_{oc} is the photovoltage conversion efficiency, ω is the circular frequency of the light modulation, α is the light absorption coefficient of the cell, I_o is the incident photon flux, ϕ accounts for reflection losses, $\gamma = \sqrt{(k_r + i\omega)/(mD_{cb})}$ is the charge-injection efficiency, L is the thickness of the anode, k_r is the rate constant for recombination, D_{cb} is the diffusion coefficient of electrons in the conduction band, m is n_{cb}/n, and n is the sum of the photoinduced electron concentration in the conduction band, n_{cb}, and in trap states. The cell series resistance and photoelectrode capacitance of the DSSCs can be extracted from EIS measurements. To determine electron transport times (τ_{sc}) and electron lifetimes (τ_{oc}) from ΔJ_{sc} and ΔV_{oc}, the real and imaginary parts of ΔJ_{sc} and ΔV_{oc} are fitted with the expressions [35]:

$$\text{Re}(\Delta J_{sc}) = \frac{X_1}{1 + (\omega\tau_{sc}^{re})^2} \tag{2.9a}$$

$$\text{Im}(\Delta J_{sc}) = -\frac{X_2\omega\tau_{sc}^{im}}{1 + (\omega\tau_{sc}^{im})^2} \tag{2.9b}$$

$$\text{Re}(\Delta V_{oc}) = \frac{X_3}{1 + (\omega\tau_{oc}^{re})^2} \tag{2.10a}$$

$$\text{Im}(\Delta V_{oc}) = -\frac{X_4\omega\tau_{oc}^{im}}{1 + (\omega\tau_{oc}^{im})^2} \tag{2.10b}$$

where X_1, X_2, X_3, X_4, τ_{sc}^{re}, τ_{sc}^{im}, τ_{oc}^{re}, and τ_{oc}^{im} are fit parameters. The electron transit time and lifetime in the anodes are obtained from the fitting results of $\tau_{sc} = \tau_{sc}^{re} \approx \tau_{sc}^{im}$ and $\tau_{oc} = \tau_{oc}^{re} \approx \tau_{oc}^{im}$, respectively.

2.3 ZnO-Based DSSCs

Zinc oxide which possesses superior performances in electronics, optics, photonics, and optoelectronics has received broad attention for its applications to varistors, transparent high power electronics, surface acoustic wave devices, piezoelectric transducers, UV light emitters, sensors, and solar cells [35−38]. The synthesis of one-dimensional (1D) ZnO nanostructures, e.g., NWs and nanorods (NRs), is of growing interest owing to their novel electrical, mechanical, chemical, and optical properties as well as the promising applications to many devices [39−43]. The merit of the vertical 1D nanostructures for use in many devices is of providing direct electron conducting channels to the electrodes. Vapor phase methods have been extensively used to prepare high-quality and well-aligned 1D ZnO nanostructures [44−46], however, the high process temperature ($\sim 500−1100°C$) limits the substrates available for ZnO growth. On the other hand, aligned 1D ZnO nanostructures have been successfully synthesized on various substrates using a low-temperature ($<100°C$) and low-cost chemical bath deposition (CBD) method [47,48].

With an energy band gap similar to that of the ordinarily used TiO_2 but possesses higher electron mobility, ZnO is an alternative anode material for DSSCs [49]. ZnO nanostructures attract considerable attention for DSSC application because of its low crystallization temperature and anisotropic growth behavior in addition to superior electron transport properties. Law et al. [18] employed polyethylenimine (PEI) to enhance the aspect ratios of the ZnO NW anodes during the CBD growth. High-density NWs with a length of $20-25$ μm have been obtained by multibath growth and the best performance of the ZnO NW DSSC was characterized by an efficiency of 1.5% under AM 1.5 irradiation (100 mW/cm^2). They suggested that the performance of the ZnO NW DSSC was still limited by inferior surface area of NWs for dye loading although ZnO NW electrodes increased the rate of electron transport. Enrichment of the surface area of the 1D nanostructures is necessary for further enhancing the DSSC performances. In addition to single-crystalline NW array, superior electron transport properties and therefore charge collection efficiencies have been demonstrated in ZnO-based DSSCs with various anode structures, such as porous crystalline film [26], NW/NP composite film [27,28], and ND/NP composite film [29].

Although ZnO possesses the advantages of rapid electron transferring and collecting, the highest power conversion efficiency of 7.5% for ZnO-based DSSC reported recently [50] is still lower than the record efficiency of 12.3% for TiO_2-based DSSC [14]. The main reason may be that DSSC dyes are always designed for TiO_2 NP film anode, whereas there is no efficient dye available for ZnO anode [26,51]. In addition to electron transport rate in the photoanode, a superior efficiency in DSSCs is also determined by electron injection rate and dye regeneration rate. The LUMO energy level and the HOMO energy level of the dye molecules have to appropriately match with the conduction band minimum of the photoanode and the redox potential of the redox mediator in the electrolyte, respectively. With most of the common dyes, the electron injection efficiency in the ZnO-based DSSCs is inferior to that in the TiO_2-based DSSCs, which may be mainly attributed to both low dielectric constant and low density of states in the conduction band of ZnO [51]. On the other hand, the poor chemical stabilities of the ZnO electrode in the acidic dye solution [52] and in the presence of complexing agents of the dyes [53] result in ZnO deterioration and the difficulty of dye uptake, which has been reported as the most probable reason for the inferior performance of ZnO DSSC [51].

It has been demonstrated that the structure of the ZnO crystals is destroyed after loading the Ru-complex dyes. Protons releasing from the dye molecules in the ethanolic solution will dissolve ZnO to generate Ru-complex dye-Zn^{2+} aggregates [52]. Compared to the Ru-complex dye, mercurochrome dye has been demonstrated to be more suitable for ZnO anode although the mercurochrome dye possesses a narrower absorptive range for light harvesting [52]. More abundant electron interfacial recombination occurs in the Ru dye-sensitized ZnO NW DSSC, compared to mercurochrome dye-sensitized one, due to the higher surface trap density in the ZnO NW photoanode after Ru dye adsorption. Nevertheless, with an optimized dye sensitization process, e.g., a shorter dye sensitization period and a lower dye concentration, recent results show that Ru dye-sensitized ZnO-based DSSCs also demonstrate encouraging efficiencies due to relatively high photocurrent density [50].

Indoline dyes have been found a comparatively good match with ZnO because of its lower acidity and the lack of complexing agent [26,51]. With the high extinction coefficient, the efficiencies of the ZnO-based DSSC higher than ~5% can be attained using the indoline dye coded D149 [26,54]. However, the absorption wavelengths of D149 dye are limited within the visible light range which is much narrower than that of ruthenium dye [26]. The narrow absorption range may restrict the efficiency of the D149-sensitized ZnO DSSCs. With the absence of dyes specially designed for ZnO anodes, the enhancement of the performances of the ZnO-based DSSC can be realized by utilizing the variety of ZnO nanostructures [27−29,55−57].

In order to increase the light harvesting efficiency of the ZnO NW DSSC, the strategies of modifications of the ZnO NW arrays have been reported by enlarging the surface area of photoanode without sacrificing fast electron transport of ZnO NW [27−29,55,57] and adding a light scattering layer for reflecting unabsorbed photons back into the NW anode [56,57]. In the following sections, a review is given on the formations of hierarchical nanostructures on the ZnO NW arrays for use in DSSCs. The photovoltaic performances of the ZnO NW-based DSSCs and the charge transport/recombination behaviors in these DSSCs will be discussed.

2.4   Wet Chemical Route to ZnO NW-Layered Basic Zinc Acetate/ZnO NP Composite Film for Use in DSSCs

For improving the surface area without sacrificing electron transport efficiency, composite films composed of NPs and the NW array have been considered to be the anode of DSSCs [27,28,43,58−61]. Baxter and Aydil [43] first hybridized ZnO NWs and NPs by filling the interstices of the NW array with NPs having diameters of 50−200 nm from an ethanol solution containing binder and surfactant. It revealed that the efficiency has been enhanced from 0.3% of the NW DSSC to 1.3% of the hybrid DSSC by loading the NPs. Nevertheless, a significant enhancement of the overall efficiency of the hybrid DSSC is not achieved, which may result from the low-density ZnO NWs grown by chemical vapor deposition and the deficient filling of the interstitial voids of the NWs by the mixing approach [43].

Formation of flower-like LBZA (layered basic zinc acetate) nanostructures consisted of LBZA nanosheets (NSs) self-assembled on the substrates has been reported using nonbasic solution routes [62,63]. The features of the self-assembled LBZA NSs are associated with the surface conditions of the substrate [62]. The LBZA NSs were found to be a two-dimensional (2D) agglomeration of NPs and were able to be transformed to ZnO NSs through pyrolysis [62−66]. In addition, the LBZA which is a derivative of α-form $Zn(OH)_2$ [63] is a high bandgap material. With the promising characteristics, the LBZA NSs have been integrated into the ZnO-nanostructured arrays for use in DSSC [27–29,58,59].

A simple wet chemical route has been developed to construct the ZnO NW-LBZA/ZnO NP composite (named as the ZnO NW–NP composite hereafter) anode [27,28], that is, formation of the aligned ZnO NW array first via an aqueous CBD followed by the heterogeneous nucleation and growth of the LBZA/ZnO NPs on the surface of the ZnO NWs using the base-free CBD. The ZnO NPs as well as insulating LBZA NPs were bottom-up grown within the interstices of NW array for increasing surface area and suppressing electron recombination simultaneously. The features of the composites from three-dimensional (3D) NW–NS networks, dense NW–NP composite film to NW–NP composite film with flower-like particles can be obtained easily by varying the growth period of base-free CBD.

2.4.1 Formation and Structural Characterizations of ZnO NW-LBZA/ZnO NP Composite Film

High-density ZnO NWs were first grown on the seeded FTO substrates in a 0.02 M aqueous solution of zinc acetate dihydrate ($Zn(CH_3COO)_2 \cdot 2H_2O$, $ZnAc \cdot 2H_2O$) and hexamethylenetetramine ($C_6H_{12}N_4$, HMTA) at 95°C for 3 h. Formation of the seed layer on FTO was conducted by the dip-coating method using a 0.05 M aqueous solution of $ZnAc \cdot 2H_2O$ and HMTA at room temperature (RT) and the layer was annealed at 350°C. Another bath growth in a fresh 0.01 M aqueous solution of $ZnAc \cdot 2H_2O$ and HMTA at 95°C for 3 h was carried out to lengthen the NWs followed by being annealed at 400°C further prior to NP growth. As shown in Figure 2.5A and B, ZnO NW arrays which possess a length of ~ 5.5 μm, a density of $\sim 6.6 \times 10^9$ cm^{-2}, and a diameter in the range of 30–300 nm were thus obtained [58]. Transmission electron microscopy (TEM) characterizations of an

Figure 2.5 (A) Top-view and (B) cross-sectional SEM images of the ZnO NW array grown by CBD. (C) HRTEM image, low-magnification TEM image (inset) and the corresponding SAED pattern (inset) of an individual ZnO NW.
Source: Reprinted with permission from Ref. [58]. Copyright 2008, American Chemical Society.

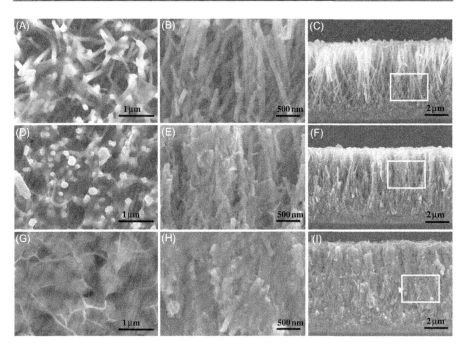

Figure 2.6 Top-view and cross-sectional SEM images of the NW−NP composite films with different NP growth periods. (A−C) 14 h; (D−F) 15 h; and (G−I) 16 h.
Reprinted with permission from Ref. [58]. Copyright 2008, American Chemical Society.

individual NW indicate that the NW possesses the single-crystal structure and the lattice spacing of around 0.52 nm along the longitudinal axis direction corresponds to the d-spacing of ZnO (001) crystal planes, as shown in Figure 2.5C.

The ZnO NW arrays were immersed in a base-free solution, that is, a 100 mL of 0.15 M methanolic solution of $ZnAc \cdot 2H_2O$, at 60°C for 14−24 h for forming various amounts of NPs within the interstices of ZnO NW arrays. The SEM images of the NW−NP films with various growth periods for base-free CBD are shown in Figures 2.6 and 2.7 [58]. They reveal that the interstices of the ZnO NWs are not fully occupied by NPs until the growth period is over 16 h. It is worth to address that the base-free solution for NP growth was still limpid after 15-h growth whereas the NSs were already grown on the surface of the NWs to form the 3D NW−NS networks after 14 h as shown in Figure 2.6A−C, suggesting that the heterogeneous nucleation of NPs on the surface of ZnO NWs is dominant in the early stage of NPs growth. On the other hand, Figure 2.6G−I shows that the NPs fully occupy the room among the NWs, resulting in the formation of the dense NW−NP composite film after 16-h NP growth. When the growth period is prolonged over 16 h, more particles are grown within the interstices of the NW array forming a denser NW−NP composite film and the composite film cracks finally as shown in Figure 2.7A−F. Moreover, some flower-like particles with sizes of 5−10 μm are

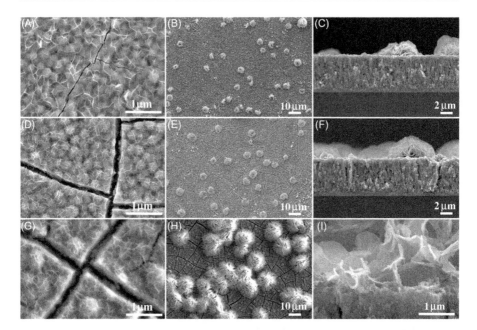

Figure 2.7 Top-view and cross-sectional SEM images of the NW−NP composite films with different NP growth periods. (A−C) 17 h; (D−F) 18 h; and (G−I) 24 h.
Reprinted with permission from Ref. [58]. Copyright 2008, American Chemical Society.

observed on the surface of the NW−NP composite film after 17−h growth, and the sizes of particles are enlarged to 20−30 μm when the growth period is extended to 24 h. As shown in Figure 2.7G−I, the particles above the NW−NP composite film are the assemblies of the NSs, which are consistent with the growth behavior of the LBZA NPs without any space confinement [60].

Figure 2.8 shows the XRD patterns of the NW−NP films wherein the NPs formed in the base-free chemical bath for 16, 19, and 24 h [58]. Since the ZnO NW array shows its preferential growth in the c-axis direction (Figure 2.5B and C), the intensities of all peaks in each pattern of Figure 2.8 are normalized to that of ZnO (002) for comparison. Apart from the diffraction peaks of FTO and ZnO, the peak at 2θ of 6.44° appears in all patterns of Figure 2.8A. This peak corresponding to the lattice spacing of 1.37 nm is indexed to be the contribution of the Zn $(OH)_{1.6}(CH_3COO)_{0.4} \cdot 0.4H_2O$ [62−65], and its normalized intensity increases significantly with the NP growth period as illustrated in Figure 2.8A. LBZA (100) and (110) peaks are also observed in the 24-h NP grown composite film. They indicate that the fraction of the LBZA in the NP portion of the composite film increases at the same time with the amount of NP occupying the interstices of the NW array when the base-free CBD period for NP growth is prolonged. On the other hand, in comparison with the pattern of the ZnO NW array, the ZnO (002) peaks of the composite films all become broadened and shift of the ZnO (002) diffraction peak

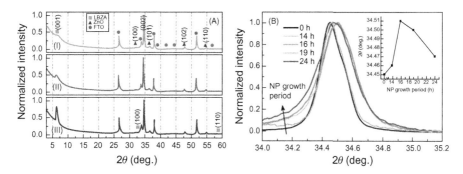

Figure 2.8 XRD patterns of the NW−NP composite films after NP growth for 16 h (pattern I), 19 h (pattern II), and 24 h (pattern III). (A) $2\theta = 3-60°$, (B) $2\theta = 34-35.2°$. The patterns of ZnO NW (0 h) and 14-h NP-grown NW−NP composite were also presented in (B) for comparison and the inset of (B) is the variation in ZnO (002) diffraction peak position of the NW−NP composite films after NP growth for 0−24 h.
Reprinted with permission from Ref. [58]. Copyright 2008, American Chemical Society.

with the NP growth period is observed in the XRD patterns, as shown in Figure 2.8B. It indicates the formation of ZnO NPs in addition to LBZA NPs in the interstices of ZnO NWs.

Typical cross-sectional TEM bright-field image and the corresponding selected area electron diffraction (SAED) pattern of the NW−NP composite are shown in Figure 2.9A and its inset [58]. The d-spacing calculated from the observable SAED rings in the inset is consistent with those of hexagonal ZnO structure. Figure 2.9B shows the corresponding dark-field image of Figure 2.9A revealing the ZnO NPs with diameters in the range of 5−30 nm. Figure 2.9C and D shows another set of bright-field and dark-field images at the interfacial region of the NW and NPs. The d-spacings calculated from the three diffraction spots closest to direct beam in the corresponding SAED pattern (inset) are indexed to be the lattice spacings of LBZA (001), (002) and (003), respectively. The dark-field image shown in Figure 2.9D was obtained by selecting the diffraction spot of LBZA (002), which illustrates the LBZA NPs appearing nearby or on the ZnO NWs. The diffraction spots of the LBZA were rarely observed during TEM analyses of the NW−NP composite film and would disappear rapidly under the irradiation of electron beam. The LBZA NPs would be gradually converted to ZnO NPs under the irradiation of electron beam owing to the easy conversion of LBZA to ZnO [62−66].

Figure 2.9E shows the typical high-resolution (HR) TEM image of the NW−NP composite at the interfacial region of the NW and NPs. It reveals that NPs with a diameter of ∼15 nm were grown on the surface of the NW and the lattice fringes of these two NPs are aligned to each other but not aligned to the single-crystal NW. The corresponding diffraction pattern of the image obtained by fast Fourier transform (FFT) spectroscopy is illustrated in the inset. It is composed of two sets of single-crystal diffraction patterns, in which the one with white-color denotations is attributed to the ZnO NW and the other is corresponding to the NPs. The HRTEM image and

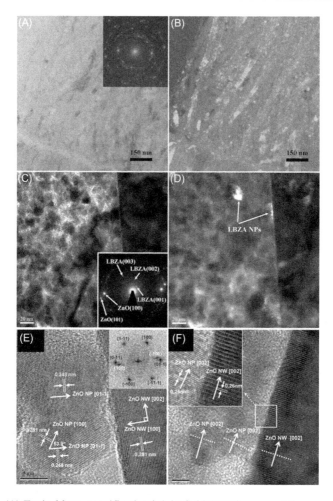

Figure 2.9 (A) Typical low-magnification bright-field TEM image and the corresponding SAED pattern (inset) of the NW−NP composite film. The NP growth period was 18 h. (B) The corresponding dark-field TEM image of (A) obtained by selecting the ZnO rings in SAED pattern. (C) Another high-magnification bright-field TEM image and the corresponding SAED pattern (inset) of the NW−NP composite film and (D) the corresponding dark-field TEM image obtained by selecting the LBZA (002) diffraction spot in (C). (E and F) Typical HRTEM images at the interfacial regions of the NW and NPs with zone axis of (011) and (010), respectively. The NP growth period was 18 h.
Reprinted with permission from Ref. [58]. Copyright 2008, American Chemical Society.

the corresponding FFT of the NPs indicate that the NPs possess the hexagonal ZnO structure as well. Another HRTEM analysis at the interfacial region of the NW and NPs (Figure 2.9F) also reveals that the (002) lattice fringes of the ZnO NW are not well-aligned to those of the NPs on it whereas only a small-angle misalignment is observed, indicating that there is no epitaxial relationship between the NW and NPs.

2.4.2   Formation Mechanism of the NW−NP Composite Film

Figure 2.10 illustrates the proposed formation mechanism of the NW−NP composite film on the FTO substrate [58]. Aligned ZnO NWs are first grown on ZnO nanocrystalline-seeded FTO substrate using the aqueous CBD method in a solution of ZnAc $\cdot$ 2H$_2$O and HMTA. The formation mechanism of ZnO NW and the role of HMTA have been reported [67−69], the following chemical reactions were involved:

$$(CH_2)_6N_4 + 6H_2O \rightarrow 6HCHO + 4NH_3 \tag{2.11}$$

$$NH_3 + H_2O \Leftrightarrow NH_4^+ + OH^- \tag{2.12}$$

$$nOH^- + Zn^{2+} \rightarrow [Zn(OH)_n]^{(n-2)-} \tag{2.13}$$

$$[Zn(OH)_n]^{(n-2)-} \rightarrow ZnO + H_2O + (n-2)OH^- \tag{2.14}$$

where $n = 2$ or 4. For synthesis of the NPs within the interstices of the ZnO NW array, the base-free CBD in a methanolic solution of ZnAc $\cdot$ 2H$_2$O was performed further. If ZnAc $\cdot$ 2H$_2$O is dissolved in methanol then a sequence of reactions took place, including solvation by methanol (Eq. 2.15), formation of the methoxyacetate complexes (Eqs. 2.16 and 2.17), hydrolysis by water originating from ZnAc $\cdot$ 2H$_2$O itself (Eq. 2.17), and polymerization of the mononuclear complexes through "olation" (Eq. 2.18) as shown below [62,63]:

$$Zn^{2+} + 6MeOH \rightarrow [Zn(MeOH)_6]^{2+} \tag{2.15}$$

$$[Zn(MeOH)_6]^{2+} + mAc^- \rightarrow [Zn(MeOH)_{6-m}Ac_m]^{2-m} + mMeOH \tag{2.16}$$

$$[Zn(MeOH)_{6-m}Ac_m]^{2-m} + (h+a)H_2O$$
$$\rightarrow [Zn(MeOH)_{6-(m+h)}(OH)_{h+a}Ac_{m-a}]^{2-(m+h)} + (h+a)H^+ + aAc^- + hMeOH \tag{2.17}$$

$$[Zn(MeOH)_{6-(m+h)}(OH)_{h+a}Ac_{m-a}]^{2-(m+h)}$$
$$+ [Zn(MeOH)_{6-(m+h)}(OH)_{h+a}Ac_{m-a}]^{2-(m+h)}$$
$$\rightarrow [Zn(MeOH)_{6-(m+h)}Ac_{m-a}(OH)_{h+a}Zn(MeOH)_{6-(m+2h+a)}(OH)_{h+a}Ac_{m-a}]^{4-2(m+h)}$$
$$+ (h+a)MeOH \tag{2.18}$$

where Ac = CH$_3$COO and Me = CH$_3$. Due to slow hydrolysis rate in a base-free solution and the large surface area of the well-aligned ZnO NWs, the

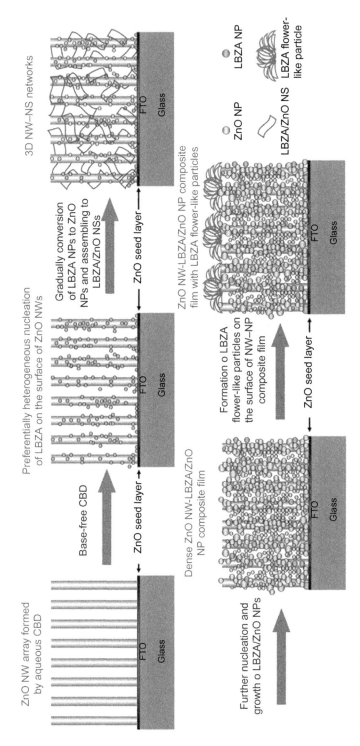

Figure 2.10 Proposed formation mechanism of the NW−NP composite.
Reprinted with permission from Ref. [58]. Copyright 2008, American Chemical Society.

heterogeneous nucleation of LBZA on the surface of the ZnO NWs would preferentially proceed in comparison with the homogeneous nucleation in the solution at such low degree of supersaturation. Further hydrolysis and dehydration reactions convert LBZA NPs into ZnO NPs.

Figure 2.6A–F show that most of the NSs are grown on the (100) surfaces of the ZnO NWs and connect neighboring ZnO NWs after base-free CBD for 14–15 h, resulting in the formation of the 3D NW–NS networks. Secondary nucleation of LBZA occurs on the surface of the NSs as the base-free CBD period is prolonged further [62]. Due to the space confinement within the interstices of the NW array, as shown in Figure 2.6G–I, the NPs fully occupy the room among the NWs after 16-h base-free CBD. The dense aligned NW–NP composite films are therefore formed on the FTO substrates.

It is noted that the water molecules for LBZA formation and further conversion to ZnO only originate from the $ZnAc \cdot 2H_2O$ precursor themselves in methanol in the base-free CBD route. An increase in LBZA fraction in the composite film with the NP growth period, as shown in Figure 2.8A, is due to the limited amount of water in the solution for the further conversion to ZnO NPs. Furthermore, as shown in Figure 2.8B, the shift of the 2θ values and the asymmetry in the low 2θ side of the ZnO (002) diffraction peak might be ascribed to the slight dissimilarity of the c-axis lattice constants of the ZnO NPs formed from a deficient conversion of LBZA to ZnO due to limited amount of water in the solution. The cracks in the NW–NP composite films observed after NP growth over 16 h are suggested to be associated with the strains accumulated within the composite films resulted from the variation of the c-axis lattice constant. Finally, LBZA flower-like particles formed on the surface of the NW–NP composite film as a result of the limited water amount and without any space confinement. HRTEM analyses, Figure 2.9, indicate that there is no epitaxial relationship between the ZnO NW and ZnO NPs. It suggests that heterogeneous nucleation of LBZA structure on the surface of the ZnO NWs is crucial for the formation of the NW–NP composite using the base-free route otherwise homoepitaxial growth on the surface of the ZnO NWs only results in enlargement of the length and the diameter of the ZnO NWs as in the case of using the aqueous solution of zinc acetate and HMTA.

2.4.3 Photovoltaic Performances of the Mercurochrome-Sensitized ZnO NW–NP DSSCs

The NW–NP composites with various amounts of NP were employed to fabricate the ZnO NW–NP composite DSSCs. The NW–NP composite anodes were immersed in an absolute methanolic dye solution refluxed at 65°C for 20 min. The concentration of mercurochrome was 1.6×10^{-4} M. The sensitized electrode and the platinized FTO counter electrode were sandwiched together with 25-μm-thick hot-melt spacers. The liquid electrolyte solution composed of 0.3 M tetra-n-propylammonium iodide and 0.03 M iodine in an ethylene carbonate and acetonitrile

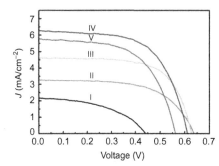

Figure 2.11 Current density−voltage characteristics of the ZnO NW DSSC (curve I) and the NW−NP composite DSSCs (curves II, III, IV, and V for NP growth periods of 15, 16, 18, and 24 h, respectively) under AM 1.5 illumination.
Reprinted with permission from Ref. [58]. Copyright 2008, American Chemical Society.

Table 2.1 Photovoltaic Parameters of the ZnO NW and Various NP Growth Period NW−NP Composite DSSCs with Photoanode Thicknesses of 5.5 μm

NP Growth Period (h)	J_{sc} (mA/cm^2)	V_{oc} (V)	FF	η (%)
0	2.15	0.44	0.47	0.45
15	3.27	0.64	0.61	1.27
16	4.61	0.63	0.67	1.93
18	6.28	0.61	0.62	2.37
24	5.79	0.56	0.59	1.94

Source: Reprinted with permission from Ref. [58]. Copyright 2008, American Chemical Society.

mixed solvent (60:40 by volume) was introduced between the sensitized and counter electrodes by capillary action. A black-painted mask was used to create an exposed area of 0.04 cm^2 for all samples.

Figure 2.11 shows the photocurrent density (J)−voltage (V) characteristics of the ZnO NW and the NW−NP composite DSSCs [58]. The photovoltaic parameters of the cells are summarized in Table 2.1 [58]. It shows that the short-circuit photocurrent density (J_{sc}) and efficiency (η) first increase with NP growth period as the period is shorter than 18 h and then decrease as the NP growth period is prolonged to 24 h. The superior performances of the NW−NP cells are ascribed to considerable enlargement of the surface area by NPs growth for dye adsorption meanwhile without sacrificing the electron transport efficiency through the vertical and single-crystalline ZnO NWs to FTO. Open-circuit voltage (V_{oc}) and fill factor (FF) of the 15-h NP-grown NW−NP composite cell are superior to those of the ZnO NW cell, indicating the LBZA/ZnO NPs grown in the interstices of the NW array are able to reduce the electron recombination. However, V_{oc} and FF slightly decrease when NP growth period is over 15 and 16 h, respectively. XRD analyses indicate that the NW−NP composite films contain more LBZA NPs as increasing the NP growth period, implying that the LBZA NP plays a crucial role for electron transport through the composite. Moreover, the NW−NP composite films formed by longer NP growth periods possess severe cracks extended to FTO substrate as shown in Figure 2.7, which may also contribute to the slight decline in performance of the 24-h NP-grown NW−NP cell. Therefore, the optimization of the LBZA content and the elimination of the crack formation in the NW−NP composite films with high-NP-occupying amount are crucial for further improvement of the performance of the NW−NP composite DSSCs. Figure 2.11 and Table 2.1 show that with an anode thickness of 5.5 μm, the highest efficiency is achieved in the 18-h NP-grown NW−NP composite DSSC. When the anode thickness is elongated to 6.2 μm, an efficiency of 3.2% is attained in the mercurochrome-sensitized ZnO NW−NP DSSC with J_{sc}, V_{oc}, and FF of 9.06 mA/cm^2, 0.60 V, and 0.58, respectively [59].

It has been reported that the performance of the TiO$_2$ NP DSSCs is able to be improved by coating a thin insulating layer, e.g., metal oxides [37,38], metal hydroxides [39], and CaCO$_3$ [40], on the surface of the anode to suppress the electron from TiO$_2$ and FTO surfaces to recombine with the oxidized species. The LBZA which is a derivative of α-form Zn(OH)$_2$ [35] may play the same role as the insulating overlays in the ZnO NW−NP DSSCs. The LBZA NPs may block the surfaces of ZnO NWs, ZnO NPs, and FTO, resulting in V_{oc} and FF of the NW−NP composite DSSC are superior to those of the ZnO NW DSSC as listed in Table 2.1. The thermal decomposition behavior of LBZA has been reported [62−66] that LBZA is converted into ZnO by the release of the intercalated and/or adsorbed water molecules at temperatures of around 70−90°C, dehydration decomposition of zinc hydroxide at around 120°C, and a slow decomposition of the acetate group in the temperature range of 130−350°C. Therefore, the influences of the LBZA fraction of the composite anode on the performance of ZnO NW−NP composite DSSC were further investigated by applying a series of postannealing conditions on the 15-h NP-grown NW−NP composite [59].

A significant enhancement of the DSSC efficiency from 1.27% (Table 2.1) to 2.77% ($J_{sc} = 8.33$ mA/cm^2, $V_{oc} = 0.58$ V, and FF = 0.58) is achieved as the 15−h NP-grown NW−NP composite was postannealed at 120°C in air for 15 min [59]. Partial conversion of LBZA to ZnO by an appropriate postannealing can increase the photoelectron injection/tunneling probability from the LUMO level of dye molecule through LBZA to the conduction band of ZnO and can suppress most trap-mediated electron leakages at the same time, resulting in a significant improvement in cell J_{sc} only at the expense of a slight decline in V_{oc} and FF [59]. This result indicates that photovoltaic performance of the NW−NP composite DSSCs can be further improved by optimizing NP occupying extent and postannealing condition.

2.4.4 Charge Transport Properties in the Mercurochrome-Sensitized ZnO NW−NP DSSCs

To examine the contribution of the vertical NW array to the electron transport in the photoanode of the ZnO NW−NP composite DSSCs, charge transport properties in the mercurochrome-sensitized ZnO NW, mercurochrome-sensitized ZnO NW−NP, and N719-sensitized TiO$_2$ NP DSSCs with an anode thickness of 5.5 μm are investigated using EIS and intensity modulation spectroscopy (IMPS and IMVS) [27,28]. The three DSSCs with the optimized performances are employed

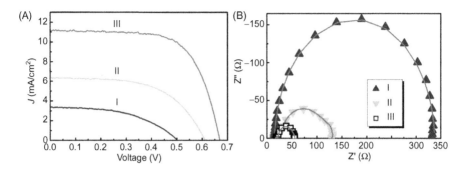

Figure 2.12 (A) Current density−voltage curves and (B) Nyquist plots of the impedance data of the mercurochrome-sensitized ZnO NW (I), mercurochrome-sensitized ZnO NW array/NP composite (II), and N719-sensitized TiO_2 NP (III) DSSCs. The solid lines in (B) are the fitting results based on the model in Figure 2.3B.
Reprinted with permission from Ref. [27]. Copyright 2007, American Institute of Physics.

Table 2.2 Performances of the Mercurochrome-Sensitized ZnO NW (I), Mercurochrome-Sensitized ZnO NW−NP Composite (II), and N719-Sensitized TiO_2 NP (III) DSSCs and the Electron Transport Properties in their 5.5-μm-Thick Photoanodes Determined by Impedance Analysis

DSSCs	J_{sc} (mA/cm^2)	V_{oc} (V)	FF	η (%)	n (cm^{-3})	D_{eff} (cm^2/s)
I	3.4	0.50	0.49	0.84	1.1×10^{18}	1.8×10^{-3}
II	6.3	0.61	0.58	2.2	1.8×10^{18}	2.1×10^{-4}
III	11.3	0.67	0.66	5.0	1.0×10^{19}	7.2×10^{-5}

Source: Reprinted with permission from Ref. [27]. Copyright 2007, American Institute of Physics.

for comparing their electron transport properties. As aforementioned dye issue, the dye employed in the ZnO and TiO_2 NP DSSCs is therefore different [27]. The photovoltaic performances of the three DSSCs are shown in Figure 2.12A and Table 2.2 [27]. Among the three cells, the N719-sensitized TiO_2 NP DSSC possesses the highest efficiency of 5.0%.

EIS measurements were carried out under the illumination of AM-1.5 100 mW/cm^2 by applying a 10 mV ac signal over the frequency range of $10^{-2}-10^5$ Hz on the top of V_{oc} of the DSSC using a potentiostat with a frequency response analyzer (FRA). IMPS and IMVS measurements were performed under a modulated green LED light (530 nm) driven by a source supply and a potentiostat with an FRA. The illumination intensity ranges from 30 to 90 mW/cm^2. The light intensity modulation was 20% of the base light intensity over the frequency range of $10^{-2}-10^4$ Hz. The amplitudes and phase shifts of the modulated short-circuit photocurrent and open-circuit voltage relative to the modulated illumination were measured by FRA through combining the signals from source supply and potentiostat for IMPS and IMVS.

The Nyquist plots of the impedance data of the three DSSCs obtained under the open-circuit condition are shown in Figure 2.12B [27]. Estimation of the electron transport parameters in the photoanode portions of the DSSCs was conducted from the Nyquist plots of the photoanode portions according to the procedure described in Section 2.2.1. The electron density (n) in the conduction band of the anode and the effective diffusion coefficient (D_{eff}) of an electron in the photoanode of the three DSSCs determined by impedance analysis are listed in Table 2.2. It reveals that electron density in the TiO$_2$ NP anode is fivefold larger than that in the ZnO NW−NP composite anode, which results from the wider absorptive range of N719 for light harvesting and the higher quantum yield of electron injection to TiO$_2$ [70] in the N719-sensitized TiO$_2$ NP DSSC. Table 2.2 also shows that the D_{eff} of an electron in the ZnO NW−NP composite anode is lower than that in the ZnO NW anode while the overall efficiency of the NW−NP composite DSSC is superior to that of the ZnO NW one. It indicates that multiple trapping/detrapping events indeed occur within the grain boundaries during electron transport in the ZnO NW−NP composite anode. The efficiency enhancement of the ZnO NW−NP DSSC is ascribed to the enrichment of the light harvesting and the reduction of the electron back reaction on the TCO surface without significant sacrificing electron transport efficiency at the same time by synthesizing dense NPs among the ZnO NW array. Moreover, a threefold enhancement of the D_{eff} of an electron in the ZnO NW−NP composite photoanode is achieved in comparison with the TiO$_2$ NP one, as listed in Table 2.2. It has been demonstrated that D_{eff} increases as more electrons are present since the deep traps are filled and electron trapping/detrapping involves shallower levels [71−73]. A lower electron density appearing in the mercurochrome-sensitized ZnO NW−NP composite compared to that of the N719-sensitized TiO$_2$ NP photoanode, as listed in Table 2.2, suggests that an even higher D_{eff} of an electron in the ZnO NW−NP composite anode should be observed when the electron density is as high as that in the TiO$_2$ NP anode. The superior D_{eff} of an electron in the NW−NP composite photoanode indicates that the ZnO NW array in the composite anode plays an important role in electron transport. The inferior efficiency of the ZnO NW−NP composite cell in comparison with the TiO$_2$ NP one is mainly limited to the issue of dye.

The charge transport/recombination dynamics in the three DSSCs were further investigated using IMPS and IMVS [28]. Light intensity dependence of dynamics of electron transport and recombination in the three DSSCs obtained, respectively, from IMPS and IMVS measurements are shown in Figure 2.13 [28]. It reveals that the electron transport rates of the three cells measured by IMPS are in order of TiO$_2$ NP < ZnO NW−NP composite < ZnO NW array, which is consistent with the order of D_{eff} in the three anodes obtained from EIS measurements. In comparison with ZnO NW anode, the electron transit times decrease by the bottom-up growth of NPs within the interstices of the ZnO NW anode. Nevertheless, the electron transport rates in the ZnO NW−NP composite anode are significantly faster than those in TiO$_2$ NP anode.

The behavior of the light intensity dependence on the transit time of the ZnO NW array anode is dissimilar to the other two photoanodes as shown in Figure 2.13.

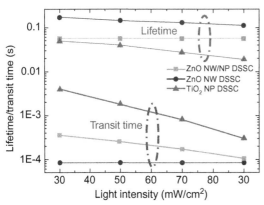

Figure 2.13 Light intensity dependence of lifetime and electron transit time in the mercurochrome-sensitized ZnO NW, mercurochrome-sensitized ZnO NW/NP composite, and N719-sensitized TiO$_2$ NP DSSCs.
Reprinted with permission from Ref. [28]. Copyright 2009, Wiley-VCH Verlag GmbH & Co. KGaA, Weinheim.

The transit time in the vertical ZnO NW array anode is insensitive to the light intensity but the transit times of the others increase as the light intensity is decreased. The constant transit time of the ZnO NW DSSC indicates that very few traps exist in the ZnO NWs [34]. On the other hand, the decrease of the electron transit time as increasing the light intensity [33,34] results from that more photoelectrons are present to fill the deep traps and electron trapping/detrapping involves shallower trap levels only [71−73]. In the case of ZnO NW array/NP composite anode, the decrease of the electron transit time is observed as well when increasing the light intensity, indicating a certain trap existing in the composite anode. Nevertheless, as illustrated in Figure 2.13, the electron transit time in the ZnO NW−NP composite anode is comparable to that in the ZnO NW array anode and is threefold shorter than that in TiO$_2$ NP anode under a high light intensity of 90 mW/cm^2. The fast electron transport in the ZnO NW−NP composite anode is ascribed to the presence of the vertical ZnO NW array and a low trap density in the composite film.

As shown in Figure 2.13, the dependences of the electron lifetime on the light intensity, the electrons in the ZnO NW array anode possess the largest lifetime. Except the ZnO NW−NP composite anode, the electron lifetimes in the other two anodes decrease as the light intensity increases. The faster recombination takes place due to local higher concentrations of injected electron and triiodide ion at higher light intensity [34]. As aforementioned, the efficiency of the ZnO NW−NP composite DSSC will be enhanced by the appearance of an appropriate fraction of the insulating LBZA in the ZnO NP portion of the composite films. The IMVS measurements demonstrate that the ZnO NW−NP composite anode still possesses a constant electron lifetime as increasing the light intensity, confirming that insulating LBZA is able to effectively suppress the recombination of the injected photoelectrons with triiodide ions and oxidized dye molecules. Consequently, the electron in the ZnO NW−NP composite anode possesses a significant larger lifetime compared to TiO$_2$ NP anode at higher light intensities.

ZnO composite anodes composed of the NW array and high-surface-area NPs have been successfully formed on TCO via a wet chemical route for use in DSSCs. A nearly threefold improvement of the efficiency of the ZnO NW DSSC is achieved by

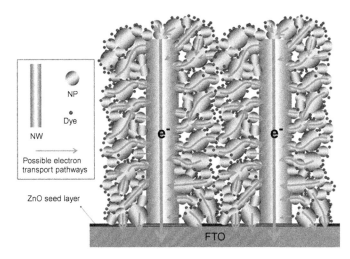

Figure 2.14 Schematic of electron transport path in the ZnO NW/NP composite DSSC.

the CBD of dense NPs within the interstices of the vertical ZnO NW array with a thickness of 5.5 μm. As illustrated in Figure 2.14, the light harvesting efficiency is enriched in the ZnO NW−NP composite anode without significantly sacrificing the electron transport efficiency. Electron transport time in the anode of the ZnO NW−NP composite DSSC is 1 order of magnitude shorter than that in the TiO$_2$ NP DSSC, which indicates that the single-crystalline ZnO NW array plays a crucial role in electron transport in the composite anode. Notably, photosensitizer employed here is mercurochrome which only absorbs photons with the wavelength in the range of 400−600 nm. By using the same wet chemical route to grow ZnO NP within the interstices of ZnO NWs, Chen and Yin [74] demonstrated that an efficiency of 7.14% can be achieved in the N719-sensitized and 25-μm-thick ZnO NW−NP DSSC.

2.5 Light Scattering Layers on 1D Nanostructured ZnO DSSC Anodes

Since there is considerable "free" space between NWs for light passing through, an alternative approach for improving the efficiency of NW DSSCs is to enhance the light harvesting efficiency by reflecting unabsorbed photons back into the NW anode. In the case of NP anode, a diffuse scattering layer composed of large colloids is utilized for this specific purpose of increasing light harvesting efficiency [75−77]. The scattering effect depends on the refractive index, size, and position of the scattering particle. It is worthy to address here that the mechanism of photocurrent improvement in aforementioned DSSCs should be carefully examined since the sensitized scattering layer also directly contributes to photocurrent.

In order to reflect the lights passing through the "free" space between NWs back to the NW electrode portion, the light scattering material in the NW DSSC

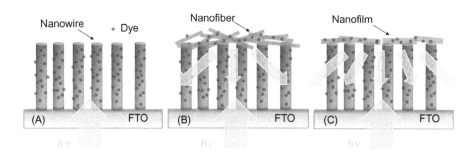

Figure 2.15 Schematics of (A) ZnO NW, (B) nanofiber/ZnO NW, and (C) nanofilm/ZnO NW anodes.
Reprinted with permission from Ref. [56]. Copyright 2008, American Chemical Society.

needs to exhibit a reflection index higher than that of the electrolyte (~ 1.4 for the conventional electrolyte) which occupies the free space between NWs. In addition, a semiconductor material with lower conduction band edge compared to that of ZnO is desired to be the scattering layer for constructing "pure" light scattering layer which enhances the light harvesting efficiency of NW DSSCs by the sole mechanism of reflecting unabsorbed photons back into the NW anode. For aforementioned criteria, SnO_2 that possesses the reflection index of ~ 2.0 [78] and the conduction band edge lower than that of ZnO [79] is the potential candidate for the light scattering material in the ZnO NW DSSCs.

Due to its continuously 1D and horizontal configuration, the electrospun SnO_2/ZnO nanofibers are employed to be the light scattering matter on the ZnO NW anode. Electrospinning of inorganic/polymer composite nanofibers has been demonstrated by mixing inorganic sol−gel precursor with water-soluble polymer solution [80,81]. A continuous gel network within the polymer matrix is produced in the electrospun composite nanofibers. Inorganic nanofibers can be subsequently obtained by calcination of the composite nanofibers at elevated temperatures in air for removing the polymer matrix [80,81]. The SnO_2/ZnO nanofibers, instead of pure SnO_2 nanofibers, are employed to improve the adhesion between the nanofibers and ZnO NWs. Moreover, by utilizing the polar solvent-soluble characteristic of polymer matrix [80,81], a continuous nanofilm on the top of ZnO NW array is obtained via methanol-vapor treatment followed by a high-temperature calcination. The schematics of the nanofiber/ZnO NW and nanofilm/ZnO NW anodes are illustrated in Figure 2.15 [56].

2.5.1 Construction of Nanocrystalline Film on ZnO NW Array via Swelling Electrospun Polyvinylpyrrolidone-Host Nanofibers

Porous nanofibers composed of SnO_2/ZnO nanocrystals can be obtained by the calcination polyvinylpyrrolidone (PVP)-hosted SnO_2/ZnO nanofibers, as shown in Figure 2.16 [56]. To prepare the nanofiber/ZnO NW anode, PVP-hosted SnO_2/ZnO nanofibers were directly electrospun onto the collector or ZnO NW array using a

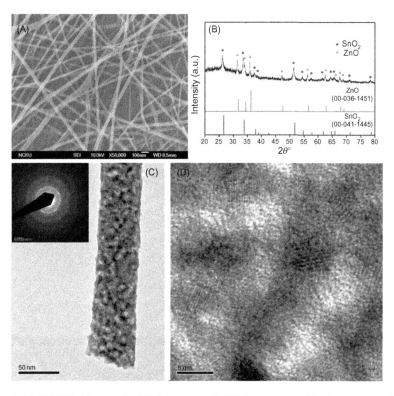

Figure 2.16 (A) SEM image, (B) XRD pattern, (C) TEM image and SEAD pattern, and (D) HRTEM mage of the porous SnO$_2$/ZnO nanofibers.
Reprinted with permission from Ref. [56]. Copyright 2008, American Chemical Society.

methanolic solution of 0.15 g/mL PVP (Mw = 1,300,000), 0.1 M ZnAc$_2$ · 2H$_2$O, 0.05 M SnCl$_4$ · 5H$_2$O, and 1.3 M *tert*-butylamine (TBA). An electrical potential of 15 kV was applied across the syringe and collector with a distance of 9 cm. SnO$_2$/ZnO nanofibers were obtained on the ZnO NW array after the calcination of PVP-hosted nanofibers at 600°C for 1 h, as shown in Figure 2.17A and B [56]. The hydrophilic characteristic of PVP is further employed to modify the feature of the light scattering matter from the PVP-hosted nanofibers on the ZnO NW photoanode. As-electrospun nanofibers on ZnO NW array were treated in a methanol-vapor ambiance at RT for 10 min followed by calcination at 600°C for 1 h. As shown in Figure 2.17C and D, a thin film instead of nanofibers is consequently formed on the top of ZnO NW array. Few particles with a size of ~1 μm also appear on the surface of the film. In addition, severe cracking of the thin film is observed and the tips of the covered NWs are embedded in the film, as shown in Figure 2.17C. Figure 2.17D shows that the thickness of the film is ~100 nm. The as-electrospun nanofibers are reconstructed to a "free-standing" thin film via absorbing methanol vapor. The "free-standing" thin film which is still composed of

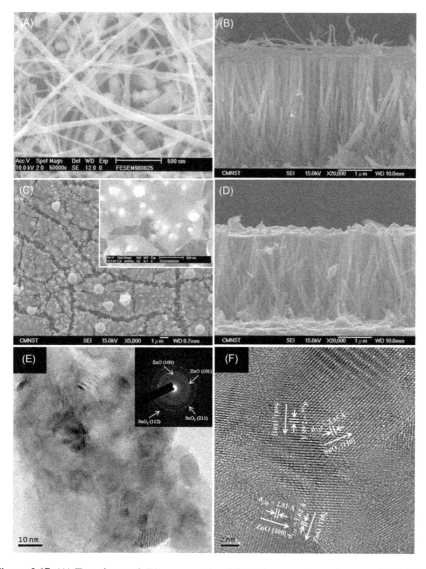

Figure 2.17 (A) Top-view and (B) cross-sectional SEM images of nanofibers on ZnO NW array; (C) top-view and (D) cross-sectional SEM images of the nanofilm on ZnO NW array, respectively. The scale bar in the inset of (C) is 500 nm. (E) TEM image and the corresponding SAED pattern (inset) of the nanofilm. (F) HRTEM image of the nanofilm. Reprinted with permission from Ref. [56]. Copyright 2008, American Chemical Society.

the inorganic sol dispersed in PVP matrix is supported by the tips of ZnO NWs. After calcination at 600°C, PVP is decomposed away and a nanofilm with many cracks is thus formed on the top of the ZnO NW array. Structural characterizations of the nanofilm, as shown in Figure 2.17E and F, reveal that the film is composed of ZnO and SnO_2 nanocrystals with a diameter of ~10 nm.

2.5.2 Photovoltaic Performances of the D149-Sensitized ZnO NW DSSCs with Electrospun Light Scattering Layers

The reflection spectra of the ZnO NW array, nanofiber/ZnO NW array, and nanofilm/ZnO NW array formed on FTO substrates are shown in Figure 2.18A [56]. It reveals that the reflections of the three nanostructures are in order of ZnO NW array < nanofiber/ZnO NW array < nano film/ZnO NW array in the wavelength range of 400–1000 nm, demonstrating the reflective abilities of the nanofilm and nanofibers on the ZnO NW arrays.

D149 dye was used to sensitize the ZnO-based electrodes. Dye adsorption was carried out by immersing the anode in a 1×10^{-5} M acetonitrile/t-butanol (1:1) solution of D149 at 80°C for 1 h. The sensitized electrode and platinized FTO counter electrodes were sandwiched together with 25-μm-thick hot-melt spacers. Liquid electrolyte solution composed of 0.5 M Pr_4NI and 50 mM I_2 in a 1:4 volume ratio of ethylene carbonate and acetonitrile was employed for the D149-sensitized DSSCs.

Figure 2.18B shows the $J-V$ characteristics of the ZnO NW DSSC, nanofiber/ZnO NW, and nanofilm/ZnO NW DSSCs under the AM-1.5 illumination at

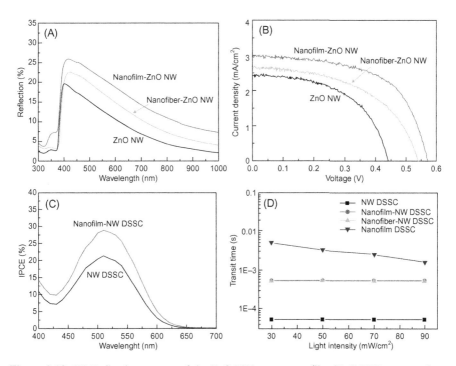

Figure 2.18 (A) Reflection spectra of the ZnO NW array, nanofiber/ZnO NW array, and nanofilm/ZnO NW array on FTO substrates. (B) Current density–voltage characteristics of the ZnO NW, nanofiber/ZnO NW, and nanofilm/ZnO NW DSSCs. (C) IPCE spectra of the nanofilm/ZnO NW and ZnO NW DSSCs. (D) Light intensity dependence of the transit times in the ZnO NW, nanofiber/ZnO NW, nanofilm/ZnO NW, and nanofilm DSSCs.
Reprinted with permission from Ref. [56]. Copyright 2008, American Chemical Society.

Table 2.3 Photovoltaic Properties of ZnO NW, Nanofiber/ZnO NW, and Nanofilm/ZnO
NW DSSCs

DSSCs	J_{sc} (mA/cm^2)	V_{oc} (V)	FF	η (%)
ZnO NW	2.46	0.44	0.53	0.58
Nanofiber/ZnO NW	2.66	0.54	0.51	0.74
Nanofilm/ZnO NW	2.99	0.57	0.59	1.01
SnO$_2$/ZnO nanofilm	1.81	0.64	0.42	0.49

Reprinted with permission from Ref. [56]. Copyright 2008, American Chemical Society.

100 mW/cm^2 [56]. The ZnO NW array is also annealed at 600°C for use in the
DSSC. The lengths of the NWs are ~3 μm for the fabrication of the three DSSCs.
The photovoltaic properties of the three cells are summarized in Table 2.3. 8 and
22% increments of the J_{sc} of ZnO NW DSSC are achieved by forming nanofibers
and nanofilm on the top of ZnO NW anode, respectively, which are ascribed to the
enhancement of the light harvesting efficiency of the photoanode. A more significant enrichment of J_{sc} is achieved in the nanofilm/ZnO NW DSSC as a result that
the coverage of nanofilm is larger than that of nanofiber layer on the ZnO NW
array as shown in Figure 2.17A and C. In addition to the enrichment of J_{sc}, the V_{oc}
and FF of ZnO NW DSSC are also improved by the formation of nanofibers or
nanofilm on the top of ZnO NW anode. Combining the enhancements of J_{sc}, V_{oc},
and FF, a 74% improvement of the efficiency of the ZnO NW is achieved by the
addition of the light scattering nanofilm. Figure 2.18C illustrates the IPCE spectra
as a function of wavelength for the nanofilm/ZnO NW and ZnO NW DSSCs. The
nanofilm/ZnO NW DSSC shows a superior photoelectrical response to ZnO NW
DSSC in the wavelengths ranging from 400 to 650 nm.

There are two possible mechanisms for the improvement of light harvesting efficiency by the nanofibers and nanofilm on the top of ZnO NW anode. One is that
the surface area of anode is increased by the addition of nanofibers or nanofilm,
which results in higher amount of dye adsorption and therefore higher J_{sc}. The other
one is that the nanofibers and nanofilm perform as the light scattering matters on
the top of ZnO NW anodes to enhance light harvesting efficiency. Since the nanofibers and nanofilm are composed of SnO$_2$/ZnO nanocrystals and the conduction
band minimum of SnO$_2$ is lower than that of ZnO [79], the photoelectrons injected
into the nanofibers and nanofilm are unable to efficiently transport to the ZnO NW
anode. Although the SnO$_2$/ZnO nanofilm DSSC demonstrates an apparent J_{sc},
as shown in Table 2.3, we suggest that J_{sc} in the nanofiber/ZnO NW and nanofilm/
ZnO NW DSSCs are mainly attributed to the photoelectrons excited from the dye
molecules adsorbed on the single-crystalline ZnO NW array. That is, the major role
of the nanofibers and nanofilm is the light scattering matter on the top of ZnO NW
anode. Further investigation of the mechanism of photocurrent enhancement was
performed by IMPS.

Light intensity dependence of dynamics of electron transport in the three DSSCs
obtained from IMPS measurements is shown in Figure 2.18D [56]. For comparison,

the electron transport properties in the SnO_2/ZnO nanofilm DSSC (in the absence of the ZnO NW array) were examined by IMPS and are also shown in Figure 2.18. The photovoltaic properties of the cell are listed in Table 2.3 as well. IMPS measurements reveal that similar to the ZnO NW DSSC, the electron transit times in the nanofiber/ZnO NW and nanofilm/ZnO NW DSSCs are insensitive to light intensity. The constant transit times indicate that electrons experience very few trapping/detrapping events as well when diffusing through the nanofiber/ZnO NW and nanofilm/ZnO NW anodes. However, many traps exist within the grain boundaries of light scattering layers since both nanofibers and nanofilm are composed of nanocrystals, as shown in Figures 2.16 and 2.17. The electron transport properties in the SnO_2/ZnO nanofilm DSSC, as shown in Figure 2.18D, reveal the significantly larger transit times than that in nanofilm/ZnO NW anode. Moreover, the decrease of the transit time in the nanofilm anode is observed as increasing the light intensity, confirming the existence of many traps in the nanofilm [33,34,71−73].

According to the IMPS results of the SnO_2/ZnO nanofilm DSSC, an increase of the transit time will be observed in the nanofiber/ZnO NW or nanofilm/ZnO NW DSSCs as decreasing the light intensity if there was significant photocurrent originated from the dye-sensitized nanofibers or nanofilm on the top of ZnO NW arrays. However, the constant electron transit times are obtained in the nanofiber/ZnO NW and nanofilm/ZnO NW DSSCs under various light intensities. In addition, the transit time in the nanofilm/ZnO NW DSSC is significantly shorter than those in the SnO_2/ZnO nanofilm DSSC. The results indicate that the light scattering layers of nanofibers and nanofilm do not contribute to significant photocurrent in the nanofiber/ZnO NW and nanofilm/ZnO NW DSSCs. The apparently longer electron transit times in nanofiber/ZnO NW and nanofilm/ZnO NW DSSCs compared to that in the ZnO NW DSSC are ascribed to reflection phase shifts of the lights reflected/ scattered by the light scattering layer during IMPS measurement [56]. Unlike the conventional light scattering layers which not only reflect light but also directly contribute photocurrent in DSSCs, the SnO_2/ZnO nanofibers and nanofilm play a major role in light scattering in the nanostructured ZnO NW DSSCs. It is therefore concluded that the considerable enhancement of the efficiency of the ZnO NW DSSC is achieved by reflecting unabsorbed photons back into the NW anode using the novel light scattering layer of nanofilm.

2.6 3D ZnO ND/NP Composite Films for Use in DSSCs

With the consideration of the single-crystalline branches also fascinating fast electron transport, branched ZnO NWs have been employed to be the anode of DSSC for enlarging the surface area of ZnO NWs [82]. Hierarchical ZnO structures constructed from rod-like units have been demonstrated via solution routes using renucleation/growth [83,84] and recoating seeds/growth [82] methods. Zhang et al. [83] reported a route of the site-specific sequential nucleation and growth to the branched ZnO micro/nanostructures. The organic structure-directing agents like diamine molecules were added into the precursor solution to synthesize branches on the primary rods. The rotorlike ZnO micro/nanostructures were therefore formed on the substrates. Sounart et al. [84] investigated the renucleation/growth of the branched ZnO micro/nanostructures using TEM. The creation of an initial layer of polycrystalline NP seeds on the surface of primary ZnO rods was observed in the presence of diaminopropane. The misalignment between the primary crystal and the seeds was also demonstrated. In addition, the densities of the primary ZnO rods on substrates shown in these two papers were rather low, implying that the branched ZnO micro/nanostructures are only successfully synthesized in the limiting condition of low-density feature by using the renucleation/growth method. On the other hand, Cheng et al. [82] reported the formation of branched ZnO NWs by dip-coating seed layers on the surface of primary ZnO NWs for the followed branch growth. The density of the primary NWs was about $\sim 10^8 \, cm^{-2}$. The branches were formed on the tops and sidewalls of primary NWs with different radial angles. TEM characterization revealed that the branches originated from the seeds recoated on the surface of the primary ZnO NWs. Nevertheless, the efficiency of the branched ZnO NW DSSC is twice as high as the bare ZnO NW cell with an anode thickness of $\sim 8 \, \mu m$. The effective electron diffusivity in the branched ZnO NW anode is $\sim 30\%$ decreased compared to that in the ZnO NW one.

In addition to branched ZnO NWs, composite anodes composed of the NW array and high-surface-area NPs have been reported to improve the efficiency of ZnO NW DSSCs [27,28,58,59]. As mentioned in Section 2.4.4, a nearly threefold improvement of the efficiency of the ZnO NW DSSC has been achieved by the CBD of dense NPs within the interstices of the vertical ZnO NW array with a thickness of 5.5 μm [27]. The light harvesting efficiency is enriched without significantly sacrificing the electron transport efficiency. Electron transport time in the anode of the ZnO NW/NP composite DSSC is 1 order of magnitude shorter than that in the TiO_2 NP DSSC, which indicates that the single-crystalline ZnO NW array plays a crucial role in electron transport in the composite anode [28]. To further enhance electron transport properties in the ZnO-based photoanodes, ZnO ND arrays which possess the feature of 3D and quasi-single-crystalline network are synthesized using a simple wet chemical route [29]. The ZnO ND array constructs a denser fast electron transport pathway compared to the NW array for electron diffusion through the photoanode. The NPs are further formed within the interstices of the NDs to enlarge the surface area for dye adsorption.

2.6.1 Formation and Structural Characterizations of the ZnO ND Arrays and ZnO ND/NP Composite Films

ZnO ND arrays were synthesized on FTO substrates using a simple wet chemical route. Aligned ZnO NW arrays were first grown on the seeded FTO substrates by CBD as described in Section 2.4.1. The branches of the ZnO NDs were further constructed by another CBD process. The ZnO NW array/FTO substrate was immersed

Figure 2.19 (A) Top-view and (B) cross-sectional SEM images of the ZnO ND array on FTO substrate. High-magnification SEM images of (C) top and (D) bottom portions of the ZnO ND array.
Reproduced from Ref. [29].

in an aqueous solution of 0.057 M ZnAc · 2H$_2$O and 0.5 M NaOH at 100°C for 3.5 h [29]. The limpid solution was obtained by preparing a solution of 0.57 M ZnAc · 2H$_2$O and 5 M NaOH and then further diluting by a factor of 10. The ND array was rinsed by DI water and was annealed at 600°C for 30 min. Base-free CBD, as described in Section 2.4.1, was employed to synthesize ZnO NPs within the interstices of ZnO ND arrays for the formation of ZnO ND/NP composite films.

As shown in Figure 2.19, in the absence of seeds and organic structure-directing agents on the surface of NWs, the branches are directly formed on the ZnO NWs after CBD at 100°C for 3.5 h [29]. The ZnO ND array with high-density 3D frameworks is therefore formed by the simple wet chemical route. The lengths and diameters of the branches are ~200 nm and in the range of 20−50 nm, respectively. On the other hand, the diameters and the lengths of the trunks (primary NWs) are also, respectively, increased to 150−250 nm and ~3.5 μm after the secondary CBD process. Moreover, as shown in Figure 2.19C and D, the SEM images of the top and bottom portions of the ZnO ND array reveal that the parallel branches are formed on the *m*-planes of the trunks.

Figure 2.20A shows a typical cross-sectional TEM image of the ZnO ND array, revealing the formation of high-density 3D frameworks in the ZnO ND array [29]. The microstructure relationship between the trunk and branch examined by

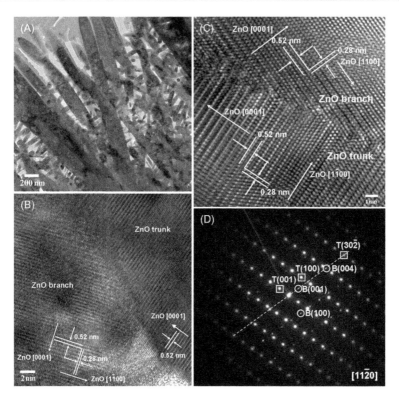

Figure 2.20 (A) Typical cross-sectional TEM image of the ZnO ND array. (B and C) HRTEM images of the trunk and branch. (D) The corresponding SAED pattern in (C). Reproduced from Ref. [29].

HRTEM is shown in Figure 2.20B and C, which was taken with the same zone axis of (11$\bar{2}$0) for both trunk and branch. The HRTEM image in Figure 2.20B reveals that NW sustains the original structure (single-crystal structure with the growth direction of (0001)) after the CBD of branches and the branch is single crystalline with the growth direction along (0001). Figure 2.20B also shows that the angle between (0001) directions of the trunk and branch is ∼110°. The interfacial region of another trunk and branch shown in Figure 2.20C reveals that a rather clear zigzag interface between the trunk and branch is observed although some disorders also appear there. Moreover, the corresponding SAED pattern, shown in Figure 2.20D, indicates that the (001) plane of the branch is parallel to the (30$\bar{2}$) plane of the trunk. Since the angle between the (001) and (30$\bar{2}$) planes is 109.82° in the wurtzite structure of ZnO crystal, it is therefore suggested that the (001) planes of the branch develop from the (30$\bar{2}$) planes of the trunk with the common zone axis of (11$\bar{2}$0), which results in the specific angle of ∼110° between the primary NWs and branches.

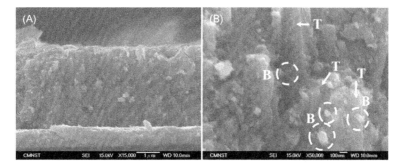

Figure 2.21 (A) Cross-sectional SEM image of the ZnO NP/ND composite film on FTO substrate. (B) High-magnification and cross-sectional SEM image of the ND/NP composite film.
Reproduced from Ref. [29].

Synthesis of NPs within the interstices of the NDs was further conducted using a base-free CBD in order to enlarge surface area for use in DSSC. Figure 2.21 shows that the interstices of the ZnO ND are fully occupied by NPs after a 20.5-h deposition [29]. Figure 2.22A displays a typical cross-sectional TEM image of the ND/NP composite film, revealing that the ND feature is sustained after NP growth. The HRTEM image of the interfacial region of the trunk and branch (region A in Figure 2.22A) is shown in Figure 2.22B and the corresponding SAED patterns of the trunk and branch are, respectively, illustrated in Figure 2.22C and D [29]. They demonstrate that both trunk and branch maintain single-crystalline structure in the ND/NP composite film. Figure 2.22E shows a typical SAED pattern of the ND/NP composite film. In addition to the dot patterns ascribed to the ZnO trunk and branch, discrete ring patterns which are pertaining to the ZnO NPs are obtained in the SAED pattern as well. Figure 2.22F and G displays the bright-field and corresponding dark-field images of region B in Figure 2.22A. The dark-field image reveals that the diameters of NPs in the composite film are in the range of 5−30 nm.

2.6.2 Formation Mechanism of ZnO ND

It was found that the solution for branch growth became opaque after staying at RT for ∼5 min, indicating that the solution is supersaturated at RT. It obeys the phase stability diagram of $ZnO_{(s)}$-H_2O [85]. Moreover, formation of the branches of ZnO NDs is achieved if the ZnO NWs/FTO substrate is initially immersed into the limpid solution before the CBD process. In the case that the ZnO NWs/FTO substrate is immersed into the already opaque solution, only ZnO NWs with slightly enlarged diameters are obtained after the CBD process. It indicates that the development of branches has to start from a supersaturated solution in a metastable state without precipitation. In the absence of seeds and organic structure-directing agents, NaOH in the solution may play a crucial role in the growth of the branches on the surface of NWs.

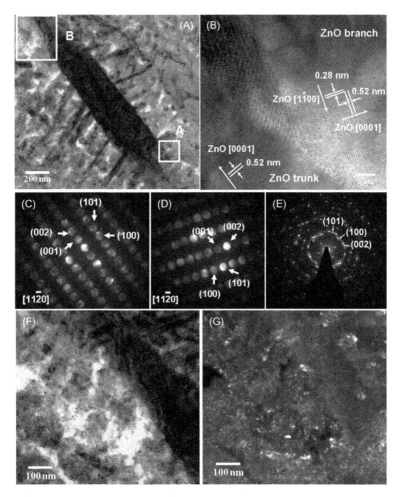

Figure 2.22 (A) Typical cross-sectional TEM image of the NP/ND composite film. (B) HRTEM image of the interfacial region of the trunk and branch (region A in (A)). (C and D) The corresponding SAED patterns of the trunk and branch in (B), respectively. (E) Typical SAED pattern of the NP/ND composite film. (F and G) The bright-field and corresponding dark-field images of region B in (A), respectively.
Reproduced from Ref. [29].

TEM characterizations of the the ZnO NWs after being immersed in the limpid solution at RT for 2 min were further conducted to investigate the development of the branches. As shown in Figure 2.23, compared to the as-prepared ZnO NW possessing a smooth surface, it is obvious that the etch pits are formed on the surface of ZnO NW after being immersed in the limpid solution which is in a metastable state with a high concentration of NaOH [29]. Due to the high surface

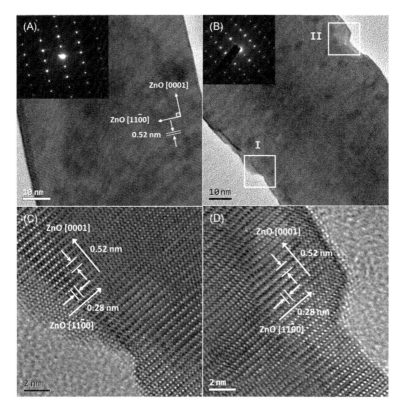

Figure 2.23 (A) HRTEM image and the corresponding SAED of the as-prepared ZnO NW. (B) HRTEM image and the corresponding SAED of the ZnO NW after immersed in the limpid solution of 0.057 M ZnAc · 2H₂O and 0.5 M NaOH at RT for 2 min. (C and D) HRTEM images of regions I and II in (B), respectively.
Reproduced from Ref. [29].

energy of the etch pits, the branches may subsequently grow from the etch pits of the trunk during the CBD process. Surface reconstruction of the exposed surfaces of the etch pits maybe needed for developing (001) planes of the branches from them, leading to the zigzag interface between the trunk and branch as shown in Figure 2.20C.

2.6.3   Photovoltaic Performances and Charge Transport Properties of the D149-Sensitized ZnO ND and ZnO ND/NP Composite DSSCs

D149-sensitized ZnO ND arrays and ZnO ND/NP composite films were employed as photoanodes to fabricate DSSCs. The photovoltaic performances and electron transport properties of the two DSSC were compared to those of the D149-sensitized ZnO NW and N719-sensitized TiO₂ NP DSSCs. The NW and ND arrays

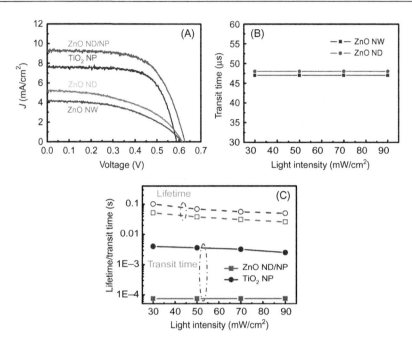

Figure 2.24 (A) Current density−voltage characteristics of the ZnO NW, ZnO ND, ZnO ND/NP, and TiO$_2$ NP DSSCs. (B) Light intensity dependence of the transit times in the ZnO NW and ZnO ND DSSCs. (C) Light intensity dependence of the transit times and lifetimes in the ZnO ND/NP and TiO$_2$ NP DSSCs.
Reproduced from Ref. [29].

were annealed at 600°C for 30 min before dye soaking process. Dye adsorption was carried out by immersing the anode in a 5×10^{-4} M acetonitrile/t-butanol (1:1) solution of D149 at 80°C for 1 h. The DSSCs were assembled using the same procedure described in the previous section.

The current−voltage characteristics of the four DSSCs are shown in Figure 2.24A and the photovoltaic parameters of the cells are summarized in Table 2.4 [29]. With an anode thickness of 3.5 μm, the J_{sc} of the ZnO ND DSSC is enhanced compared to that of the ZnO NW cell, which is ascribed to the higher surface area of ND anode. A higher efficiency of the ZnO ND DSSC is therefore obtained. Moreover, dynamics of charge transport in the two DSSCs, as shown in Figure 2.24B, indicate that the electron transit time in the ND anode is almost identical to that in the NW cell. In addition, the same as the NW anode, the transit times in the ZnO ND anode are insensitive to the light intensity, indicating that very few traps exist in the NDs as well [33,34,71−73]. The slightly higher constant electron transit time in the ND anode is ascribed to its longer electron transport pathway compared to NW anode. The optical modulation measurements confirm that quasi-single-crystalline 3D frameworks are successfully constructed in the ZnO ND electrode for fast electron transport.

Table 2.4 Photovoltaic Properties of D149-Sensitized (I) ZnO NW, (II) ZnO ND DSSCs, (III) ZnO ND/NP Composite, and (IV) N719-Sensitized TiO_2 NP DSSCs

DSSCs	V_{oc} (V)	J_{sc} (mA/cm^2)	FF	η (%)
(I)	0.61	4.18	0.48	1.24
(II)	0.62	5.24	0.48	1.55
(III)	0.63	9.22	0.65	3.74
(IV)	0.59	7.66	0.69	3.09

Reproduced from Ref. [29].

When the NPs were grown within the interstices of ZnO ND array, as shown in Figure 2.24A and Table 2.4, the D149-sensitized ZnO ND/NP composite DSSC has an efficiency of 3.74%, which is higher than that of TiO_2 NP cell. J_{sc} and V_{oc} of the ZnO ND/NP composite DSSC are superior to those of TiO_2 NP cell although the FF of ZnO composite DSSC is worse than that of the TiO_2 NP cell. Since the absorption range of the D149 dye is narrower than that of N719 dye, the superior photovoltaic properties of the D149-sensitized ZnO ND/NP composite DSSC may be attributed to the sufficient surface area for dye adsorption and fast electron transport in the 3D single-crystalline network of ZnO NDs.

Dynamics of charge transport and recombination in the D149-sensitized ZnO ND/NP composite and N719-sensitized TiO_2 NP cell DSSCs are shown in Figure 2.24C. IMVS measurements show that electron lifetimes in the ZnO ND/NP composite anode are shorter than those in the TiO_2 NP one. Unlike N719 dye for the TiO_2 anodes, D149 dye is not specially designed for ZnO anodes. In addition, there is no coadsorber, such as tertbutylpyridine, in the electrolyte of D149-sensitized ZnO ND/NP composite DSSC. Therefore, the worse FF and electron lifetimes in the ZnO ND/NP composite DSSC may be mainly attributed to the poor coverage of D149 dye adsorbed on ZnO ND/NP composite anode. On the other hand, as shown in Figure 2.24C, electron transport rate in the ZnO ND/NP composite anode is 30-fold enhanced compared to that in the TiO_2 NP anode. The transit time in the TiO_2 NP anode decreases significantly as the light intensity is increased, which is attributed to the existence of trap defects in the anode [33,34,71–73]. Surprisingly, however, the dynamics of electron transport in the ZnO ND/NP composite DSSC are insensitive to light intensity. The ZnO ND/NP composite anode behaves like the single-crystalline ZnO NW anode in the electron transport aspect although there are NPs within the composite anode as shown in Figure 2.21. The IMPS results indicate that trap density in the ZnO ND/NP composite anode is significantly lower than that in TiO_2 NP anode. Moreover, all traps in the ZnO ND/NP composite anode can be filled even at the lowest light intensity during IMPS measurement. The high-density ND array in the ND/NP composite film provides a 3D and quasi-single-crystalline framework for fast electron transport, as illustrated in Figure 2.25 [29]. In comparison with multiple trapping/detrapping events significantly occurring in the pure TiO_2 NP anode, photoelectrons injected into ZnO NPs

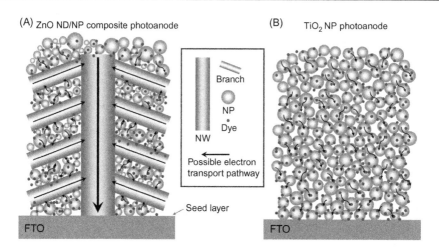

Figure 2.25 Proposed electron transport mechanisms in (A) ZnO NP/ND and (B) TiO$_2$ NP photoanodes.
Reproduced from Ref. [29].

within the interstices of the ZnO ND array can easily diffuse through much fewer traps to the fast electron transport pathway of the high-density ND and subsequently to the FTO electrode. Superior electron transport property in the ZnO ND/NP composite DSSC compared to that in the TiO$_2$ NP cell is mainly due to the 3D quasi-single-crystalline frameworks of ZnO NDs in the composite anode.

2.7 Room-Temperature Fast Construction of ZnO Nanoarchitectures on 1D Array Templates for DSSCs

When working on the formation mechanism of ZnO ND demonstrated in Section 2.6.2, it was found that a novel hierarchical ZnO nanostructure, so called nanocactus (NC) array, can be constructed on the ZnO NW array at RT by using the same chemical solution for the growth of ZnO branches [55]. With the development of the facile RTCBD route, ZnO nanoarchitectures with various morphologies can be constructed on 1D nanostructural templates simply by using this process. In addition to ZnO NC arrays, ZnO NS frameworks [86] and NC−NS arrays [57] are also successfully constructed on ZnO nanoneedle and NW arrays via this route. They have been used as the electron transporter/collector in dye-sensitized and hybrid polymer solar cells. Significantly enhancements of the photovoltaic performances are attained through the modifications of the primary ZnO nanostructures simply by RTCBD. The formation of the ZnO NC array and its application to DSSCs is described in this section.

2.7.1 Formation and Structural Characterizations of the ZnO NC Arrays

For construction of the hierarchical nanostructures on ZnO NWs, the ZnO NW array/FTO substrate was immersed in a stirred aqueous solution of 0.057 M ZnAc $\cdot$ 2H$_2$O and 0.5 M NaOH at RT. As mentioned in Section 2.6.1, the limpid solution was obtained by preparing a solution of 0.57 M ZnAc $\cdot$ 2H$_2$O and 5 M NaOH and then further diluting by a factor of 10. Figure 2.26 shows that the spines are directly constructed on the ZnO NWs to form NC array at RT for 5 min [55]. The diameters of the NWs are also slightly increased after the RTCBD. That is, the hierarchical nanostructures of shells and spines on the ZnO NWs are successfully synthesized at RT without any assistance of seeds and organic structure-directing agents. Moreover, the diameters of the NWs and the dimensions of the spines are increased with the period of RTCBD.

The limpid solution for the RTCBD, which is obtained by preparing a solution of 0.57 M ZnAc $\cdot$ 2H$_2$O and 5 M NaOH and then further diluting by a factor of 10, will become opaque after staying at RT for 5 min. The limpid solution is supersaturated at RT [29,55] and possesses a high concentration of NaOH. It is worth addressing that formation of the hierarchical nanostructures on ZnO NWs is attained if the ZnO NWs/FTO substrate is initially immersed into the limpid solution for the RTCBD process. As long as the ZnO NWs/FTO substrate is immersed into the already opaque solution, the feature of ZnO NCs is not achievable.

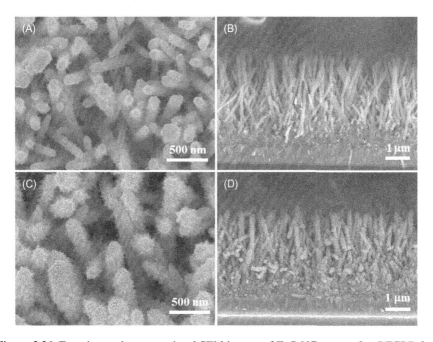

Figure 2.26 Top-view and cross-sectional SEM images of ZnO NC arrays after RTCBD for 5 min (A and B) and for 20 min (C and D).
Reproduced from Ref. [55].

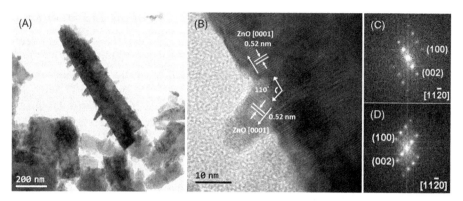

Figure 2.27 (A) Typical TEM image of the ZnO NCs scratched from ZnO NC-20. (B) HRTEM image of the NW and spine of the ZnO NC. (C and D) SAED patterns of the NW and spine portions in (B), respectively, taken from FFT.
Reproduced from Ref. [55].

As shown in the TEM images of Figure 2.23 in Section 2.6.2, the single-crystalline structure is sustained but many etch pits are formed on the surface of ZnO NW after staying in the limpid solution for 2 min. Formation of the etch pits is ascribed to the high concentration of NaOH in the limpid solution. Typical TEM image of the scratched ZnO NCs, which is shown in Figure 2.27A, reveals that the parallel spines are formed on the NWs [55]. Figure 2.27B shows a HRTEM image of the NW and spine, which were taken with the same zone axis of $(11\bar{2}0)$ for both NW and spine. It evidently reveals that the spine is developed from an etch pit on the NW. The spine is single crystalline and grows along (0001) as well. Combining with the SAED patterns of the NW and spine portions taken from FFT, as respectively shown in Figure 2.27C and D, it is concluded that the primary NW sustains the original structure and the shell of the NC is epitaxially grown on surface of primary NW after the RTCBD. With the same zone axis of $(11\bar{2}0)$, the angle between (0001) directions of the NW and spine is $\sim 110°$, which is the same as that of NW and branch in ZnO ND described in the previous section. Since the angle between the (001) and $(30\bar{2})$ planes is 109.82° in the wurtzite structure of ZnO crystal, it is therefore suggested that the (001) planes of the spine develop from the $(30\bar{2})$ planes of the etch pit on NW with the common zone axis of $(11\bar{2}0)$ [29,55]. In addition, unlike the m-plane terminated NW with a hexagonal faceted feature and a constant diameter through whole NW, as shown in Figure 2.27B, the needle-like feature of spines and the rough surface of NW indicate that the surfaces of NCs are semipolar-plane terminated.

2.7.2 Photovoltaic Performances and Charge Transport Properties of the D149-Sensitized ZnO NC DSSCs

The as-prepared ZnO NC arrays were employed as anodes to fabricate DSSCs. For simplicity, the ZnO NC arrays formed from the 3.3-μm-thick NW arrays through

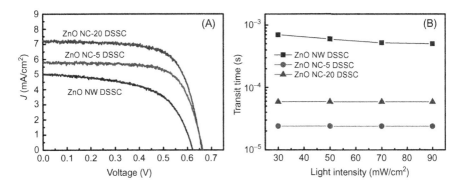

Figure 2.28 (A) Current density−voltage characteristics of the ZnO NW, ZnO NC-5, and ZnO NC-20 DSSCs. (B) Dynamics of charge transport in the three DSSCs. Reproduced from Ref. [55].

Table 2.5 Photovoltaic Properties of D149-Sensitized ZnO NW and ZnO NC DSSCs with an Anode Thickness of 3.3 μm

DSSCs	V_{oc} (V)	J_{sc} (mA/cm^2)	FF	η (%)	The Amounts of Dye Adsorption (nmol/cm^2)	Shunt Resistance (Ωcm^2)
ZnO NW	0.62	5.07	0.58	1.82	8	∼500
ZnO NC-5	0.66	5.80	0.71	2.73	–	∼2500
ZnO NC-20	0.66	7.20	0.70	3.32	12	∼2500

Reproduced from Ref. [55].

the RTCBD for 5 min (Figure 2.26A and B) and 20 min (Figure 2.26C and D) are named as ZnO NC-5 and ZnO NC-20 anodes, respectively. The photovoltaic and electron transport properties of the ZnO NC DSSCs were compared to those of ZnO NW DSSC. D149 dye is employed to be the sensitizer for the ZnO-based cells here. It should be addressed that, unlike the procedure described in the previous sections, the as-prepared NW and NC arrays were not annealed at a high temperature before dye soaking process.

The current density−voltage characteristics of the ZnO NW and ZnO NC DSSCs with an anode thickness of 3.3 μm are shown in Figure 2.28A [55]. The photovoltaic properties of these cells are listed in Table 2.5. It shows that the efficiencies of the ZnO NC DSSCs are significantly superior to that of the ZnO NW DSSC. The efficiencies of the ZnO NC DSSCs increase with the period of RTCBD for hierarchical nanostructure growth and a DSSC efficiency of 3.32% is obtained using the ZnO NC-20 anode. Nearly twofold enhancement of the efficiency of the ZnO NW DSSC is achieved simply by the formation of the hierarchical nanostructures at RT for 20 min. The J_{sc}, V_{oc}, and FF of the ZnO NC DSSCs are all superior to those of the ZnO NW DSSC. The amounts of dye adsorption on the ZnO NW and ZnO NC-20 photoanodes are $8 \, nmol/cm^2$ and $12 \, nmol/cm^2$, respectively, resulting in that a $\sim 40\%$ enrichment of the J_{sc} is achieved in the ZnO NC-20 DCCS compared to that of ZnO NW cell. In addition, the FFs of the D149-sensitized ZnO NC DSSCs are ~ 0.7 which is significantly larger than that of the ZnO NW DSSC.

By analyzing the $J-V$ characteristics shown in Figure 2.28A, the shunt resistances of ZnO NC and ZnO NW DSSCs are ~ 2500 and $\sim 500 \, \Omega cm^2$, respectively. Considerable improvement of the shunt resistance of NW DSSC is achieved by constructing the hierarchical nanostructures on the ZnO NW anode. The semipolar surface of ZnO NCs may be more fitting for D149 dye and possesses higher dye adsorption coverage than the m-plane surface of ZnO NWs does. The high dye adsorption coverage on the ZnO NC anode results in not only the increase of injected photoelectrons in the ZnO NC anode but also the diminution of the back reaction of photoelectron in the ZnO NC anode with I_3^- in the electrolyte in the absence of coadsorber. Consequently, FFs of the ZnO NC DSSCs are superior to that of the ZnO NW DSSC. V_{oc} is fundamentally corresponding to the difference of the Fermi level of anode and the redox potential of electrolyte [87]. The higher concentration of photoelectron on the conduction band, that is, larger photocurrent as shown in Figure 2.28A, elevates the quasi-Fermi level of ZnO NC anode which results in a larger V_{oc} in the ZnO NC DSSCs. Moreover, the increase of shunt resistance in the ZnO NC DSSCs also enhances the V_{oc} [88] compared to the ZnO NW DSSC.

It should be addressed here again that the ZnO NW and ZnO NC DSSCs were fabricated using as-prepared NW and NC anodes which were not annealed at high temperatures. Dynamics of the electron transport in the three cells were investigated using IMPS. As shown in Figure 2.28B, the dependence of dynamics of the electron transport on light intensity is observed in the ZnO NW DSSC, indicating that traps exist in the unannealed NW anode. The transit time in the ZnO NW anode decreases as the light intensity is increased. It is attributed to that more photoelectrons are generated at higher light intensity to fill the deep traps as a result that electron trapping/detrapping involves shallower trap levels only [33,34,71−73]. Surprisingly, however, the electron transit times in the ZnO NC anodes are significantly shorter than those in the ZnO NW anode. Moreover, the dynamics of electron transport in the ZnO NC DSSC are insensitive to light intensity. The interesting results suggest that the shells and spines of the ZnO NCs possess superior electron transport properties to the primary NWs. Fewer traps exist in the RT-synthesized hierarchical nanostructures on the NWs. As a result, in the ZnO NC anode, the electrons diffuse to TCO via the electron transport pathway of the hierarchical nanostructures instead of the primary NWs. On the other hand, the constant electron transit time increases with the period of RTCBD, which is pertaining to the elongated electron transport pathway of the longer spines.

A thick ZnO NC array was prepared for further increasing the efficiency of the ZnO NC DSSC. $J-V$ characteristic of the D149-sensitized ZnO NC DSSC with an anode thickness of $10 \, \mu m$ is shown in Figure 2.29. J_{sc}, V_{oc}, and FF of the ZnO NC

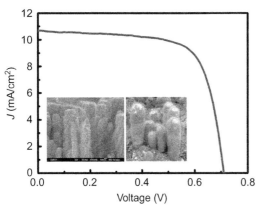

Figure 2.29 Current density—voltage characteristic of the D149-sensitized ZnO NC DSSC with an anode thickness of 10 μm. The insets are a SEM image of the ZnO NC array and a photo of cactuses.

DSSC are 10.69 mA/cm^2, 0.71 V, and 0.71, respectively. An efficiency of 5.37% is achieved simply by the formation of the RT hierarchical nanostructures on the ZnO NW array.

2.8 Concluding Remarks

As demonstrated in this chapter, the formations of hierarchical nanostructures on the ZnO NW arrays are able to improve the light harvest efficiencies and therefore the DSSC efficiencies by the strategies of increasing the surface area of photoanode without sacrificing fast electron transport of ZnO NW and adding a light scattering layer for reflecting unabsorbed photons back into the NW anode. Especially, the NC architectures, which possess the characteristics of large surface area and fast electron transport rate, have been successfully synthesized on the ZnO NW array template at RT for a short period. A notable efficiency of 5.4% is therefore attained in the D149-sensitized ZnO NC DSSC. However, compared to the typical Ru dye-sensitized TiO_2 NP DSSC, the limitation of light harvesting wavelength range is a significant issue for further enhancement of the efficiency of the D149-sensitized ZnO NC DSSC. We suggest that dyes specially designed for ZnO anode with wider absorption range and larger electron injection efficiency are needed to further increase the efficiency of ZnO NC DSSC.

There has been a great deal of interests in flexible and light-weight DSSCs since they are potentially suitable to be the renewable and mobile power sources for portable electronic devices. Fabrication of DSSCs on TCO-coated plastic substrates is therefore an important undergoing research topic. However, the formation of high efficient TiO_2 NP DSSCs on plastic substrates is a significant challenge since most plastic substrates are only able to be sustained at a limited temperature of ~ 150°C whereas the thermal treatment at a higher temperature is usually necessary for TiO_2 NP film anodes. On the other hand, ZnO possesses high electron mobility, low crystallization temperature, and anisotropic growth behavior, which make it being a very promising candidate for the flexible DSSC anode. With the low-temperature wet chemical route ($<100°C$), the hierarchical ZnO NC nanostructures demonstrated in this chapter are potentially suitable for use in flexible DSSC. Moreover, complex nanoarchitectural ZnO anodes, for example, the configuration composed of the ZnO NC arrays and ZnO NPs, may able to be chemically integrated on plastic substrates using the facile RTCBD method. By taking the advantages of low crystallization temperature and anisotropic growth behavior, ZnO-based flexible DSSCs are anticipated to exhibit encouraging performances once the specially designed dyes for ZnO are realized.

Apart from the application to DSSCs, the hierarchically nanostructured ZnO arrays described in this chapter, which possesses good electron transport properties and tunable surface area, may also perform as efficient electrodes for potential use in other applications, such as electrochromic windows, electroluminescent devices, and chemical and biological sensors.

References

[1] Centi G, Perathoner S. The role of nanostructure in improving the performance of electrodes for energy storage and conversion. Eur J Inorg Chem 2009;26:3851−78.

[2] Kim SW, Seo DH, Ma X, Ceder G, Kang K. Electrode materials for rechargeable sodium-ion batteries: potential alternatives to current lithium-ion batteries. Adv Energy Mater 2012;2(7):710−21.

[3] Su L, Jing Y, Zhou Z. Li ion battery materials with core−shell nanostructures. Nanoscale 2011;3(10):3967−83.

[4] Manthiram A, Vadivel Murugan A, Sarkar A, Muraliganth T. Nanostructured electrode materials for electrochemical energy storage and conversion. Energy Environ Sci 2008;1(6):621−38.

[5] Chen HY, Kuang DB, Su CY. Hierarchically micro/nanostructured photoanode materials for dye-sensitized solar cells. J Mater Chem 2012;22(31):15475−89.

[6] Xu F, Sun L. Solution-derived ZnO nanostructures for photoanodes of dye-sensitized solar cells. Energy Environ Sci 2011;4(3):818−41.

[7] Zhang Q, Cao G. Nanostructured photoelectrodes for dye-sensitized solar cells. Nano Today 2011;6(1):91−109.

[8] Zheng H, Ou JZ, Strano MS, Kaner RB, Mitchell A, Kalantar-zadeh K. Nanostructured tungsten oxide-properties, synthesis, and applications. Adv Funct Mater 2011;21 (12):2175−96.

[9] Granqvist CG. Electrochromic tungsten oxide films: review of progress 1993−1998. Sol Energy Mater Sol Cells 2000;60(3):201−62.

[10] Li Z, Luo W, Zhang M, Feng J, Zou Z. Photoelectrochemical cells for solar hydrogen production: current state of promising photoelectrodes, methods to improve their properties, and outlook. Energy Environ Sci 2013;6(2):347−70.

[11] Osterloh FE. Inorganic nanostructures for photoelectrochemical and photocatalytic water splitting. Chem Soc Rev 2013;42(6):2294−320.

[12] Sivula K, Le Formal F, Gratzel M. Solar water splitting: progress using hematite (alpha-Fe_2O_3) photoelectrodes. ChemSusChem 2011;4(4):432−49.

[13] Oregan B, Gratzel M. A low-cost, high-efficiency solar-cell based on dye-sensitized colloidal TiO_2 films. Nature 1991;353(6346):737−40.

[14] Yella A, Lee HW, Tsao HN, Yi C, Chandiran AK, Nazeeruddin MK, et al. Porphyrin-sensitized solar cells with cobalt (II/III)-based redox electrolyte exceed 12 percent efficiency. Science 2011;334(6056):629−34.

[15] Fisher AC, Peter LM, Ponomarev EA, Walker AB, Wijayantha KGU. Intensity dependence of the back reaction and transport of electrons in dye-sensitized nanacrystalline TiO_2 solar cells. J Phys Chem B 2000;104(5):949−58.

[16] Kopidakis N, Schiff EA, Park NG, van de Lagemaat J, Frank AJ. Ambipolar diffusion of photocarriers in electrolyte-filled, nanoporous TiO_2. J Phys Chem B 2000;104 (16):3930−6.

[17] van de Lagemaat J, Frank AJ. Nonthermalized electron transport in dye-sensitized nanocrystalline TiO_2 films: transient photocurrent and random-walk modeling studies. J Phys Chem B 2001;105(45):11194−205.

[18] Law M, Greene LE, Johnson JC, Saykally R, Yang P. Nanowire dye-sensitized solar cells. Nat Mater 2005;4(6):455−9.

[19] Liu B, Aydil ES. Growth of oriented single-crystalline rutile TiO_2 nanorods on transparent conducting substrates for dye-sensitized solar cells. J Am Chem Soc 2009;131 (11):3985−90.

[20] Feng X, Shankar K, Varghese OK, Paulose M, Latempa TJ, Grimes CA. Vertically aligned single crystal TiO_2 nanowire arrays grown directly on transparent conducting oxide coated glass: synthesis details and applications. Nano Lett 2008;8(11):3781−6.

[21] Liao WP, Wu JJ. Wet chemical route to hierarchical TiO_2 nanodendrite/nanoparticle composite anodes for dye-sensitized solar cells. J Mater Chem 2011;21(25):9255−62.

[22] Enache-Pommer E, Liu B, Aydil ES. Electron transport and recombination in dye-sensitized solar cells made from single-crystal rutile TiO_2 nanowires. Phys Chem Chem Phys 2009;11(42):9648−52.

[23] Schlichthorl G, Park NG, Frank AJ. Evaluation of the charge-collection efficiency of dye-sensitized nanocrystalline TiO_2 solar cells. J Phys Chem B 1999;103(5):782−91.

[24] Fabregat-Santiago F, Bisquert J, Garcia-Belmonte G, Boschloo G, Hagfeldt A. Influence of electrolyte in transport and recombination in dye-sensitized solar cells studied by impedance spectroscopy. Sol Energy Mater Sol Cells 2005;87(1−4):117−31.

[25] Fabregat-Santiago F, Bisquert J, Palomares E, Otero L, Kuang D, Zakeeruddin SM, et al. Correlation between photovoltaic performance and impedance spectroscopy of dye-sensitized solar cells based on ionic liquids. J Phys Chem C 2007;111(17):6550−60.

[26] Jiang X, Chen HY, Galvan G, Yoshida M, Lahann J. Vapor-based initiator coatings for atom transfer radical polymerization. Adv Funct Mater 2008;18(1):27−35.

[27] Ku CH, Wu JJ. Electron transport properties in ZnO nanowire array/nanoparticle composite dye-sensitized solar cells. Appl Phys Lett 2007;91(9):093117.

[28] Wong DK, Ku CH, Chen YR, Chen GR, Wu JJ. Enhancing electron collection efficiency and effective diffusion length in dye-sensitized solar cells. Chemphyschem 2009;10(15):2698−702.

[29] Wu CT, Liao WP, Wu JJ. Three-dimensional ZnO nanodendrite/nanoparticle composite solar cells. J Mater Chem 2011;21(9):2871−6.

[30] Bisquert J. Theory of the impedance of electron diffusion and recombination in a thin layer. J Phys Chem B 2002;106(2):325−33.

[31] Adachi M, Sakamoto M, Jiu J, Ogata Y, Isoda S. Determination of parameters of electron transport in dye-sensitized solar cells using electrochemical impedance spectroscopy. J Phys Chem B 2006;110(28):13872−80.

[32] Kern R, Sastrawan R, Ferber J, Stangl R, Luther J. Modeling and interpretation of electrical impedance spectra of dye solar cells operated under open-circuit conditions. Electrochim Acta 2002;47(26):4213−25.

[33] Oekermann T, Yoshida T, Minoura H, Wijayantha KGU, Peter LM. Electron transport and back reaction in electrochemically self-assembled nanoporous ZnO/dye hybrid films. J Phys Chem B 2004;108(24):8364−70.

[34] Martinson AB, McGarrah JE, Parpia MO, Hupp JT. Dynamics of charge transport and recombination in ZnO nanorod array dye-sensitized solar cells. Phys Chem Chem Phys 2006;8(40):4655−9.

[35] Pearton SJ, Norton DP, Ip K, Heo YW, Steiner T. Recent advances in processing of ZnO. J Vac Sci Technol B 2004;22(3):932−48.

[36] Özgür U, Alivov YI, Liu C, Teke A, Reshchikov MA, Doğan S, et al. A comprehensive review of ZnO materials and devices. J Appl Phys 2005;98(4):041301.

[37] Wang ZL. Zinc oxide nanostructures: growth, properties and applications. J Phys-Condens Mater 2004;16(25):R829−58.

[38] Pearton S. Recent progress in processing and properties of ZnO. J Mater Sci 2005;50 (3):293−340.

[39] Könenkamp R, Word RC, Schlegel C. Vertical nanowire light-emitting diode. Appl Phys Lett 2004;85(24):6004.

[40] Huang MH, Mao S, Feick H, Yan H, Wu Y, Kind H, et al. Room-temperature ultraviolet nanowire nanolasers. Science 2001;292(5523):1897−9.

[41] Li QH, Wan Q, Liang YX, Wang TH. Electronic transport through individual ZnO nanowires. Appl Phys Lett 2004;84(22):4556.

[42] Goldberger J, Sirbuly DJ, Law M, Yang P. ZnO nanowire transistors. J Phys Chem B 2005;109(1):9−14.

[43] Baxter JB, Aydil ES. Nanowire-based dye-sensitized solar cells. Appl Phys Lett 2005;86(5):053114.

[44] Yang PD, Yan HQ, Mao S, Russo R, Johnson J, Saykally R, et al. Controlled growth of ZnO nanowires and their optical properties. Adv Funct Mater 2002;12(5):323−31.

[45] Pan ZW, Dai ZR, Wang ZL. Nanobelts of semiconducting oxides. Science 2001;291 (5510):1947−9.

[46] Wu JJ, Liu SC. Low-temperature growth of well-aligned ZnO nanorods by chemical vapor deposition. Adv Mater 2002;14(3):215−8.

[47] Govender K, Boyle DS, O'Brien P, Binks D, West D, Coleman D. Room-temperature lasing observed from ZnO nanocolumns grown by aqueous solution deposition. Adv Mater 2002;14(17):1221−4.

[48] Choy JH, Jang ES, Won JH, Chung JH, Jang DJ, Kim YW. Soft solution route to directionally grown ZnO nanorod arrays on Si wafer; room-temperature ultraviolet laser. Adv Mater 2003;15(22):1911−4.

[49] Zhang Q, Dandeneau CS, Zhou X, Cao G. ZnO nanostructures for dye-sensitized solar cells. Adv Mater 2009;21(41):4087−108.

[50] Memarian N, Concina I, Braga A, Rozati SM, Vomiero A, Sberveglieri G. Hierarchically assembled ZnO nanocrystallites for high-efficiency dye-sensitized solar cells. Angew Chem Int Ed 2011;50(51):12321−5.

[51] Anta JA, Guillén E, Tena-Zaera R. ZnO-based dye-sensitized solar cells. J Phys Chem C 2012;116(21):11413−25.

[52] Wu JJ, Chen GR, Yang HH, Ku CH, Lai JY. Effects of dye adsorption on the electron transport properties in ZnO-nanowire dye-sensitized solar cells. Appl Phys Lett 2007;90(21):213109.

[53] González-Moreno R, Cook PL, Zegkinoglou I, Liu X, Johnson PS, Yang W, et al. Attachment of protoporphyrin dyes to nanostructured ZnO surfaces: characterization by near edge X-ray absorption fine structure spectroscopy. J Phys Chem C 2011;115 (37):18195−201.

[54] Lin CY, Lai YH, Chen HW, Chen JG, Kung CW, Vittal R, et al. Highly efficient dye-sensitized solar cell with a ZnO nanosheet-based photoanode. Energy Environ Sci 2011;4(9):3448−55.

[55] Wu CT, Wu JJ. Room-temperature synthesis of hierarchical nanostructures on ZnO nanowire anodes for dye-sensitized solar cells. J Mater Chem 2011;21(35):13605−10.

[56] Wu JJ, Chen YR, Liao WP, Wu CT, Chen CY. Construction of nanocrystalline film on nanowire array via swelling electrospun polyvinylpyrrolidone-hosted nanofibers for use in dye-sensitized solar cells. ACS Nano 2010;4(10):5679−84.

[57] Jiang WT, Wu CT, Sung YH, Wu JJ. Room-temperature fast construction of outper-formed ZnO nanoarchitectures on nanowire-array templates for dye-sensitized solar cells. ACS Appl Mater Interfaces 2013;5(3):911−7.

[58] Ku CH, Yang HH, Chen GR, Wu JJ. Wet-chemical route to ZnO nanowire-layered basic zinc acetate/ZnO nanoparticle composite film. Cryst Growth Des 2008;8 (1):283−90.

[59] Ku CH, Wu JJ. Chemical bath deposition of ZnO nanowire−nanoparticle composite electrodes for use in dye-sensitized solar cells. Nanotechnology 2007;18 (50):505706.

[60] Tan B, Wu Y. Dye-sensitized solar cells based on anatase TiO_2 nanoparticle/nanowire composites. J Phys Chem B 2006;110(32):15932−8.

[61] Wu JJ, Chen GR, Lu CC, Wu WT, Chen JS. Performance and electron transport prop-erties of TiO_2 nanocomposite dye-sensitized solar cells. Nanotechnology 2008;19 (10):105702.

[62] Hosono E, Fujihara S, Kimura T, Imai H. Growth of layered basic zinc acetate in methanolic solutions and its pyrolytic transformation into porous zinc oxide films. J Colloid Interface Sci 2004;272(2):391−8.

[63] Hosono E, Fujihara S, Kimura T, Imai H. Non-basic solution routes to prepare ZnO nanoparticles. J Sol−Gel Sci Technol 2004;29(2):71−9.

[64] Morioka H, Tagaya H, Karasu M, Kadokawa J, Chiba K. Effects of zinc on the new preparation method of hydroxy double salts. Inorg Chem 1999;38(19):4211−6.

[65] Song RQ, Xu AW, Deng B, Li Q, Chen GY. From layered basic zinc acetate nanobelts to hierarchical zinc oxide nanostructures and porous zinc oxide nanobelts. Adv Funct Mater 2007;17(2):296−306.

[66] Poul L, Jouini N, Fiévet F. Layered hydroxide metal acetates (metal = zinc, cobalt, and nickel): elaboration via hydrolysis in polyol medium and comparative study. Chem Mater 2000;12(10):3123−32.

[67] Vayssieres L. Growth of arrayed nanorods and nanowires of ZnO from aqueous solu-tions. Adv Mater 2003;15(5):464−6.

[68] Govender K, Boyle DS, Kenway PB, O'Brien P. Understanding the factors that govern the deposition and morphology of thin films of ZnO from aqueous solution. J Mater Chem 2004;14(16):2575−91.

[69] Li QC, Kumar V, Li Y, Zhang HT, Marks TJ, Chang RPH. Fabrication of ZnO nanor-ods and nanotubes in aqueous solutions. Chem Mater 2005;17(5):1001−6.

[70] Redmond G, Fitzmaurice D, Graetzel M. Visible-light sensitization by cis-bis(thiocya-nato)bis(2,2′-bipyridyl-4,4′-dicarboxylato)ruthenium(II) of a transparent nanocrystalline ZnO film prepared by sol-gel techniques. Chem Mater 1994;6(5):686−91.

[71] Peter LM, Wijayantha KGU. Intensity dependence of the electron diffusion length in dye-sensitised nanocrystalline TiO$_2$ photovoltaic cells. Electrochem Commun 1999;1 (12):576−80.

[72] van der Zanden B, Goossens A. The nature of electron migration in dye-sensitized nanostructured TiO$_2$. J Phys Chem B 2000;104(30):7171−8.

[73] Nakade S, Saito Y, Kubo W, Kanzaki T, Kitamura T, Wada Y, et al. Enhancement of electron transport in nano-porous TiO$_2$ electrodes by dye adsorption. Electrochem Commun 2003;5(9):804−8.

[74] Chen LY, Yin YT. Hierarchically assembled ZnO nanoparticles on high diffusion coefficient ZnO nanowire arrays for high efficiency dye-sensitized solar cells. Nanoscale 2013;5(5):1777−80.

[75] Rothenberger G, Comte P, Gratzel M. A contribution to the optical design of dye-sensitized nanocrystalline solar cells. Sol Energy Mater Sol Cells 1999;58(3):321−36.

[76] Hore S, Vetter C, Kern R, Smit H, Hinsch A. Influence of scattering layers on efficiency of dye-sensitized solar cells. Sol Energy Mater Sol Cells 2006;90(9):1176−88.

[77] Zheng YZ, Tao X, Wang LX, Xu H, Hou Q, Zhou WL, et al. Novel ZnO-based film with double light-scattering layers as photoelectrodes for enhanced efficiency in dye-sensitized solar cells. Chem Mater 2010;22(3):928−34.

[78] Lide DR. CRC handbook of chemistry and physics. 76th ed. 1995. p. 4−133.

[79] Thavasi V, Renugopalakrishnan V, Jose R, Ramakrishna S. Controlled electron injection and transport at materials interfaces in dye sensitized solar cells. Mater Sci Eng R 2009;63(3):81−99.

[80] Li D, Xia YN. Electrospinning of nanofibers: reinventing the wheel? Adv Mater 2004;16(14):1151−70.

[81] Greiner A, Wendorff JH. Electrospinning: a fascinating method for the preparation of ultrathin fibers. Angew Chem Int Ed 2007;46(30):5670−703.

[82] Cheng HM, Chiu WH, Lee CH, Tsai SY, Hsieh WF. Formation of branched ZnO nanowires from solvothermal method and dye-sensitized solar cells applications. J Phys Chem C 2008;112(42):16359−64.

[83] Zhang T, Dong W, Keeter-Brewer M, Konar S, Njabon RN, Tian ZR. Site-specific nucleation and growth kinetics in hierarchical nanosyntheses of branched ZnO crystallites. J Am Chem Soc 2006;128(33):10960−8.

[84] Sounart TL, Liu J, Voigt JA, Hsu JWP, Spoerke ED, Tian Z, et al. Sequential nucleation and growth of complex nanostructured films. Adv Funct Mater 2006;16 (3):335−44.

[85] Yamabi S, Imai H. Growth conditions for wurtzite zinc oxide films in aqueous solutions. J Mater Chem 2002;12(12):3773−8.

[86] Sung YH, Frolov VD, Pimenov SM, Wu JJ. Investigation of charge transfer in Au nanoparticle−ZnO nanosheet composite photocatalysts. Phys Chem Chem Phys 2012;14(42):14492−4.

[87] Peter L. "Sticky electrons" transport and interfacial transfer of electrons in the dye-sensitized solar cell. Accounts Chem Res 2009;42(11):1839−47.

[88] Luque A, Hegedus S. Handbook of photovoltaic science and engineering. 2003. p. 102−4.

3 One-Dimensional Nanomaterials for Energy Applications

Ning Han and Johnny C. Ho

Department of Physics and Materials Science, and Centre for Functional Photonics (CFP), City University of Hong Kong, Tat Chee Avenue, Kowloon, Hong Kong SAR, PR China

3.1 Introduction

Nowadays, the world is running on the production peak of traditional fossil energy resources, such as coal, petroleum, and natural gas, which would be depleted in a foreseeable future. However, at the same time, the power needed to sustain the development of human society never stops increasing due to the ever expanding population as well as the average consumption per capita. Meanwhile, the greenhouse gases and toxic pollutants are degrading our living environment and threatening people's health. In this regard, exploring renewable and more environmental-friendly energy sources are the consensus of the whole world. Among many investigations, the transformation of solar, mechanical, and thermal energies into electricity is the most promising. The key to this transformation is the realization of active materials with excellent optical, electrical, mechanical, and thermal properties, enabling the cost-effective production of electricity, which is cost-competitive with the conventional energy sources. On the other hand, flexible generators are also required for the integration with self-powered electronics, facilitating the smart, wearable, and advanced applications.

In this context, one-dimensional (1D) nanomaterials are widely explored with the aim to enhance the energy harvesting efficiency and to enable the flexible nanogenerators, due to their superior light absorption, excellent electronic conduction, high mechanical strength, and ultra-low thermal conductivity. In this chapter, the fundamentals as well as the most recent advances of 1D nanomaterial preparation technologies, extraordinary physical properties, technological applications in the solar, mechanical, thermal energy harvesting and storage are reviewed.

Nanocrystalline Materials. DOI: http://dx.doi.org/10.1016/B978-0-12-407796-6.00003-8

3.2   Synthesis, Characterization, and Properties of 1D Nanomaterials

3.2.1   Synthesis of 1D Nanomaterials

In general, nanomaterials can be prepared by two distinct methods namely "top-down" and "bottom-up" technologies. The former utilizes a bulk material as the precursor, and reduces its dimension to the nanosize, while the latter adopts atoms or molecules to assemble materials in the nanoscale. Similarly, the typical synthesis of 1D nanomaterials also follows these two strategies, where anisotropic properties of the crystal are essential to etch the bulk material (top-down scheme) or induce atomic precursors assembled in the preferred direction (bottom-up scheme), in order to produce the 1D nanostructured materials, including nanorods (NRs), nanowires (NWs), nanotubes (NTs), etc.

3.2.1.1   Top-Down Methods

The key advantage of this top-down method is the inheritance of bulk properties in the synthesized nanomaterials, such as the doping concentration and crystal orientation [1]. However, the production throughput is relatively low and the production cost is expected to be higher as it sacrifices majority of the materials in etching during the fabrication.

Reactive Ion Etching

By utilizing a well-defined masking step, lateral and vertical NR arrays can be obtained by the reactive ion etching (RIE) from their bulk materials. For example, lateral Si NWs can be etched from the silicon on insulator (SOI) structure by a well-controlled electron beam lithography (EBL) and the subsequent XF_2 etching, enabling the direct electrical and thermal property measurement [1]. Also, as shown in Figure 3.1 [2], vertical InP NR arrays can be obtained by RIE of the bulk crystal with the silica sphere mask in order to enhance the light absorption and thus increase the energy conversion efficiency of the solar cell underneath [2,3].

Anodic Oxidation

In another case, metals such as Al and Ti can be anodically oxidized in acid electrolytes forming a nanopore or NT surface layer, as typically presented in Figure 3.2 [4]. As exemplified by the TiO_2 NT array preparation, the Ti surface is first oxidized by OH^- and converted into TiO_2, which is later dissolved into the F^- containing electrolyte at corrosive pits $(TiO_2 + 6F^- + 4H^+ \leftrightarrow [TiF_6]^{2-} + 2H_2O)$. The oxidants OH^- and O^{2-} are then migrated or diffused through the oxide layer to further oxidize the metallic Ti at the TiO_2/Ti interface. As the dissolution and oxidation processes continue in the pore sites, the nanopores grow stepwise until the growth rate equals to the wall etching rate [5]. Some key parameters should be well considered in this anodic oxidation: (a) there must exist an oxide layer (e.g., the native oxide layer) first before the anodic oxidation initiates, (b) water is needed to provide as the oxidant even in organic electrolytes, and (c) organic

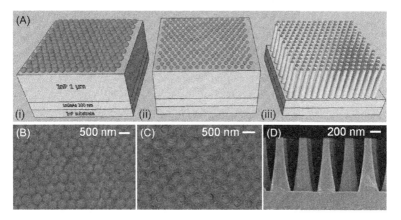

Figure 3.1 RIE method for the preparation of InP NR arrays. (A) Schematics of the fabrication process, (i) SiO$_2$ colloidal nanoparticles are dispersed by spin coating onto the oxygen plasma pretreated InP crystal, (ii) the size of SiO$_2$ particles is reduced by RIE (CHF$_3$ etchant gas) to define the diameter of intended NRs, and (iii) the exposed InP crystal is then selectively etched by RIE (Cl$_2$/H$_2$/CH$_4$ etchant gas; etching rate of InP over SiO$_2$ > 12:1) to obtain the NR arrays. (B and D) The SEM images of (i)−(iii) process steps, respectively. *Source*: Reprinted with permission from Ref. [2]. Copyright 2013, American Chemical Society.

electrolytes with the well-controlled oxidant concentration can yield super long TiO$_2$ NTs up to hundreds of micrometers [6].

Metal-Assisted Chemical Etching

Metal-assisted chemical etching is another common technique for the fabrication of 1D semiconductor nanostructures with the ability to control various parameters, such as cross-sectional shape, diameter, length, orientation, doping type, and doping level [7−9]. Taking Si as an illustrative example (Figure 3.3), the bulk Si < 100 > crystal can be chemically etched by Ag catalyst particles in HF and H$_2$O$_2$ solution along <100 > direction. Specifically, the etching process is generally understood as follows: (a) the oxidant is preferentially reduced at the surface of Ag particles due to the catalytic activity of Ag on the reduction of the oxidant. (b) The holes (electrons) generated due to the reduction of the oxidant diffuse through Ag and are injected into (extracted from) Si that is in contact with the Ag particles. (c) Si is then oxidized by the injected holes and dissolved at the Si/Ag interface by HF. The reactant (HF) and the byproducts next diffuse along the interface between the Si and Ag particles. (d) Once the hole concentration reaches its maximum at the Si/Ag interface, Si in contact with Ag would be etched much faster by HF than the bare Si surface. (e) Eventually, holes would diffuse from Si under the Ag particles to off-metal areas or to the wall of the pore if the rate of hole consumption at the Si/Ag interface is smaller than the rate of hole injection. As a result, Si < 100 > NWs can be obtained in this case. It is also noted that this electrochemical etching only occurs at <100 > direction, and if other orientated Si

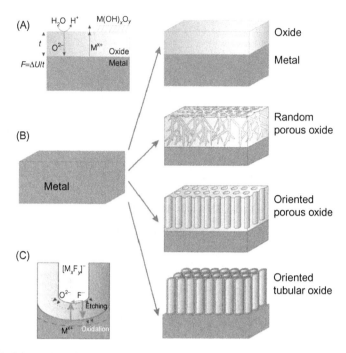

Figure 3.2 Schematics of the anodic oxidation process. (A) Oxide formation mechanism on metals, (B) various morphologies can be obtained by the electrochemical oxidation of metallic titanium, namely, a compact oxide film, a disordered nanoporous layer, a self-ordered nanopore array layer, and a self-ordered NT array layer, and (C) NT formation mechanism.

Source: Reprinted with permission from Ref. [4]. Copyright 2013, Royal Society of Chemistry.

wafers, such as $<110>$ and $<111>$, are utilized, the as-obtained NWs would tilt with respect to the wafer at the crystallographic angle between $<100>$ and $<110>$ or $<100>$ and $<111>$ [7].

3.2.1.2 Bottom-Up Approaches

Vapor Phase Growth
In addition to the above-mentioned techniques, 1D semiconductor nanomaterials, such as Si and Ge NWs, can also be assembled from vapor phase by using metal nanoparticles as catalysts (e.g., Au), that is, the vapor−liquid−solid (VLS) growth [10]. In this process, the growth mainly relies on the fact that Si or Ge precursor atoms can be dissolved in the Au nanoparticles in a high supersaturation, due to the increased solubility in nanoscale as inferred from the Gibbs−Thomson effect [11]. This way, Si or Ge atoms would precipitate from the supersaturation of metal−semiconductor alloy with a continuous supply of

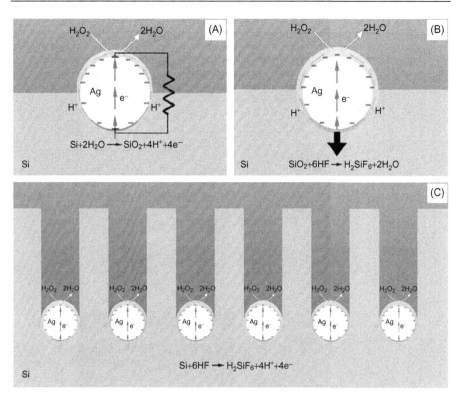

Figure 3.3 Schematics of metal-assisted chemical etching of Si NW array from bulk Si. (A) An electric field is generated from H^+ gradient across the Ag particle catalyst, (B) self-electrophoresis driven motion of Ag particles, and (C) Si NW array formation by the extended tunneling motion of the Ag catalyst particles.
Source: Reprinted with permission from Ref. [7]. Copyright 2013, Wiley.

the precursor flux, and then the NWs would grow along the axial direction. Notably, the Au—Si and Au—Ge alloys were believed to exist in the liquid state above the eutectic temperature and continue the NW growth in the VLS growth mode, but until recently *in situ* TEM showed that Si NWs can also grow well below the eutectic temperature of Al—Si, inferring a vapor—solid—solid (VSS) growth mechanism, as depicted in the binary phase diagram in Figure 3.4 [15]. At the same time, the vapor phase growth of other compound semiconductors is similar, except that the catalyst alloy would be Au-III alloy (III stands for Al, Ga, In, etc.) or Au-II alloy (II stands for Zn, Cd, etc.), while the VLS versus VSS growth mechanisms are still under an extensive debate in the research community [12,16]. Depending on the precursor types, there are several different growth systems commonly utilized in this vapor phase synthesis, such as molecular beam epitaxy (MBE), metal-organic chemical vapor deposition (MOCVD), plasma-enhanced chemical vapor deposition (PECVD), and solid-source chemical vapor deposition (SSCVD).

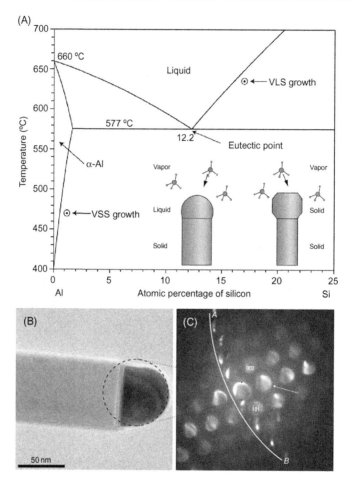

Figure 3.4 VLS and VSS growth mechanisms of NWs. (A) Growth temperature zones in the binary phase diagram of Al−Si, inset is the growth schematics with a hemispherical and polygonal catalytic tip inferring the VLS and VSS growth modes. (B and C) *In situ* TEM image and electron diffraction pattern of the GaAs NW and the corresponding Au−Ga catalyst tip at 540°C. The electron diffraction pattern shows a solid phase of the Au−Ga catalytic alloy even though existed in hemispherical shape.
Source: Reprinted with permission from Refs. [12−14]. Copyright 2013, Nature Publishing Group and Royal Society of Chemistry.

Since catalysts are employed in the synthesis, the catalyst residues may exist in the NW body and behave as dopants or trap states, which would influence the electrical and optical properties of the grown NWs significantly [17,18]. In this case, the precursor group II or III metals are then adopted as the catalyst in the III−V and II−VI semiconductor NW growth, which is known as the self-catalyzed growth mechanism. As the catalyst contributes to one of the

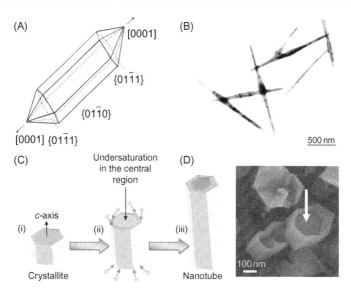

Figure 3.5 VS growth mechanism of 1D nanomaterials. (A and B) Schematics and TEM image of ZnO NRs grown along the [0001] direction, (C and D) schematics and SEM image of the grown Te NTs.
Source: Reprinted with permission from Refs. [19,20]. Copyright 2013, American Chemical Society and Nature Publishing Group.

constituents of NWs, such as the case for Ga-catalyzed GaAs NWs, there would be no catalyst residual-related problem and thus the resulting electrical properties can be greatly enhanced [17].

It should be noted that if there are neither homogenous nor heterogeneous catalyst particles, 1D nanostructures would also be grown noncatalytically due to the large difference in the surface energy of crystal planes. The NW growth direction would follow the stacking planes which have the lowest surface energy. For example, since ZnO (0001) planes are known to have the lowest surface energy, then ZnO NRs can be prepared by the vapor−solid (VS) growth mode along the [0001] direction as shown in Figure 3.5A and B [19]. Recently, Te NTs are also successfully demonstrated to synthesize by the VS growth mechanism due to the large difference in surface energy, as illustrated in Figure 3.5C and D [20]. More importantly, self-catalyzed NWs also have no catalytic seed existed on the NW tips, which is typically consumed in the cooling down step by gaseous precursors [21]; therefore, it is challenging to distinguish explicitly between the VS and self-catalyzed growth mechanisms in this case.

Liquid Phase Growth

Hydrothermal/solvothermal utilizes the crystallization of nanomaterials from the supersaturated aqueous (hydrothermal) or organic (solvothermal) solutions. Due to the surface energy difference of different crystal planes, high energy surface

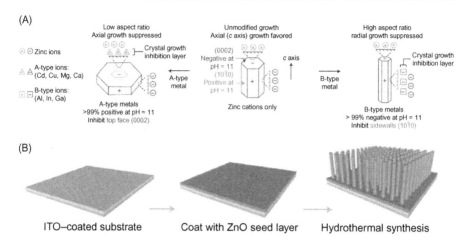

Figure 3.6 Hydrothermal synthesis of ZnO NW and NR arrays. (A) Growth mechanism of ZnO NW by controlling the crystal planes occupation. A-type cations (positive at pH 11, Cd, Cu, etc.) would prohibit Zn accessing onto the negative (0002) plane, leading to planar structure, while B-type ions (negative at pH 11, Al, In, etc.) would occupy the positive sidewalls leading to NW growth. (B) Schematic illustration of the seed-assisted hydrothermal synthesis of ZnO NR arrays.
Source: Reprinted with permission from Ref. [22]. Copyright 2013, Nature Publishing Group.

would grow faster and thus the crystallization occurs along the low energy surface plane forming a 1D nanomaterial as depicted in Figure 3.6A. This synthesis technique is highly versatile because the preferred orientation can also be tailored by adding other coordinates (e.g., simply pH, surfactants) to occupy the low energy surface and prevent the growth; as a result, NWs can then grow on other low energy directions to form various nanostructures, such as nanocubes and nanocages. As shown in Figure 3.6B, addition of the seed layer would accelerate the growth of ZnO NRs and make the diameter more uniform, since the nucleation process without the seeds would be relatively slow and random in supersaturated solutions contributing to this irregularity. This seeding process is usually adopted to prepare metal oxide semiconductor, such as ZnO and TiO$_2$, and other 1D nanomaterials like GaAs NWs can also be prepared in the solvothermal process [23].

Electrospinning

Another common technique employs the electrostatic force to stretch a viscous conductive fluid through a thin jet of nanometer to micrometer scale, then the fluid containing semiconductor precursors can be solidified to form semiconductor NWs or nanofibers, with length up to centimeter range. By introducing another one to two fluids, the coaxial electrospinning (ES) can also be achieved in preparing multistructures (e.g., core/shell NWs or NWs in microtubes) as presented in Figure 3.7 [24]. This technique is simple, cost-effective, and more

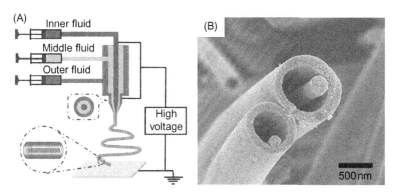

Figure 3.7 ES of 1D nanomaterials. (A) Schematic of the multifluidic coaxial ES setup and (B) SEM image of the NW fabricated in microtube.
Source: Reprinted with permission from Ref. [24]. Copyright 2013, American Chemical Society.

importantly it can prepare complex compound materials which are challenging to be realized in other conventional methods. For example, thermoelectric PZT (Pb $(Zr,Ti)O_3$) NWs can be synthesized by this ES method [25], which is seldom reported by other synthesis technique due to the complex material structures. In this case, the as-spun nanomaterial is often polycrystalline, as a result of the random condensation at low temperature or even at a slightly higher annealing temperature. This polycrystalline material has relatively low electron mobility because of the severe electron scattering at the grain boundary. However, these PZT NWs are still very attractive to various technological applications as the thermal conductance is significantly reduced by the large surface areas, and thus can be used in thermoelectric generators [24].

Template-Assisted Synthesis
Using anodic aluminum oxide (AAO) as the template (similarly prepared as TiO_2 NT array in Figure 3.2), NW materials like Bi NWs can be prepared by electrode-position in the nanopores followed by the removal of the AAO template [26]. On another hand, utilizing the polycarbonate template, $Bi_2Te_3/(Bi_{0.3}Sb_{0.7})_2Te_3$ super-lattice NWs can also be fabricated by the electrodeposition [27]. Moreover, NTs can also be achieved by the careful control of the template and process parameters as depicted in Figure 3.8 [28]. It is obvious that this method would suffer from the complex processes including careful preparation of the template, electrode incorpo-ration for the electrodeposition, removal of the barrier layer, proper electrolyte selection, and the final template removal step; however, the NT diameter distribu-tion can be well controlled in accordance with the template, the NT composition can be tuned in a wide range depending on the precursors in different steps of the electrodeposition, and also the orientation can be controllable by adjusting the potential in certain cases [29]. In recent years, more and more template-assisted

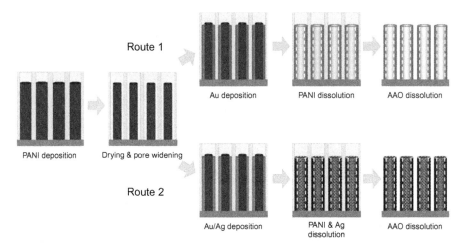

Figure 3.8 Schematics of the electrodeposition of smooth and porous Au NTs into the AAO template.
Source: Reprinted with permission from Ref. [28]. Copyright 2013, Royal Society of Chemistry.

syntheses are explored for the fabrication of 1D nanostructures, such as virus [30] and carbon fiber [31].

3.2.2 Characterization

3.2.2.1 Morphology, Composition, and Crystallography

Scanning electron microscopy (SEM) and transmission electron microscopy (TEM) are always used to observe the morphology of synthesized 1D nanostructures. Aspect ratio, tubular structure, core/shell structure, tapering and surface coating, etc. can be all well determined by the SEM/TEM technologies. The elemental analysis can be carried out by energy-dispersive X-ray spectroscopy (EDS) in SEM/TEM mode, and the surface elements within ∼5 nm can be detected by X-ray photoelectron spectroscopy (XPS) and ultraviolet photoelectron spectroscopy (UPS). Radial and axial chemical composition can also be assessed by EDS or electron energy loss spectroscopy (EELS) attached with TEM. Nowadays all these characterization technologies are very common in the routine processes and will not be discussed herein.

Meanwhile, the crystal structure of 1D nanomaterials is rather complex except for the top-down fabricated nanostructures which inherit their bulk material properties. For the bottom-up synthesized 1D nanomaterials, the crystal structure, including crystal phase and orientation, should be carefully determined because the unstable crystal structure in the bulk would transform and become more stable when existed in nanostructures due to the large surface-to-volume ratio [32,33]. For example, the cubic zinc blende (ZB) phase is the most stable crystal structure for most of the III−V semiconductor bulk materials, but the hexagonal wurtzite (WZ) phase is found to be more stable for the NWs down to <10 nm diameter [34] and a mixed phase is existed in the range of 10−100 nm in NW diameter [35]. Furthermore, the NW growth orientations are also mixed (known as inversion domains) even in the same ZB phase, because the unstable surfaces such as (100) and (110) would become thermally favorable in the nanoscale, and thus the corresponding mixed growth directions are usually found for the NWs [35].

In addition, X-ray diffraction (XRD) is another classical technique in the determination of crystal phase, but it can only give the average collective information and has limited resolution in assessing the mixed crystal phases. Also, it cannot determine the NW orientation except for the well-aligned vertical NW arrays. For example, XRD cannot detect small amount of WZ GaAs NWs while it can only illustrate the typical ZB crystal structure in the measured spectrum [35,36]. At the same time, Raman spectroscopy can also be employed to distinguish different crystal structures by determining different vibration modes. Specifically, cubic $BaTiO_3$ inherently has no active Raman modes while the active modes are only expected for the noncentrosymmetric structure [37]. In any case, the conventional Raman spectroscopy has the same disadvantages as XRD of giving the average collective information of the bulk crystal and lacks information regarding crystal orientations. Therefore, single NW characterization is necessary and essential in order to differentiate the wire to wire variations, such as performing micro-Raman spectroscopy for the crystal phase determination, and selected area electron diffraction (SAED) and high-resolution (HR)TEM for both phase and orientation determination for the specific individual NW.

Other than the above-mentioned postgrowth characterization, the *in situ* TEM imaging is more versatile in determining the growth mechanism by detecting and tracing the growth process of one specific nanostructure [38]. It is especially utilized in the vapor phase growth of NWs via CVD method, and is extremely helpful in the VLS/VSS mechanism assessment and the catalytic behavior observation during the NW growth.

3.2.2.2   Bandgap and Point Defects

Photoluminescence (PL) and cathodeluminescence (CL) are the common technologies which can determine the electronic bandgap and crystal defect energy level in nanomaterials. Due to the effect of quantum confinement, the bandgap energy of nanomaterials would become larger, and a blue-shifted in PL or CL is usually observed. For instance, PL peaks of 750−820 nm are obtained for the 10−50 nm diameter of GaAs NWs [39,40]. At the same time, crystal defect-related luminescence can also be observed in PL and CL. Taking ZnO NRs as the example, various crystal defects induced luminescence are detected in the PL peaks which can correlate well with their electrical properties [41]. It should be noted that because of the abundant crystal defects and/or surface states contributing to the nonradioactive electron/hole recombination, some nanomaterials have no luminescence emission, and a subsequent annealing or passivation process is needed to enhance to radioacitve recombination [42]. Therefore, PL or CL can also be used to dedicate the crystal quality of nanostructures.

UV−visible absorption/reflection/transmission spectroscopy is another typical technique to determine the bandgap of nanomaterials [34,43], especially useful for the cases when PL or CL is not detectable. For large bandgap semiconductors like Ga_2O_3 (4.9 eV) which has the UV emission, no suitable excitation can be used to generate a detectable emission [44]; therefore, it is always easier for large bandgap nanomaterials to be assessed by the absorption spectroscopy for the bandgap determination. In specific, the bandgap energy is determined by deducing the absorption edge in an $(ahv)^2-hv$ plot, where a is the absorption coefficient of the measured sample and hv is the photon energy with all the details available in the literature. It should also be noted that small bandgap semiconductors, such as InAs (0.34 eV) and InSb (0.17 eV), have the emission located in the far infrared region, which is not detectable by the conventional detectors. In this case, there is not suitable technology to determine the bandgap of these materials till now.

3.2.2.3   Trace Dopants

In order to tune the electrical and optical properties of 1D nanomaterials, doping by other elements is a common method to achieve this modulation. In the top-down synthesis scheme, materials inherit their bulk properties including the doping concentration. But for the bottom-up route, the doping concentration, if any, is not always the same as that in the precursor, which thus needs the well quantification. On the other hand, since there may be residual dopants introduced in the growth, either coming from the catalyst or process ambient, it is essential to investigate and understand the effect of such dopants on the physical properties of fabricated materials. Large amount of dopants can be determined by EDS or EELS in TEM, while it is a great challenge to assess the small amount of doping concentration (<0.01%).

Although the conventional secondary ion mass spectrometry (SIMS) has very high dynamic range for the doping concentration determination, its spatial resolution is somewhat lowered and restricted for nanomaterials due to the relatively large ion beam size. Therefore, novel probing technologies in the resolution of one single dopant atom are being developed, such as local electrode atom probe (LEAP) [45] and high-angle annular dark-field (HAADF) scanning transmission electron microscopy (STEM) [46,47]. Taking the VLS grown GaAs NWs as an illustration, the Au catalyst may reside certain amount of atoms in the as-grown NW body, which forms deep level trap states and they are detrimental to the electrical and optical properties [17,18]. Since the Au atoms are at a relatively low concentration (<0.01%), far beyond the detection limit of typical EDS or XPS technologies. In this regard, the HAADF STEM technology is developed and successfully utilized of direct imaging the Au atoms in GaAs NW body, which account for $10^{17} - 10^{18}$ cm^{-3} (ppm level) concentration [46].

3.2.3   Fundamental Properties

3.2.3.1   Electrical Property

Due to the quantum confinement of 1D nanostructures, electrons can flow efficiently in the axial direction without much scattering in the radial direction; in this case, the electrical properties should get improved theoretically as compared with their bulk counterparts. However, there are still several key factors which can practically affect the carrier mobility, such as crystal defects, diameter scaling, surface states, surface roughness, and configuration of the heterostructure utilized [48−50].

Specifically, when the concentration of crystal defects increases and the diameter of the nanostructure decreases, it is commonly observed that the resulting carrier mobility would be reduced due to the higher carrier scattering by those defects or surface states [49,50]. Similarly, higher surface roughness would also lead to the enhanced carrier scattering which induces the degradation of corresponding carrier mobility. In general, these surface states contribute to the surface carrier recombination as well as lead to the lowering of carrier mobility, which can be tackled by the suitable surface passivation [51]. Also, the proper design of core/shell heterostructure can align the band structures appropriately such that utilizing the shell material with a larger bandgap can confine carriers in the core region for the enhanced electrical transport properties. For example, InP shell is utilized and found to discriminate the Fermi level pinning of InAs NW core, resulting to achieve the p-type InAs NW, as it is challenging to be realized due to its well-known surface Fermi level pinning above the conduction band (CB) [52]. Moreover, gallium oxide shell can also be employed to deplete electrons in the GaAs NW core, leading to a "p-to-n" type switching of the NWs through the manipulation of surface oxide depletion layer by enlarging the NW diameters [53].

3.2.3.2   Optical Property

Similarly, because of the quantum confinement, optical properties of 1D nanostructures can also be easily tuned by controlling the diameter and shape, which are especially important in the application of solar cells. In photovoltaics, 1D nanostructure-based antireflection layers are usually fabricated on the top of the active layer in order to boost up the photon collection efficiency. In brief, thin vertical NR or NW arrays are shown to be more effective in absorbing the short-wavelength photons while the thicker ones are more suitable for the long-wavelength photons. This way, by forming a dual diameter NR array, a nearly complete photon absorption is observed in the 300−900 nm range [54]. Besides the geometrical effect, tuning the composition of nanostructures as well as integrating nanomaterials with different bandgaps are also demonstrated to improve the resultant optical properties of the nanostructure. For example, by modulating the In/Ga ratio in $In_xGa_{1-x}As$ NWs, the bandgap can be reliably tuned from the pure GaAs phase (1.42 eV) to the InAs phase (0.34 eV) for broadband photon absorption [43]. Likewise, by using CdTe/CdSe double layers, the corresponding light absorption efficiency is greatly enhanced due to the fact that CdTe can absorb short-wavelength photons and CdSe would capture long-wavelength ones [55]. All these unique optical properties arising from the geometries, compositional manipulation, and integration of different nanostructures have illustrated great technological potency for the future high-performance, low-cost third-generation solar cells.

3.2.3.3 Thermal Property

Recently, fundamental thermal properties of 1D nanomaterials are extensively explored for thermoelectric and pyroelectric generator applications in which an electrical potential can be generated by the temperature difference applied in the nanostructures. Also, their thermal properties are essential to understand for electrical devices since Joule heating is unavoidable and needed to be minimized for the low-standby-power electronics. For most thermal properties, nanomaterials show significant variation from their bulk counterparts. For instance, although both Bi NWs and bulk films have the anisotropic thermal expansion, the NW format is illustrated to have the diameter-dependent thermal expansion coefficient and even the transition from positive to negative coefficients for the change from low to high temperatures, while the bulk films have only the positive coefficient [29,56]. Another well-known difference in the thermal properties of nanomaterials is the reduction of their melting points due to the high surface-to-volume ratio. When the Bi NWs are scaled in their diameters from 60 to 12 nm, the NWs would exhibit a continuous lowering of their melting points from 542 to 536 K, as compared with the one of 544 K in the bulk material [29].

In addition to the melting point reduction, the thermal conductance of NW materials is also found to decrease with the reducing diameters, including Si, Ge, GaAs, and the alloying superlattices [1,57,58]. Generally, thermal conductance in the solid materials is mainly composed of the electron contribution and atomic lattice vibration (phonon). As phonons have a relatively shorter wavelength, they can be effectively scattered by the larger NW surface effect due to scaled NWs and thus resulting in a decreased thermal conductance. Interestingly, when Si NWs are miniaturized down to ~ 50 nm, they would display the insulating thermal property which is comparable with SiO_2 [1]. Furthermore, with the increasing NW surface roughness, the thermal conductivity is observed to decrease further for the enhanced surface effect [58]. All these modulation in the thermal conductivity of NWs can be utilized for a wide range of applications in the thermoelectric generators.

3.2.3.4 Mechanical Property

On another hand, mechanical properties of 1D nanostructures, such as the Young's moduli (E) and resonance frequency, can be assessed accurately by using the resonance atomic force microscopy (AFM) or STEM. Because of the high surface-to-volume ratio, small diameter NWs would have higher surface tensions, which increase the elastic modulus drastically for the cases of Ag and Pb NWs (diameter <70 nm) as compared with their bulk counterparts [59]. Similarly, since small diameter WO_3 NWs (<30 nm) are shown to have the near-perfect crystallinity as proved in HRTEM images, the resulting Young's modulus exhibits a $3 \times$ increase when compared to the thicker NWs and bulk materials where the abundant crystal defects are believed to dissipate energies for the lower moduli [60]. However, the enhanced modulus might also be material dependent, as Au NWs show the diameter-independent modulus in the 40−200 nm range but the yield strength is increased by nearly 100 times due to the significant strain-hardening effect [61]. At the same time, ZnO NWs also display the diameter-independent modulus while the strength is increased 40 times more than their bulk counterparts [62]. It is as well interesting that short Cu NWs usually illustrate the ductile failure via the dislocation of multiple slip planes, while long NWs show the brittle failure by an unstable localized shear along a single slip plane [63].

All these conflicting information regarding the mechanical properties of 1D NWs suggest the effort for continuous investigation here, especially needed for the development of piezoelectric cells as the electricity can be generated by the lattice deformation. For example, short ZnO NRs are difficult to be deformed and thus only a small piezoelectric potential is detected [64]. Also, mechanical properties of different plastic matrix can be reinforced by incorporating inorganic nanoscale additives [65]. This way, it is important to further evaluate those mechanical properties with the aim to properly select appropriate nanomaterials, composites, and their geometrical dimensions for the flexible and wearable piezoelectric generators.

3.3 Solar Energy Harvesting

It is roughly estimated that the total amount of sunlight received by the earth for 1 h would be enough for the energy consumption of the entire world for 1 year, together with its cleanliness. However, the grand challenge is how we can harvest this solar energy effectively, due to the following difficulties: (a) one need to cover a large area of the surface to absorb the low density of solar energy (AM 1.5G, ~ 1 kW/m^2); (b) sunlight varies largely depending on different areas and different time in the world; and (c) one have to store the converted electricity in an efficient way in case when there is little sunlight available (e.g., night, rainy, winter). Consequently, the cost of the solar energy is relatively higher than that of the fossil energy, making it account for a very small fraction ($\sim 0.015\%$) in the current energy consumption in the world [66].

Due to the advent of various advanced semiconductor technologies, the solar energy can now be harvested with an energy conversion efficiency of 10−20% depending on the device types, including the photovoltaic (PV) cells, photoelectrochemical cells (PECs), and solar hydrogen production. As abovementioned in the fundamental properties, the advantages of using 1D nanomaterials in the solar energy harvesting are the improved photon absorption as well as the enhanced carrier mobility for photocarriers collection, resulting in the higher solar energy conversion efficiency. Also, the amount of active materials utilized can be greatly reduced because of the higher photon absorbability, and thus the material cost would be reduced. All these would contribute significantly to the next-generation high-efficiency solar cells, as well as the high-performance flexible and wearable generators for the residential use.

3.3.1 Photovoltaic Cells

A PV cell can be simply configured as a potential barrier with the use to sweep photogenerated carriers once it receives the sunlight. The cell can be varied from the form of a semiconductor p−n junction to a metal−semiconductor Schottky contact as shown in Figure 3.9A and B [67,69]. Specifically, when the incident sunlight shines on the active device materials (with the bandgap energies less than the visible photon energy of 2−3 eV), the photogenerated electrons would get excited from the valence band (VB) to CB leading the holes reside in the VB. Then the excited electrons and holes will be separated by the potential barrier and collected to the external circuit generating a photocurrent. It should also be noted that photons with the energy lower than the bandgap of the active materials will not be absorbed, while higher energy photons would produce hot electrons, leading to the energy loss. Theoretically, the bandgap-dependent PV efficiency can be calculated as shown in Figure 3.9C [68].

In general, the PV cell performance can be characterized by the output parameters, such as open-circuit voltage (V_{oc}), short-circuit current (I_{sc}), fill factor (FF),

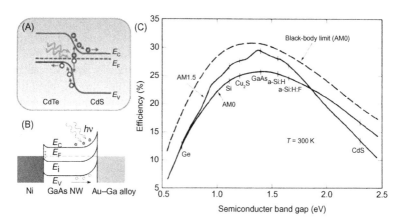

Figure 3.9 Operating mechanism of the PV cell. (A and B) The p−n junctions of CdTe/CdS heterostructures and Au−Ga alloy tip/GaAs NW metal−semiconductor Schottky contact-based PC cells. (C) The theoretical bandgap-dependent photon to electricity conversion efficiency.
Source: Reprinted with permission from Refs. [67−69]. Copyright 2013, Nature Publishing Group, American Institute of Physics, and Elsevier.

and the energy conversion efficiency ($\eta = V_{oc} \times I_{sc} \times FF/P_{in}$, where P_{in} is the incident solar power of 100 mW/cm^2 intensity on the active area). For a given semiconductor material, V_{oc} is determined mainly by the potential barrier height in the p−n junction or Schottky contact, while increases slightly (logarithmically) with the photocurrent. On the other side, I_{sc} increases drastically with the enhanced photocurrent by higher photon absorption and larger collection area. From the operating mechanism, one would find the way to improve the solar cell efficiency: (1) selection of semiconductor materials with the suitable bandgap should absorb more sunlight to generate more photoelectrons in p-type material and photoholes in n-type material (i.e., the minorities) and (2) the photoinduced carriers should have long enough lifetime and diffusion length to be separated and collected to the external circuit before the recombination.

Considering (1), it is predicted that 1D nanostructure-based antireflection layers can be fabricated on top of the commercial bulk Si PV cells to boost up the energy conversion efficiency. In this regard, several novel hierarchical 1D nanostructures have been recently developed in order to absorb more photons, including the dual diameter nanopillar arrays [54], nanospike arrays [70], integrated NW/nanowell arrays [71], etc. as presented in Figure 3.10. The unique advantage lies in the effect that the thin tips can absorb the short-wavelength photons while the scattered long-wavelength photons can be absorbed by the deep thicker bottoms. Notably, through the systematic optimization, all these nanostructures can achieve nearly 100% photon absorption in the solar spectra of 300−900 nm, inferring a promising enhancement in the solar cell efficiency.

Consider (2), the electron and/or hole mobility in most III−V and II−VI semiconductors are typically higher than those of Si, meaning a faster photocarrier separation and collection before the recombination. Meanwhile, most III−V and II−VI semiconductors have the direct bandgap, inferring a thinner active layer is needed to absorb the incident sunlight than Si with the indirect bandgap. This way, more III−V and II−VI semiconductor-based PV cells are being explored in the recent years. For example, the radial p−i−n structured GaAs single NW solar cell holds the record of >10% efficiency for the single NW PVs [72], in comparison with the 3.4% efficiency of Si single NW PV utilizing the same architecture [73]. In order to have a larger output, 3D NW array-based PVs are developed with respectably high efficiency. For example, InP n−i−p NW arrays were grown on the p-InP substrate and the as-fabricated PV cell has the outstanding performance with an efficiency up to 13.8% [74]. Also, by embedding n-CdS nanopillars into the p-CdSe polycrystalline thin film, a large junction area was obtained and thus the efficiency is observed as high as 6% even fabricated on the flexible substrate as depicted in Figure 3.11, showing the prospect in the wearable solar cells [67]. However, it should be noted that enlarging the junction area (i.e., the photocurrent collection area) by the 3D radial junction is not always helpful to enhance the performance due to the issue of larger area for the junction leakage. In specific, the radial 3D InP NW arrays are demonstrated with only a relatively low efficiency of 3.37% [75] as compared with the above-mentioned 13.8%, while the radial GaAs p−n junction NW arrays have even a surprisingly low

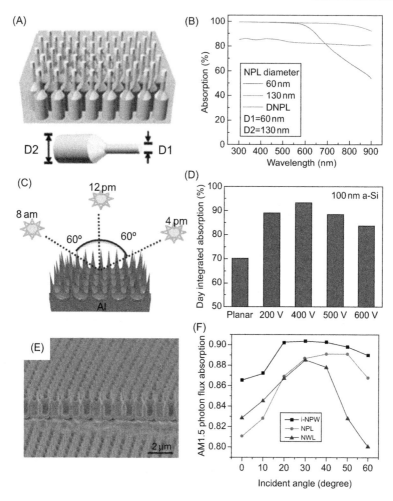

Figure 3.10 Integrated 1D nanostructure-based antireflective layers. (A and B) The schematic of dual diameter nanopillar arrays and the corresponding light absorption comparison with the single diameter nanopillar arrays. (C and D) The schematic of nanospike arrays and the corresponding light absorption comparison with the nanospikes prepared at different potentials. (E and F) The SEM image of the integrated NW/nanowell arrays and the corresponding light absorption comparison with NW and nanowell arrays alone.
Source: Reprinted with permission from Refs. [54,70,71]. Copyright 2013, American Chemical Society.

efficiency of 0.83% [76]. All these indicate the dominant effect of the large-area junction leakage, arising from the interface states induced recombination of photocarriers, which requires a further investigation in the compromisation between large junction area for the photon absorption and the carrier recombination.

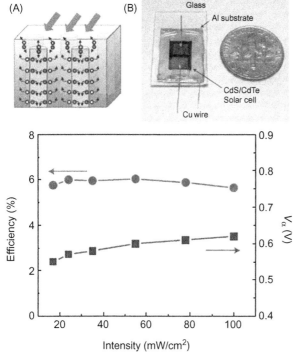

Figure 3.11 3D nanopillar array-based solar cells. (A) Schematic of the 3D p-CdTe/n-CdS heterojunction PV device, (B) optical image of one 3D cell bonded on a glass substrate, and (C) efficiency and V_{oc} of the PV cell under different illumination intensities. *Source*: Reprinted with permission from Ref. [67]. Copyright 2013, Nature Publishing Group.

3.3.2 Photoelectrochemical Cells

The dye-sensitized photoelectrochemical solar cells (DSSCs) are typical PECs which were first reported by Gratzel's group in 1990s [77,78], with the working mechanism as shown in Figure 3.12. Typically, it comprises of a dye-loaded TiO_2 anode and an inert cathode, with the I_3^-/I^- reduction/oxidation pair in between for the electron transformation. Sunlight is absorbed by the dye molecules producing photogenerated electrons, which are then collected by the TiO_2 electrode into the external circuit. The working principle is different from PVs because the photocarrier generation and collection processes are separately designated to the dyes and semiconductors, respectively. One advantage of using this PEC process is that the dyes would have the higher photon absorption efficiency as they have a larger overlap with the solar spectra as compared with the large bandgap semiconductors [80,81]. At the same time, as the semiconductors do not take the role of absorbers, their bandgap is not so important as in PVs and thus more low-cost large bandgap, chemically more stable metal-oxide semiconductors can be chosen as the active electrode materials. However, in this PEC scheme, an extra electron injection process step from the dye to semiconductor is needed, this way, the overall incident photon to current conversion efficiency (IPCE) is then determined by the light harvesting efficiency (LHE), electron injection efficiency ($\varphi_{injection}$), and electron collection efficiency ($\eta_{collection}$) by the equation of $IPCE = LHE \times \varphi_{injection} \times \eta_{collection}$. Similar to the role of 1D

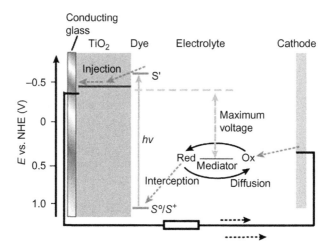

Figure 3.12 Operating mechanism of the PEC. The TiO$_2$ photoanode receives electrons from photoexcited dyes. The positively charged dyes are then reduced by a mediator red/ox pair dissolved in the electrolyte. The mediator redox pair is next regenerated by reduction near the cathode by the electrons.
Source: Reprinted with permission from Ref. [79]. Copyright 2013, Nature Publishing Group.

nanomaterials in the PV cells for the improved photon absorption, the 1D nanostructures in the DSSCs can also scatter the incident sunlight effectively for the subsequent absorption by the dye molecules. In this case, LHE can be enhanced by the use of 1D nanomaterials.

In general, after the photocarriers are generated in the dye molecules, two processes of the electron-hole recombination and electron injection into the semiconductor materials would compete against each other. It is well accepted that the electron injection velocity (time) should be at least 100 times higher (lower) than the recombination velocity (time) [82]. As 1D nanomaterials can provide the electron expressway in the axial direction, as compared with the random diffusion and scattering in the nanoparticle aggregates, the electrons can inject and be collected into semiconductors efficiently. For example, the recombination time of electrons and holes is 150 times longer than the injection time, and thus results in a near 100% collection efficiency [83]. Also, since electrons can flow easily in axial direction, the reduced resistance here would contribute to an enhanced fill factor and thus a higher power output [84]. Furthermore, by well designing the band diagram of the core/shell 1D nanostructures, there would be a potential drop from the shell to the core region, which would favor for the electron injection and confines them in the core reducing the surface recombination [85,86].

However, when compared to the anode configuration utilizing nanoparticles, the use of 1D nanomaterials may lead to the lower dye molecule loading due to the reduced surface area. As a result, the pure 1D nanostructured TiO$_2$ anode can only yield the highest energy conversion efficiency of <8%, as compared with the one employing TiO_2 nanoparticles [84,87,88]. In this regard, integrating 1D nanomaterials with nanoparticles would help to improve the efficiency. As shown in Figure 3.13, the TiO_2 nanoparticle/NR hybrid anode is optimized to have a ratio of 8:2 in order to get a high efficiency of 8.6%. Similar results are also found by Suzuki et al. that the TiO_2 nanoparticle anode with a load of 5−10 wt% of NWs would result the highest efficiency [89].

Although DSSCs have a high energy conversion efficiency of ∼8% in their first appearance in 1990s, only there is a slight improvement of up to 11% after 20 years of investigation [82]. Apart from the dye selection, one bottleneck would be the recombination loss in the large surface area of nanomaterials, which infers a continuous effort to further develop high electron mobility materials for the higher electron collection efficiency. In this case, ZnO [90], SnO_2 [85], GaN [91] NWs attracted extensive attention here, because of their excellent chemical stability in addition to the high electron mobility. Special care should be paid to the adhesion and band alignment of the anode materials contacting with the dyes, which might have the significant effect in the photon absorbability and the consequent electron injection efficiency [91]. Anyway, due to the low-cost and relatively high energy conversion efficiency, the overall operation of DSSCs would be practically promising in the near future.

3.3.3 Solar Hydrogen Production

One main drawback of the PV and PEC cells is that there is no electricity generated when no sunlight is available, such as at night and in rainy days; therefore, the electricity produced has to be stored for the real-time use. The energy storage in batteries and in hydrogen produced by electrolyzing water with electricity would be discussed in the following energy storage section. However, there is another choice for PECs before the electricity is generated, in which water is reduced/oxidized to H_2 and O_2 by the photosensitive anode and cathode, as shown in Figure 3.14A [79,92,93]. In the photosensitive anode, the photoexcited holes would diffuse into contact with water and oxidize it into O_2, while the electrons would flow through the external circuit to the cathode to reduce H^+ into H_2. It should be noted that the photogenerated electrons and holes should have the suitable potentials as compared with H_2/H_2O and H_2O/O_2 in order to make the effective reduction and oxidation, which are summarized in Figure 3.14B with the common semiconductor electrodes.

Again, the use of 1D nanostructured materials would enhance the photon absorption efficiency and electron transport velocity, which helps to increase the hydrogen production yield. Due to the suitable band diagram and the stability in electrolytes, ZnO and TiO_2 nanomaterials are commonly used as photoanodes in solar hydrogen production [94−96]. However, because of their relatively large bandgap (∼3.3 eV), only a small portion of photons in the ultraviolet range can be utilized leading to an overall low light to hydrogen production efficiency. For example, even though TiO_2 NT arrays show a respectably high conversion

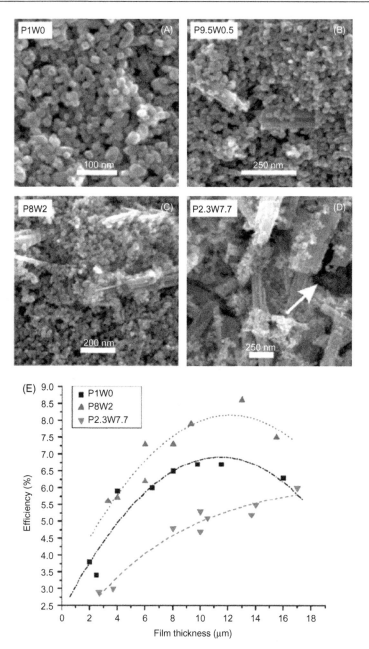

Figure 3.13 Advantages of utilizing NW/nanoparticle composite anodes. Cross-sectional SEM images of (A) TiO$_2$ nanoparticles anode (P1W0), (B) hybrid anode with TiO$_2$ NW load 5% (P9.5W0.5), (C) hybrid anode with TiO$_2$ NW load 20% (P8W2), (D) hybrid anode with TiO$_2$ NW load 77% (P2.3W7.7), and (E) thickness dependence of the energy conversion efficiencies of the photoanodes under AM 1.5G sunlight. The anode with TiO$_2$ NWs loaded at 20% (P8W2) has the highest solar energy harvesting efficiency.
Source: Reprinted with permission from Ref. [89]. Copyright 2013, American Chemical Society.

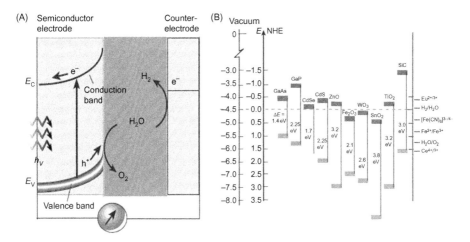

Figure 3.14 Operating mechanism of solar hydrogen production. (A) Separation, transport, and reaction of photogenerated electron-hole pairs. (B) Band positions of several semiconductors in contact with aqueous electrolyte at pH 1, and the standard potentials of several redox couples against the standard hydrogen electrode potential. Theoretically, photoexcited electrons with energy higher than the H_2/H_2O potential will reduce H_2O to H_2, and photoexcited holes with energy lower than the H_2O/O_2 potential will oxidize H_2O to O_2. *Source*: Reprinted with permission from Ref. [79]. Copyright 2013, Nature Publishing Group.

efficiency of up to 16.25% in the 320−400 nm range under AM 1.5G illumination, the overall energy conversion efficiency is still relatively low because of the small weight of UV energy in the normal solar energy distribution [97], and so does for ZnO material [98].

To circumvent this problem, two strategies are often adopted. One is decorating small bandgap quantum dots (QDs) onto ZnO and TiO_2 anodes and the other one is using small bandgap semiconductor materials directly as the active electrodes. For example, double-sided ZnO NR arrays are prepared on the ITO glass, with the top layer decorated with CdS QDs and bottom layer decorated with CdSe QDs. The medium bandgap CdS (~ 2.3 eV) helps absorbing photons in addition to the UV absorption of ZnO, and the smaller bandgap CdSe (1.7 eV) enhances the long-wavelength photon absorption, which increases the overall photon absorption efficiency significantly. Moreover, the band alignment of CdS and CdSe with ZnO favors the electron injection into ZnO layer by providing an energy gradient. As a result, the ultimate photon to hydrogen conversion efficiency is demonstrated as high as 45% in the 300−600 nm range, as illustrated in Figure 3.15 [55]. 1D nano-materials with the smaller bandgap are also explored to be the active electrode, such as Fe_2O_3 (1.9−2.1 eV) [99], CuO/Cu_2O (1.2−1.9 eV) [100−102], W_2N (2.2 eV) [103], and other III−V and II−VI compound semiconductors [104−106]. But special care should be paid to the energy band diagram of these semiconductors with the potential of H^+/H_2 and OH^-/O_2, as an external bias would be needed if the electrons (holes) have not enough energy to reduce H^+ (OH^-). Also, the

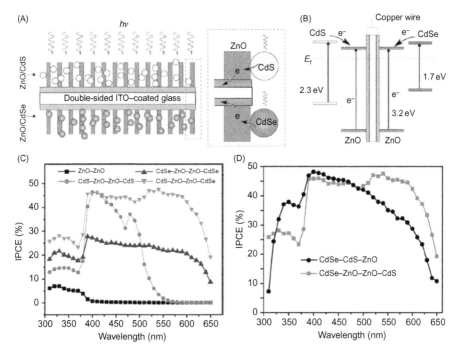

Figure 3.15 Advantages of the double-sided CdS−ZnO−ZnO−CdSe photoanode.
(A) Schematic view, (B) band energy alignment, (C) IPCE spectra in the range of
310−650 nm at 0 V versus Ag/AgCl, and (D) IPCE comparison with the single-sided
CdSe−CdS−ZnO photoanode. The higher IPCE of double-sided CdS−ZnO−ZnO−CdSe
photoanode in (C) is due to the higher light absorbance, with CdS and CdSe absorbing the
short-wavelength and long-wavelength photons, respectively, and better electron collection
efficiency due to the band energy alignment in (B). The higher IPCE of double-sided
CdS−ZnO−ZnO−CdSe photoanode in (D) is due to the more efficient electron collection
than single-sided one.
Source: Reprinted with permission from Ref. [55]. Copyright 2013, American Chemical
Society.

chemical stability is another important issue for the long-term production of hydrogen, otherwise an anticorrosive layer should be used to protect the active materials
[100,107,108].

3.4   Piezoelectric Energy Transformation

3.4.1   Piezoelectric Effect

Piezoelectric phenomenon is a coupling effect of material's mechanical and electrical properties. The phenomenon would exist for the materials, which do not have a

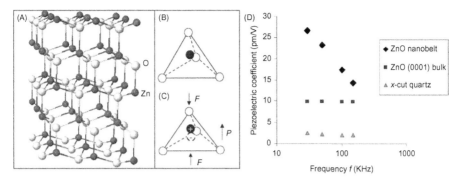

Figure 3.16 Piezoelectric effect of ZnO. (A) The WZ structure of ZnO, (B) tetragonal-coordinated Zn and O atoms in ZnO, (C) the positive and negative charge distortion by an external force *F* while a molecular dipole *P* forms, and (D) the piezoelectric coefficient comparison of the bulk ZnO and ZnO nanobelt.
Source: Reprinted with permission from Refs. [109,110]. Copyright 2013, Wiley and American Chemical Society.

center of symmetry in their crystal lattice (e.g., hexagonal or tetragonal structure) such that when the lattice is deformed by an external force, the positive and negative charge centers are then distorted as shown in Figure 3.16A−C. Another requirement is that the material should have a high degree of crystallinity as to make all the molecular dipoles aligned in the same direction and forms a macroscopic electric field. Polycrystalline crystal usually would not have any piezoelectric effect. On the contrary, when a piezoelectric material is subjected to a potential difference, the lattice would deform mechanically [110]. Although this piezoelectric effect has long been discovered and utilized in our daily life, including the usage of bulk piezoelectric materials such as PZT for various technological applications, the nanostructured configuration of piezoelectric materials has not been actively explored till recent years.

Among many piezoelectric materials, ZnO has a relatively low piezoelectric coefficient as a bulk material and is not widely studied before. However, the nanoscale ZnO materials, e.g., ZnO NWs, NRs, and nanobelts, etc., all having significant larger piezoelectric coefficients are presented in Figure 3.16D [109]. More importantly, due to the flexibility of synthesis in various nanostructures, piezoelectric nanogenerators were firstly realized by the employment of ZnO NWs.

3.4.2 Piezoelectric Nanogenerators

The first ZnO NW-based nanogenerator was reported by ZL Wang's group in 2006 [111,112], with the working principle as depicted in Figure 3.17. The ZnO NW was prepared by CVD via the VLS mechanism. The bottom end of the vertical ZnO NW is connected to the Ag electrode while the top end was scanned by an AFM tip (Pt-coated Si) in the contact mode at a normal force of 5 nN as shown in Figure 3.17A and B. Due to the work function difference, Ag (4.2 eV) forms ohmic

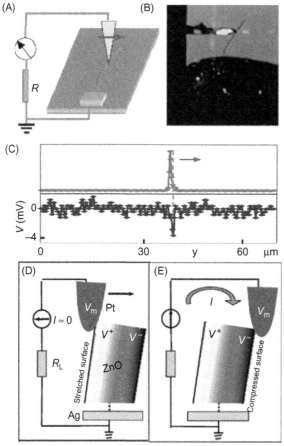

Figure 3.17 Schematics and working principle of the ZnO nanogenerator. (A) Schematic and (B) photography of the ZnO nanobelt bent by the Pt-coated Si AFM tip, (C) the topography and the output voltage of the ZnO NW generator, (D and E) schematics of the reverse- and forward-biased Schottky contact of AFM tip and ZnO nanobelt.
Source: Reprinted with permission from Ref. [112]. Copyright 2013, American Chemical Society.

contact to ZnO (4.5 eV) and Pt (6.1 eV) forms a Schottky contact with ZnO. Once the NW is bent by the AFM tip, ZnO would generate a potential difference between the compressed and stretched sides. When the Pt tip contacts the stretched positive side, the Pt/ZnO diode is reversely biased and thus no current generates. However, when the Pt tip comes to the compressed negative potential side, the diode is forward biased, this way, a forward current is generated, as presented in Figure 3.17C−E.

In this configuration, ZnO NW acts as a "capacitor," which stores electrons in the reverse bias case from the external circuit and discharges in the forward bias connection. The charging process is relatively slow as the NW has to be bent back by the Pt tip gradually, and thus the charging current is not detectable. While the discharge process occurs suddenly when the Pt tip contacts the negative compressed side, the discharge current is high enough to be detected. In this case, a direct current (DC) is generated here instead of the switching alternative current (AC). Furthermore, when the scanning speed is higher, the output power of the ZnO nanogenerators would become higher [113,114]. It should also be noted that the potential is generated by the distorted ZnO positive and negative charge centers in the lattice which is not mobile; as a result, this effect has no correlation with the free carrier concentration within the crystal. However, when the free carrier concentration is too high, these carriers would prevent the potential neutralization by the carrier diffusion. On another hand, if the carrier concentration is too low, the crystal would then have a high inner resistance and yield a low output current [115,116]. In fact, there exists an optimal carrier concentration of $\sim 10^{15}$ cm^{-3} to balance the above two aspects [117,118].

The Schottky contact of Pt with ZnO is another key factor in realizing the piezoelectric current, as the Schottky diode can prevent current leakage in the reverse bias and accumulate charges, and then discharges in the subsequent forward bias step. If not, the charges stored in the ZnO capacitor would leak gradually, leading to no detectable current generation, which can be further proved by the symmetric Ag ohmic contact on both ZnO NW ends.

As mentioned above, the charging current is too low to be detected in the nanogenerator driven by a single AFM tip, as a result of gradually bending process. However, the charging current can be produced and detected reasonably well in the laterally configured nanogenerator as the bending process is fast as depicted in Figure 3.18. As a result, an AC current in charging and discharging process is generated [114,119].

When configuring them into the parallel NW array device configuration, several parameters can be further optimized in order to enhance the piezoelectric effect. For example, NWs with the small Young's modulus should not be too short and the NW density should not be too high as to be easily bent to produce the potential [64,115,116]. Also, the Schottky barrier height can be tuned by adopting electrodes with higher work function difference like indium zinc tin oxide (IZTO) to produce a higher potential. The carrier density should neither be too high nor too low with the aim to generate a high current output. The nanogenerators can be connected in series or in parallel as to scale up the output voltage or current [114,120,121]. If needed, multilayered structure is also developed to further increase the current density for more portable applications [120−122].

It is worth noting that ZnO material has the relatively low piezoelectric coefficient of 12 pC/N, and thus has a relatively low piezoelectric potential (e.g., 4 mV in Figure 3.17). Therefore, materials with the higher piezoelectric coefficient should be adopted in order to design high output nanogenerators. Under this circumstance, other piezoelectric materials, such as InN [115,123], GaN [116,124], CdSe [125], and CdTe [126] NWs, are all fabricated and configured into nanogenerators, with an improved potential and current (e.g., 60 mV for InN [115], 137 mV for CdSe, and 0.3 V for CdTe [126]).

The conventional perovskite materials have the far larger piezoelectric coefficients, such as BaTiO$_3$ (100 pC/N) and PZT (200 pC/N). In this regard, perovskite-structured NWs are prepared by the ES method and fabricated into the high output nanogenerators. For example, the output voltage and current of the fabricated BaTiO$_3$ nanogenerators are observed up to 5.5 V and 350 nA under a stress of 1 MPa, respectively, which are much larger than their thin film counterparts [37]. ZnSnO$_3$ triangular belts based nanogenerators can also be demonstrated to generate

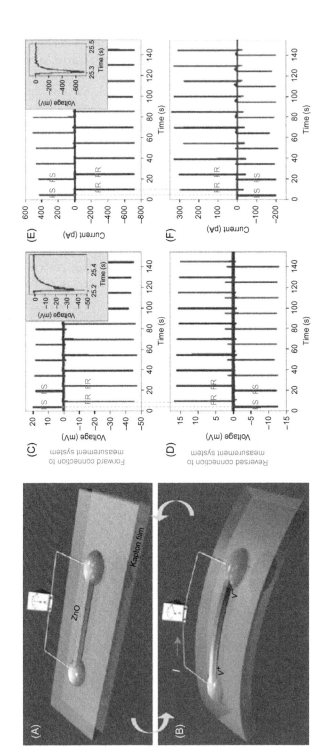

Figure 3.18 An AC current nanogenerator fabricated on the plastic substrate. (A and B) Schematics of the ZnO NW piezoelectric device fabricated on the Kapton substrate and bent by an external force, (C and E) open-circuit voltage and short-circuit current of the generator when subject to repeated fast stretching (FS) and fast release (FR) processes at the forward connection, (D and F) the reverse connection of (C) and (E). *Source:* Reprinted with permission from Ref. [114]. Copyright 2013, Nature Publishing Group.

a current and voltage of $0.13\,\mu A$ and 5.3 V accordingly [127]. Also, the lead zirconate titanate NW generators can produce a 6 V and 45 nA output voltage and current, that is large enough to power a liquid crystal display and a UV sensor [25]. And recently, a vertically aligned ultra-long $Pb(Zr_{0.52}Ti_{0.48})O_3$ NW has an ultra-high output voltage of 209 V and current density of $23.5\,\mu A/cm^2$, which are 3.6 times and 2.9 times of the previous record values, respectively [128]. Consequently, using flexible substrates, such as Al foils [120,129], plastics [64,121], and papers, flexible and wearable generators with the acceptable output power by the industrial demand would be produced in the near future.

3.5 Thermal Energy Conversion

It is estimated that $>50\%$ of the energy consumption is ultimately wasted to the environment in the form of heat [130,131]. Therefore, it would be ideal if we can recycle this heat energy in order to improve the energy utilization efficiency, which is beneficial to solve the energy crisis and global warming. However, this heat energy is existed at the lowest intensity among all energy forms, which is a grand challenge to efficiently transform back into electricity. Two kinds of methods are under the current investigation: one utilizes the thermoelectric response and the other adopts the pyroelectric effect.

3.5.1 Thermoelectric Generators

Typically, thermoelectric generators utilize the Seebeck effect of semiconductor materials as shown in Figure 3.19. When there is a temperature difference between the two ends of the semiconductor connected in parallel (a pair of p-type and n-type semiconductors), free carriers will diffuse from the hot side to cold side. This way, a potential difference is generated between the two ends, which is dependent on the Seebeck coefficient (α) of the material and the temperature difference by equation $V = \alpha(T_{hot} - T_{cold})$. The reverse process is called the Peltier effect, where the forced current is given by a potential difference between two ends of the semiconductor, would carry heat from the cold to hot side, namely a solid-state refrigerator. It should be noted that both Seebeck and Peltier effects have a relatively low efficiency where they are only used for the case where other technologies are not available (e.g., space satellites) [130].

For practical applications, a thermoelectric figure of merit (z) is usually adopted to characterize the performance, $z = \alpha^2\sigma/\kappa$, where σ and κ are electrical and thermal conductivities of the material. A high-performance thermoelectric generator should have a zT (T is temperature) of $>3-4$ for the competitive usage [133]. It is also easy to find how to further improve Z from the equation. First, high Z materials are the top choice, such as Bi-, Te-, and Pb-based materials rather than the commonly employed Si and III–V, II–VI semiconductors, as illustrated in Figure 3.20 [134]. Second, the electrical conductivity of materials should be tuned

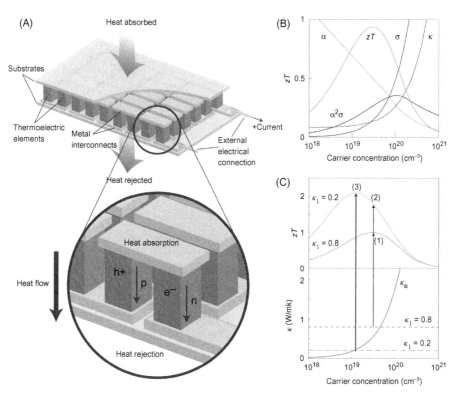

Figure 3.19 Thermoelectric mechanism and relating key parameters. (A) Schematic module showing the charge and heat flow, (B) influence of the increasing free carrier density on the electric conductivity σ, Seebeck coefficient α, and thermal conductivity κ, and figure of merit zT, (C) reducing the lattice contribution to the thermal conductivity from 0.8 to 0.2 W/mK helps increasing the figure of merit from 1 (point 1) to 2 (point 3).
Source: Reprinted with permission from Ref. [132]. Copyright 2013, Nature Publishing Group.

higher. And finally, it should have the lower κ (i.e., thermal insulating) as to maintain the temperature difference. To sum all these up, the ideal candidate should be a phonon-glass electron-crystal material. However, the bottleneck for this thermoelectric technology is that when the electrical conductivity is increased, the thermal conductivity would increase at the same time, since the hot electrons also carry heat to cold side as well.

One obvious advantage in utilizing semiconductor 1D nanomaterials as the active materials for thermoelectric generators is that the electrical conductivity can be maintained the same as high as the bulk materials, while the thermal conductivity can be modulated far lower than the bulk counterparts. It is due to the fact that the thermal conductivity can be decomposed into the electron contribution (κ_e) and phonon (lattice vibration) contribution (κ_L), where κ_L plays a main role in the semiconductors. As a result of the reduced dimension, the phonon is drastically

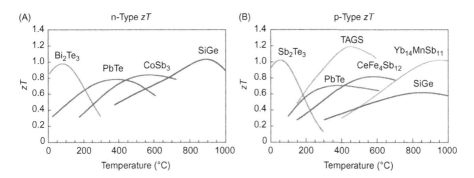

Figure 3.20 The thermoelectric figure of merit of common semiconductor materials: (A) n-type and (B) p-type semiconductors.
Source: Reprinted with permission from Ref. [132]. Copyright 2013, Nature Publishing Group.

scattered by the NW surface (large surface-to-volume ratio in NWs); this way, the thermal conductivity can be reduced far lower than their bulk equivalents, and thus a higher figure of merit can be obtained in the 1D nanomaterial-based thermoelectric generators. For example, it is theoretically predicted that 5-nm-diameter PbSe/PbS and PbTe/PbSe NW alloys would have a zT of 4−6 at 77 K [135]. Also, p-type PbSe NWs are observed to have a twofold lower thermal conductivity than the bulk materials [136]. Similarly, tubular materials are illustrated to have a far lower thermal conductivity than the NW format, due to the enhanced surface area for more effective phonon scattering [137]. Furthermore, it is found that when the NWs have rougher surfaces, the phonon would be scattered more significantly and result a lower thermal conductivity [26]. Notably, the electrical conductivity of these 1D nanostructures should also be well considered in order to balance the contribution of positive-charged carriers versus negative-charged carriers to the thermal conductivity to maintain the high thermoelectric response for applications [138,139].

Till now, a more successful example can be demonstrated with the Si NW-based thermoelectric generators, which can be easily fabricated as compared to the complicated material structures described above [1,58,140]. Due to the good thermal conductivity, bulk Si material has a low zT of ~0.01 which is usually not considered for thermoelectric applications. However, the rough Si NW is experimentally shown to exhibit a thermal insulating property (i.e., comparable with SiO_2) down to the diameter of <50 nm, and thus enhance the zT to 0.6 as presented in Figure 3.21 [58]. When the Si NW is further scaled to 20 nm in the diameter, it shows a high zT of ~1 at 200 K, which is 2 orders of magnitude higher than that of the bulk Si [1]. Although this zT improvement is believed to be attributed by the enhanced surface phonon scattering on the rough NW surface, the detailed underlying physical mechanism is still being explored extensively in order to clarify this effect. In any case, it is another technological breakthrough utilizing 1D nanostructures to achieve high-performance thermoelectric generators which are competitive for the future practical usage.

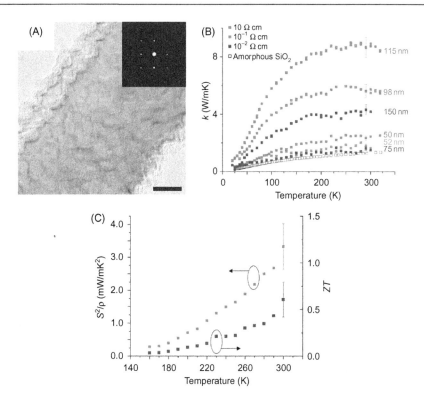

Figure 3.21 Si NW with significant rough surfaces as thermoelectric generators. (A) TEM image and SAED pattern of the wet chemical etched rough Si NW (scale bar = 20 nm), (B) temperature-dependent thermal conductivity of Si NWs with different resistivity (the open squares stand for the thermal conductivity of bulk SiO_2 material as comparison), and (C) single NW power factor and the calculated zT using the 52-nm-thick NW described in (B). *Source*: Reprinted with permission from Ref. [58]. Copyright 2013, Nature Publishing Group.

3.5.2 Pyroelectric Generators

Pyroelectric effect describes a temporary voltage generated by certain materials when the temperature applied on the materials is changed with time ($dT/dt \neq 0$). It is different from the thermoelectric effect where the temperature gradient is varied in the space, instead of the time ($dT/dx \neq 0$). In specific, when the temperature applied to the pyroelectric material changes, the atomic positions would be changed slightly such that a molecular dipole generates due to the anisotropic property of the crystal. On a macroscopic view, a voltage drop would also be obtained across the crystal along the dipole direction, which is similar with the thermoelectric potential.

The first nanogenerator based on pyroelectric effect was reported in 2012 by ZL Wang's group as shown in Figure 3.22 [131]. The ZnO NW arrays were prepared by a solution-based growth, and a Schottky contact is formed between the Ag

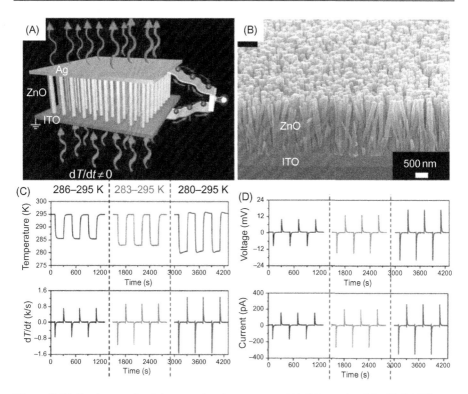

Figure 3.22 Fundamentals of the pyroelectric generator. (A) Schematic of the ZnO NW array pyroelectric generator, (B) SEM image of ZnO NW arrays grown on ITO predeposited glass, (C) temperature changes as a function of time, and (D) the generated voltage and current output.
Source: Reprinted with permission from Ref. [131]. Copyright 2013, American Chemical Society.

electrode and ZnO. It is obvious that once there is a temperature change (either increase or decrease with time), there would be an electric output. Although the power generated is relatively low at this prototyping stage, it provides another feasible route to harvest waste thermal energy, especially when there is no temperature gradient in a spatially equilibrium environment.

3.6 Energy Storage

It should be noted that the new energy sources (sunlight, mechanical, heat, etc.) are not stable in both time and spatial domains, which cannot satisfy the energy consumption requirement of our society around the clock. As a result, the generated electric power should be stored for the later use when the production exceeds the consumption and this energy storage is essential for the stability of electricity supply and quality of the power output. Generally, the electricity can be stored in many forms, such as hydroenergy and compressed air. Among all, the energy storage in the chemical bonds in H_2 fuels can provide one of the highest energy density [141,142]. Also, H_2 is a clean fuel as only water is the byproduct; in this case, storing electricity in H_2 can be considered to be the best choice in which the energy can be later retransformed into the electricity by H_2 fuel cells [142]. H_2 can be produced by the direct electrolysis of water, but the efficiency is relatively low considering the electricity generating efficiency of <20%, electrolysis efficiency of 50−70%, and fuel cell efficiency of 50−60% [141]. In addition, H_2 can also be produced *in situ* by PECs as detailed in the solar hydrogen production section. Another promising energy storage device is the Li-ion battery, which has similar power and energy density as fuel cells and is being widely explored nowadays. In any case, the power density provided by both fuel cells and batteries is still not high enough for daily applications due to the relatively slow red/ox speed as compared with the direct electron transport. In this regard, 1D nanostructured materials have been recently under the vast investigation as active materials for the above energy storage technologies. Supercapacitors are also actively studied for the energy storage with higher power density.

3.6.1 H_2 Fuel Cells

The H_2 fuel cell is typically composed of two catalytic electrodes, one of which oxidizes hydrogen into H^+ (anode). The H^+ ions would then penetrate through a proton exchange membrane (PEM) to the cathode, where O_2 (pure or in air) is reduced to combine with H^+ forming clean water and the current would generate in these two electrode reactions as shown in Figure 3.23A [143,145]. In general, a highly active catalytic electrode (e.g., Pt- and Pd-based noble metals) is needed to effectively accelerate these two reactions in order to get a higher energy density [143,146]. As the catalytic property is greatly dependent on the surface area, porosity as well as different crystal planes, 1D nanostructured materials can be the ideal candidates for their large surface-to-volume ratio. Successful examples include the Pt and Pt/Pb NTs [147] and freestanding Pt NWs membrane [148], which possess all the advantages of large surface area, porosity, and preferential exposure of the highly active planes. In order to reduce the expensive Pt load, Pt nanoparticles can be dispersed onto carbon NTs, to achieve another highly conductive and high-surface-area 1D material and exhibit the high catalytic activity as presented in Figure 3.23B and C.

3.6.2 Li-Ion Battery

In typical Li-ion batteries, the cathode is constructed by the Li-containing material that provides Li^+ to the anode in the charging process, making the anode material lithiated. Then the reduced metallic Li atoms inserted into the anode would lose their electrons in the subsequent discharging process to generate the current, and the anode material would be delithiated. This way, the lithium intercalation and

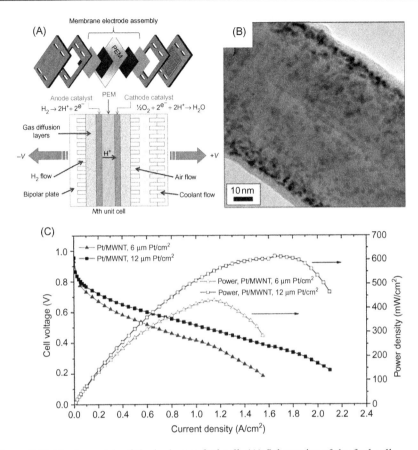

Figure 3.23 Fundamentals of the hydrogen fuel cell. (A) Schematics of the fuel cell components, (B) TEM image of Pt nanoparticles catalyst dispersed on multiwall carbon nanotubes (MWCNTs), and (C) power density of the Pt/MWCNT cathode.
Source: Reprinted with permission from Refs. [143,144]. Copyright 2013, Nature Publishing Group and American Chemical Society.

deintercalation reversible cycles enable the operation of the Li-ion battery as depicted in Figure 3.24A [149]. One main problem is that the anode material inflates significantly in its volume to house the high Li-ion load and then shrinks back after the delithiation [151]. For example, Si has the theoretically largest capacity of ~4200 mAh/g, but the volume changes by 400% in the Li$^+$ insertion and extraction process, making the anode Si pulverized by the stress, which induces the operational instability and capacity loss. The well-aligned Si NW arrays with suitable filling density would delimitate this problem since there would be enough interspace for the volume change and thus release stress as shown in Figure 3.24B. This way, a high capacity of >3000 mAh/g can be maintained at long circles [150], far higher than the commercial graphite-based anode (~372 mAh/g). Recently, Cui's group designed a double-walled silicon NT anode material by using

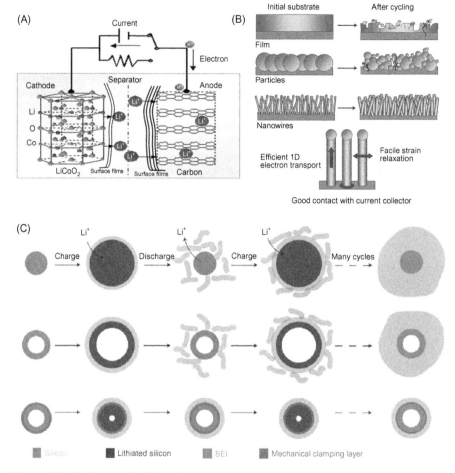

Figure 3.24 Fundamentals of the Li-ion battery. (A) Schematic, (B) the stability advantage of Si NW arrays on the conductive substrate over the thin film and nanoparticle counterparts, due to the stress relaxation in the space, and (C) the stability advantage of double-walled Si NT over the NW and normal NT counterparts, due to the mechanical clamping layer.
Source: Reprinted with permission from Refs. [31,149,150]. Copyright 2013, Nature Publishing Group and Royal Society of Chemistry.

the carbon fiber as a sacrificing template as illustrated in Figure 3.24C [31]. The NT prevents the pulverization effect greatly, as the Si wall would expand into the inner space as constrained by the outmost oxide shell. Also, the oxide shell would prohibit the direct contact of electrolyte with Si, and helps to form a stable outer solid—electrolyte interface (SEI). All these make the NT anode very stable in 6000 times charge/discharge testing with a high capacity >1000 mAh/g, holding the promise for industrial applications.

Another advantage of using 1D nanostructures is the high charge mobility in the confined axial direction as mentioned before, which helps to increase the capacity as well as the stability. This enhanced mobility has been shown in many anode NW and NT materials such as Co_3O_4 NWs [30,152,153]. Furthermore, highly conductive carbon NTs can also be added to increase the conductivity and thus improve the capacity [152]. However, special attention should be paid to the direct contact of the NW anode with the conductive charge-collecting substrate, as it has been proved that reduced contact resistance by the direct connection would help to increase capacity and stability greatly in Co_3O_4 [153] and Ge [154] NW arrays. Also, the increased surface area by making porous NWs and NTs would also be favorable [155].

3.6.3 Supercapacitors

Recently, supercapacitors (or ultracapacitors) are attracting more and more interests because they have higher power density, long lifecycle and bridging the energy gap between a battery/cell and a traditional capacitor, which are very important when there is a sudden increase in the load [156]. Basically, they store the electrical energy in the positive and negative electrodes separated by an insulating dielectric, as depicted in Figure 3.25A [157]. As capacitors store electrons or adsorbed ions instead of absorbed ions [158], there is not any expansion or mechanically overstress problem in the cathode materials. Also, the conductance of 1D nanomaterials is of high importance as in the Li-ion batteries, since the well-modulated conductance can result in a fast response and a higher power for the energy storage. For example, although carbon NTs have high conductivity, they only exhibit high capacitance and power output (48 kW/kg) when they are epitaxially grown on the anchored Au catalyst in the conductive substrate as shown in Figure 3.25B and C [157].

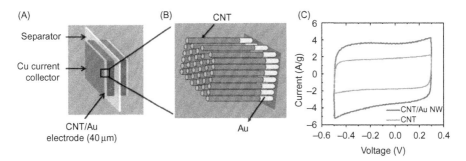

Figure 3.25 CNT/Au electrode-based supercapacitors. (A and B) Schematics of the supercapacitor configuration and the CNT/Au electrode, (C) current−voltage characteristics of the supercapacitor at a scan rate of 100 mV/s in the 6 M KOH electrolyte, demonstrating the advantage of the CNT/Au over pure CNT electrode.
Source: Reprinted with permission from Ref. [157]. Copyright 2013, Royal Society of Chemistry.

3.7   Summary and Outlook

All in all, finding reliable energy sources is urgent for the human society in order to ensure the sustainability in our near future. More importantly, producing smart power sources is essential for human beings to maintain the high technological development trend. Solar cells, piezoelectric generators, and waste heat harvesting generators are typical devices for the electricity generation from our daily energy resources. In these generators, 1D nanostructured materials have demonstrated their promising prospect in enhancing the efficiency while reducing the cost due to their outstanding light coupling effect, high electrical conductivity, high mechanical strength, and low thermal conductivity. Also, they are highly versatile in the electricity storage and releasing systems due to the fact of their high electrical conductivity and high catalytic activity. All these technological breakthroughs are encouraging and facilitate a wide horizon of application domains in the energy-related disciplines.

References

[1] Boukai AI, Bunimovich Y, Tahir-Kheli J, Yu J-K, Goddard Iii WA, Heath JR. Silicon nanowires as efficient thermoelectric materials. Nature 2008;451(7175):168−71.

[2] Naureen S, Sanatinia R, Shahid N, Anand S. High optical quality InP-based nanopillars fabricated by a top-down approach. Nano Lett 2011;11(11):4805−11.

[3] Cho K, Ruebusch DJ, Lee MH, Moon JH, Ford AC, Kapadia R, et al. Molecular monolayers for conformal, nanoscale doping of InP nanopillar photovoltaics. Appl Phys Lett 2011;98(20):203101.

[4] Ghicov A, Schmuki P. Self-ordering electrochemistry: a review on growth and functionality of TiO_2 nanotubes and other self-aligned MO_x structures. Chem Commun 2009;20:2791−808.

[5] Su Z, Zhou W. Formation, morphology control and applications of anodic TiO_2 nanotube arrays. J Mater Chem 2011;21(25):8955−70.

[6] Shankar K, Mor GK, Prakasam HE, Yoriya S, Paulose M, Varghese OK, et al. Highly-ordered TiO_2 nanotube arrays up to 220 μm in length: use in water photoelectrolysis and dye-sensitized solar cells. Nanotechnology 2007;18(6):065707.

[7] Peng K, Lu A, Zhang R, Lee ST. Motility of metal nanoparticles in silicon and induced anisotropic silicon etching. Adv Funct Mater 2008;18(19):3026−35.

[8] DeJarld M, Shin JC, Chern W, Chanda D, Balasundaram K, Rogers JA, et al. Formation of high aspect ratio GaAs nanostructures with metal-assisted chemical etching. Nano Lett 2011;11(12):5259−63.

[9] Foll H, Langa S, Carstensen J, Christophersen M, Tiginyanu IM. Pores in III−V semiconductors. Adv Mater 2003;15(3):183−98.

[10] Shi WS, Zheng YF, Wang N, Lee CS, Lee ST. A general synthetic route to III−V compound semiconductor nanowires. Adv Mater 2001;13(8):591−4.

[11] Han N, Wang F, Hou JJ, Yip S, Lin H, Fang M, et al. Manipulated growth of GaAs nanowires: controllable crystal quality and growth orientations via a supersaturation-controlled engineering process. Cryst Growth Des 2012;12(12):6243−9.

[12] Persson AI, Larsson MW, Stenstrom S, Ohlsson BJ, Samuelson L, Wallenberg LR. Solid-phase diffusion mechanism for GaAs nanowire growth. Nat Mater 2004;3 (10):677−81.

[13] Wang YW, Schmidt V, Senz S, Gosele U. Epitaxial growth of silicon nanowires using an aluminium catalyst. Nat Nanotechnol 2006;1(3):186−9.

[14] Lensch-Falk JL, Hemesath ER, Perea DE, Lauhon LJ. Alternative catalysts for VSS growth of silicon and germanium nanowires. J Mater Chem 2009;19(7):849−57.

[15] Kodambaka S, Tersoff J, Reuter MC, Ross FM. Germanium nanowire growth below the eutectic temperature. Science 2007;316(5825):729−32.

[16] Lin PA, Liang D, Gao XPA, Sankaran RM. Shape-controlled Au particles for InAs nanowire growth. Nano Lett 2012;12(1):315−20.

[17] Breuer S, Pfuüller C, Flissikowski T, Brandt O, Grahn HT, Geelhaar L, et al. Suitability of Au-and self-assisted GaAs nanowires for optoelectronic applications. Nano Lett 2011;11(3):1276−9.

[18] Tambe MJ, Ren S, Gradečak S. Effects of gold diffusion on n-type doping of GaAs nanowires. Nano Lett 2010;10(11):4584−9.

[19] Hu P, Yuan F, Bai L, Li J, Chen Y. Plasma synthesis of large quantities of zinc oxide nanorods. J Phys Chem C 2007;111(1):194−200.

[20] Vasileiadis T, Dracopoulos V, Kollia M, Yannopoulos SN. Laser-assisted growth of t-Te nanotubes and their controlled photo-induced unzipping to ultrathin core-Te/sheath-TeO_2 nanowires. Sci Rep 2013;3.

[21] Mandl B, Stangl J, Hilner E, Zakharov AA, Hillerich K, Dey AW, et al. Growth mechanism of self-catalyzed group III−V nanowires. Nano Lett 2010;10(11):4443−9.

[22] Joo J, Chow BY, Prakash M, Boyden ES, Jacobson JM. Face-selective electrostatic control of hydrothermal zinc oxide nanowire synthesis. Nat Mater 2011;10 (8):596−601.

[23] Yu H, Buhro WE. Solution−liquid−solid growth of soluble GaAs nanowires. Adv Mater 2003;15(5):416−9.

[24] Chen H, Wang N, Di J, Zhao Y, Song Y, Jiang L. Nanowire-in-microtube structured core/shell fibers via multifluidic coaxial electrospinning. Langmuir 2010;26 (13):11291−6.

[25] Wu W, Bai S, Yuan M, Qin Y, Wang ZL, Jing T. Lead zirconate titanate nanowire textile nanogenerator for wearable energy-harvesting and self-powered devices. ACS Nano 2012;6(7):6231−5.

[26] Zhou J, Jin C, Seol JH, Li X, Shi L. Thermoelectric properties of individual electrode-posited bismuth telluride nanowires. Appl Phys Lett 2005;87(13):133109.

[27] Yoo B, Xiao F, Bozhilov KN, Herman J, Ryan MA, Myung NV. Electrodeposition of thermoelectric superlattice nanowires. Adv Mater 2007;19(2):296−9.

[28] Shin T-Y, Yoo S-H, Park S. Gold nanotubes with a nanoporous wall: their ultrathin platinum coating and superior electrocatalytic activity toward methanol oxidation. Chem Mater 2008;20(17):5682−6.

[29] Zhu Y, Dou X, Huang X, Li L, Li G. Thermal properties of Bi nanowire arrays with different orientations and diameters. J Phys Chem B 2006;110(51):26189−93.

[30] Nam KT, Kim D-W, Yoo PJ, Chiang C-Y, Meethong N, Hammond PT, et al. Virus-enabled synthesis and assembly of nanowires for lithium ion battery electrodes. Science 2006;312(5775):885−8.

[31] Wu H, Chan G, Choi JW, Ryu I, Yao Y, McDowell MT, et al. Stable cycling of double-walled silicon nanotube battery anodes through solid−electrolyte interphase control. Nat Nanotechnol 2012;7(5):310−5.

[32] Joyce HJ, Wong-Leung J, Gao Q, Tan HH, Jagadish C. Phase perfection in zinc blende and wurtzite III−V nanowires using basic growth parameters. Nano Lett 2010;10 (3):908−15.

[33] Ikejiri K, Kitauchi Y, Tomioka K, Motohisa J, Fukui T. Zinc-blende and wurtzite crystal phase mixing and transition in indium phosphide nanowires. Nano Lett 2011;11 (10):4314−8.

[34] Han N, Hou JJ, Wang F, Yip S, Lin H, Fang M, et al. Large-scale and uniform preparation of pure-phase wurtzite GaAs NWs on non-crystalline substrates. Nanoscale Res Lett 2012;7(1):1−6.

[35] Han N, Hui AT, Wang F, Hou JJ, Xiu F, Hung TF, et al. Crystal phase and growth orientation dependence of GaAs nanowires on Ni_xGa_y seeds via vapor−solid−solid mechanism. Appl Phys Lett 2011;99(8):083114.

[36] Han N, Wang FY, Hui AT, Hou JJ, Shan GC, Xiu F, et al. Facile synthesis and growth mechanism of Ni-catalyzed GaAs nanowires on non-crystalline substrates. Nanotechnology 2011;22(28):285607.

[37] Lin ZH, Yang Y, Wu JM, Liu Y, Zhang F, Wang ZL. $BaTiO_3$ nanotubes-based flexible and transparent nanogenerators. J Phys Chem Lett 2012;3(23):3599−604.

[38] Wu YY, Yang PD. Direct observation of vapor−liquid−solid nanowire growth. J Am Chem Soc 2001;123(13):3165−6.

[39] Duan X, Wang J, Lieber C. Synthesis and optical properties of gallium arsenide nanowires. Appl Phys Lett 2000;76(9):1116.

[40] Zhang GQ, Tateno K, Sanada H, Tawara T, Gotoh H, Nakano H. Synthesis of GaAs nanowires with very small diameters and their optical properties with the radial quantum-confinement effect. Appl Phys Lett 2009;95(12):123104.

[41] Han N, Hu P, Zuo AH, Zhang DW, Tian YJ, Chen YF. Photoluminescence investigation on the gas sensing property of ZnO nanorods prepared by plasma-enhanced CVD method. Sens Actuators B 2010;145(1):114−9.

[42] Tajik N, Peng Z, Kuyanov P, LaPierre R. Sulfur passivation and contact methods for GaAs nanowire solar cells. Nanotechnology 2011;22(22):225402.

[43] Hou JJ, Wang F, Han N, Xiu F, Yip S, Fang M, et al. Stoichiometric effect on electrical, optical, and structural properties of composition-tunable $In_xGa_{1-x}As$ nanowires. ACS Nano 2012;6(10):9320−5.

[44] Chun HJ, Choi YS, Bae SY, Choi HC, Park J. Single-crystalline gallium-doped indium oxide nanowires. Appl Phys Lett 2004;85(3):461−3.

[45] Perea DE, Allen JE, Steven J, Wessels BW, Seidman DN, Lauhon LJ. Three-dimensional nanoscale composition mapping of semiconductor nanowires. Nano Lett 2006;6(2):181−5.

[46] Bar-Sadan M, Barthel J, Shtrikman H, Houben L. Direct imaging of single Au atoms within GaAs nanowires. Nano Lett 2012;12(5):2352−6.

[47] Hemesath ER, Schreiber DK, Gulsoy EB, Kisielowski CF, Petford-Long AK, Voorhees PW, et al. Catalyst incorporation at defects during nanowire growth. Nano Lett 2011;12(1):167−71.

[48] Schroer MD, Petta JR. Correlating the nanostructure and electronic properties of InAs nanowires. Nano Lett 2010;10(5):1618−22.

[49] Motayed A, Vaudin M, Davydov AV, Melngailis J, He M, Mohammad S. Diameter dependent transport properties of gallium nitride nanowire field effect transistors. Appl Phys Lett 2007;90(4):043104.

[50] Ford AC, Ho JC, Chueh YL, Tseng YC, Fan ZY, Guo J, et al. Diameter-dependent electron mobility of InAs nanowires. Nano Lett 2009;9(1):360−5.

[51] Jabeen F, Rubini S, Martelli F, Franciosi A, Kolmakov A, Gregoratti L, et al. Contactless monitoring of the diameter-dependent conductivity of GaAs nanowires. Nano Res 2010;3(10):706−13.

[52] Li HY, Wunnicke O, Borgström M, Immink W, Van Weert M, Verheijen M, et al. Remote p-doping of InAs nanowires. Nano Lett 2007;7(5):1144−8.

[53] Han N, Wang F, Hou JJ, Xiu F, Yip S, Hui AT, et al. Controllable p−n switching behaviors of GaAs nanowires *via* an interface effect. ACS Nano 2012;6 (5):4428−33.

[54] Fan ZY, Kapadia R, Leu PW, Zhang XB, Chueh YL, Takei K, et al. Ordered arrays of dual-diameter nanopillars for maximized optical absorption. Nano Lett 2010;10 (10):3823−7.

[55] Wang GM, Yang XY, Qian F, Zhang JZ, Li Y. Double-sided CdS and CdSe quantum dot co-sensitized ZnO nanowire arrays for photoelectrochemical hydrogen generation. Nano Lett 2010;10(3):1088−92.

[56] Li L, Zhang Y, Yang Y, Huang X, Li G, Zhang L. Diameter-depended thermal expansion properties of Bi nanowire arrays. Appl Phys Lett 2005;87(3):031912.

[57] Martin PN, Aksamija Z, Pop E, Ravaioli U. Reduced thermal conductivity in nanoengineered rough GE and GAAS nanowires. Nano Lett 2010;10(4):1120−4.

[58] Hochbaum AI, Chen R, Delgado RD, Liang W, Garnett EC, Najarian M, et al. Enhanced thermoelectric performance of rough silicon nanowires. Nature 2008;451 (7175):163−7.

[59] Cuenot S, Frétigny C, Demoustier-Champagne S, Nysten B. Surface tension effect on the mechanical properties of nanomaterials measured by atomic force microscopy. Phys Rev B 2004;69(16):165410.

[60] Liu K, Wang W, Xu Z, Liao L, Bai X, Wang E. *In situ* probing mechanical properties of individual tungsten oxide nanowires directly grown on tungsten tips inside transmission electron microscope. Appl Phys Lett 2006;89(22):221908.

[61] Wu B, Heidelberg A, Boland JJ. Mechanical properties of ultrahigh-strength gold nanowires. Nat Mater 2005;4(7):525−9.

[62] Wen B, Sader JE, Boland JJ. Mechanical properties of ZnO nanowires. Phys Rev Lett 2008;101(17):175502.

[63] Wu Z, Zhang Y-W, Jhon MH, Gao H, Srolovitz DJ. Nanowire failure: long = brittle and short = ductile. Nano Lett 2012;12(2):910−4.

[64] Gao PX, Song J, Liu J, Wang ZL. Nanowire piezoelectric nanogenerators on plastic substrates as flexible power sources for nanodevices. Adv Mater 2006;19(1):67−72.

[65] Vivekchand S, Ramamurty U, Rao C. Mechanical properties of inorganic nanowire reinforced polymer−matrix composites. Nanotechnology 2006;17(11):S344.

[66] Crabtree GW, Lewis NS. Solar energy conversion. Phys Today 2007;60(3):37−42.

[67] Fan ZY, Razavi H, Do JW, Moriwaki A, Ergen O, Chueh YL, et al. Three-dimensional nanopillar-array photovoltaics on low-cost and flexible substrates. Nat Mater 2009;8 (8):648−53.

[68] Goetzberger A, Hebling C, Schock H-W. Photovoltaic materials, history, status and outlook. Mater Sci Eng R 2003;40(1):1−46.

[69] Han N, Wang F, Yip S, Hou JJ, Xiu F, Shi X, et al. GaAs nanowire schottky barrier photovoltaics utilizing Au−Ga alloy catalytic tips. Appl Phys Lett 2012;101 (1):013105.

[70] Yu R, Ching K-L, Lin Q, Leung S-F, Arcrossito D, Fan Z. Strong light absorption of self-organized 3-D nanospike arrays for photovoltaic applications. ACS Nano 2011;5 (11):9291−8.

[71] Lin Q, Hua B, Leung S-F, Duan X, Fan Z. Efficient light absorption with integrated nanopillar nanowell arrays for three-dimensional thin-film photovoltaic applications. ACS Nano 2013;7(3):2725–32.

[72] Holm JV, Jørgensen HI, Krogstrup P, Nygård J, Liu H, Aagesen M. Surface-passivated GaAsP single-nanowire solar cells exceeding 10% efficiency grown on silicon. Nat Commun 2013;4:1498.

[73] Tian BZ, Zheng XL, Kempa TJ, Fang Y, Yu NF, Yu GH, et al. Coaxial silicon nanowires as solar cells and nanoelectronic power sources. Nature 2007;449 (7164):885–90.

[74] Wallentin J, Anttu N, Asoli D, Huffman M, Åberg I, Magnusson MH, et al. InP nanowire array solar cells achieving 13.8% efficiency by exceeding the ray optics limit. Science 2013;339(6123):1057–60.

[75] Goto H, Nosaki K, Tomioka K, Hara S, Hiruma K, Motohisa J, et al. Growth of core––shell InP nanowires for photovoltaic application by selective-area metal organic vapor phase epitaxy. Appl Phys Express 2009;2(3):035004.

[76] Czaban JA, Thompson DA, LaPierre RR. GaAs core–shell nanowires for photovoltaic applications. Nano Lett 2009;9(1):148–54.

[77] Oregan B, Gratzel M. A low-cost, high-efficiency solar-cell based on dye-sensitized colloidal TiO$_2$ films. Nature 1991;353(6346):737–40.

[78] Vlachopoulos N, Liska P, Augustynski J, Gratzel M. Very efficient visible-light energy harvesting and conversion by spectral sensitization of high surface-area polycrystalline titanium-dioxide films. J Am Chem Soc 1988;110(4):1216–20.

[79] Gratzel M. Photoelectrochemical cells. Nature 2001;414(6861):338–44.

[80] Wang ZS, Cui Y, Hara K, Dan-Oh Y, Kasada C, Shinpo A. A high-light-harvesting-efficiency coumarin dye for stable dye-sensitized solar cells. Adv Mater 2007;19 (8):1138–41.

[81] Meng S, Ren J, Kaxiras E. Natural dyes adsorbed on TiO$_2$ nanowire for photovoltaic applications: enhanced light absorption and ultrafast electron injection. Nano Lett 2008;8(10):3266–72.

[82] Gratzel M. Solar energy conversion by dye-sensitized photovoltaic cells. Inorg Chem 2005;44(20):6841–51.

[83] Enache-Pommer E, Boercker JE, Aydil ES. Electron transport and recombination in polycrystalline TiO$_2$ nanowire dye-sensitized solar cells. Appl Phys Lett 2007;91 (12):123116.

[84] Jiu JT, Isoda S, Wang FM, Adachi M. Dye-sensitized solar cells based on a single-crystalline TiO$_2$ nanorod film. J Phys Chem B 2006;110(5):2087–92.

[85] Gubbala S, Chakrapani V, Kumar V, Sunkara MK. Band-edge engineered hybrid structures for dye-sensitized solar cells based on SnO$_2$ nanowires. Adv Funct Mater 2008;18(16):2411–8.

[86] Law M, Greene LE, Radenovic A, Kuykendall T, Liphardt J, Yang PD. ZnO–Al$_2$O$_3$ and ZnO–TiO$_2$ core–shell nanowire dye-sensitized solar cells. J Phys Chem B 2006;110(45):22652–63.

[87] Bang JH, Kamat PV. Solar cells by design: photoelectrochemistry of TiO$_2$ nanorod arrays decorated with CdSe. Adv Funct Mater 2010;20(12):1970–6.

[88] Zhu K, Neale NR, Miedaner A, Frank AJ. Enhanced charge-collection efficiencies and light scattering in dye-sensitized solar cells using oriented TiO$_2$ nanotubes arrays. Nano Lett 2007;7(1):69–74.

[89] Tan B, Wu YY. Dye-sensitized solar cells based on anatase TiO$_2$ nanoparticle/nanowire composites. J Phys Chem B 2006;110(32):15932–8.

[90] Law M, Greene LE, Johnson JC, Saykally R, Yang PD. Nanowire dye-sensitized solar cells. Nat Mater 2005;4(6):455−9.

[91] Chen XY, Yip CT, Fung MK, Djurisic AB, Chan WK. GaN-nanowire-based dye-sensitized solar cells. Appl Phys A 2010;100(1):15−9.

[92] Fujishima A, Honda K. Electrochemical photolysis of water at a semiconductor electrode. Nature 1972;238(5358):37−8.

[93] Heller A. Hydrogen-evolving solar cells. Science 1984;223(4641):1141−8.

[94] van de Krol R, Liang YQ, Schoonman J. Solar hydrogen production with nanostructured metal oxides. J Mater Chem 2008;18(20):2311−20.

[95] Li Y, Zhang JZ. Hydrogen generation from photoelectrochemical water splitting based on nanomaterials. Laser Photonics Rev 2010;4(4):517−28.

[96] Shankar K, Basham JI, Allam NK, Varghese OK, Mor GK, Feng XJ, et al. Recent advances in the use of TiO_2 nanotube and nanowire arrays for oxidative photoelectrochemistry. J Phys Chem C 2009;113(16):6327−59.

[97] Paulose M, Shankar K, Yoriya S, Prakasam HE, Varghese OK, Mor GK, et al. Anodic growth of highly ordered TiO_2 nanotube arrays to 134 µm in length. J Phys Chem B 2006;110(33):16179−84.

[98] Yang XY, Wolcott A, Wang GM, Sobo A, Fitzmorris RC, Qian F, et al. Nitrogen-doped ZnO nanowire arrays for photoelectrochemical water splitting. Nano Lett 2009;9(6):2331−6.

[99] Vayssieres L, Sathe C, Butorin SM, Shuh DK, Nordgren J, Guo JH. One-dimensional quantum-confinement effect in a-Fe_2O_3 ultrafine nanorod arrays. Adv Mater 2005;17 (19):2320−3.

[100] Siripala W, Ivanovskaya A, Jaramillo TF, Baeck S-H, McFarland EWA. Cu_2O/TiO_2 heterojunction thin film cathode for photoelectrocatalysis. Sol Energy Mater Sol Cells 2003;77(3):229−37.

[101] Barreca D, Fornasiero P, Gasparotto A, Gombac V, Maccato C, Montini T, et al. The potential of supported Cu_2O and CuO nanosystems in photocatalytic H_2 production. Chemsuschem 2009;2(3):230−3.

[102] Zhao X, Wang P, Li B. CuO/ZnO core/shell heterostructure nanowire arrays: synthesis, optical property, and energy application. Chem Commun 2010;46(36):6768−70.

[103] Chakrapani V, Thangala J, Sunkara MK. WO_3 and W_2N nanowire arrays for photoelectrochemical hydrogen production. Int J Hydrogen Energy 2009;34(22):9050−9.

[104] Aryal K, Pantha BN, Li J, Lin JY, Jiang HX. Hydrogen generation by solar water splitting using p-InGaN photoelectrochemical cells. Appl Phys Lett 2010;96 (5):052110.

[105] Hensel J, Wang G, Li Y, Zhang JZ. Synergistic effect of CdSe quantum dot sensitization and nitrogen doping of TiO_2 nanostructures for photoelectrochemical solar hydrogen generation. Nano Lett 2010;10(2):478−83.

[106] Yu ZG, Pryor CE, Lau WH, Berding MA, MacQueen DB. Core−shell nanorods for efficient photoelectrochemical hydrogen production. J Phys Chem B 2005;109 (48):22913−9.

[107] Nowotny J, Sorrell CC, Bak T, Sheppard LR. Solar-hydrogen: unresolved problems in solid-state science. Sol Energy 2005;78(5):593−602.

[108] Hwang YJ, Boukai A, Yang PD. High density n-Si/n-TiO_2 core/shell nanowire arrays with enhanced photoactivity. Nano Lett 2009;9(1):410−5.

[109] Zhao M-H, Wang Z-L, Mao SX. Piezoelectric characterization of individual zinc oxide nanobelt probed by piezoresponse force microscope. Nano Lett 2004; 4(4):587−90.

[110] Wang ZL. Nanopiezotronics. Adv Mater 2007;19(6):889−92.

[111] Wang ZL, Song JH. Piezoelectric nanogenerators based on zinc oxide nanowire arrays. Science 2006;312(5771):242−6.

[112] Song J, Zhou J, Wang ZL. Piezoelectric and semiconducting coupled power generating process of a single ZnO belt/wire. A technology for harvesting electricity from the environment. Nano Lett 2006;6(8):1656−62.

[113] Periasamy C, Chakrabarti P. Time-dependent degradation of Pt/ZnO nanoneedle rectifying contact based piezoelectric nanogenerator. J Appl Phys 2011;109 (5):054306.

[114] Yang R, Qin Y, Dai L, Wang ZL. Power generation with laterally packaged piezoelectric fine wires. Nat Nanotechnol 2008;4(1):34−9.

[115] Wang X, Song J, Zhang F, He C, Hu Z, Wang Z. Electricity generation based on onedimensional group-III nitride nanomaterials. Adv Mater 2010;22(19):2155−8.

[116] Hung S-C. Generation of piezoelectricity by deflecting nanorods vertically on GaN template. J Electrochem Soc 2011;158(12):H1265−9.

[117] Scrymgeour DA, Hsu JW. Correlated piezoelectric and electrical properties in individual ZnO nanorods. Nano Lett 2008;8(8):2204−9.

[118] Sohn JI, Cha SN, Song BG, Lee S, Kim SM, Ku J, et al. Engineering of efficiency limiting free carriers and an interfacial energy barrier for an enhancing piezoelectric generation. Energy Environ Sci 2013;6(1):97−104.

[119] Park HK, Lee KY, Seo JS, Jeong JA, Kim HK, Choi D, et al. Charge-generating mode control in high-performance transparent flexible piezoelectric nanogenerators. Adv Funct Mater 2011;21(6):1187−93.

[120] Lee S, Hong JI, Xu C, Lee M, Kim D, Lin L, et al. Toward robust nanogenerator using aluminum substrate. Adv Mater 2012;24(32):4398−402.

[121] Xu S, Qin Y, Xu C, Wei Y, Yang R, Wang ZL. Self-powered nanowire devices. Nat Nanotechnol 2010;5(5):366−73.

[122] Xu S, Wei Y, Liu J, Yang R, Wang ZL. Integrated multilayer nanogenerator fabricated using paired nanotip-to-nanowire brushes. Nano Lett 2008;8(11):4027−32.

[123] Ku NJ, Wang CH, Huang JH, Fang HC, Huang PC, Liu CP. Energy harvesting from the obliquely aligned InN nanowire array with a surface electron-accumulation layer. Adv Mater 2013;25(6):861−6.

[124] Huang C-T, Song J, Lee W-F, Ding Y, Gao Z, Hao Y, et al. GaN nanowire arrays for high-output nanogenerators. J Am Chem Soc 2010;132(13):4766−71.

[125] Zhou YS, Wang K, Han W, Rai SC, Zhang Y, Ding Y, et al. Vertically aligned CdSe nanowire arrays for energy harvesting and piezotronic devices. ACS Nano 2012; 6(7):6478−82.

[126] Hou T-C, Yang Y, Lin Z-H, Ding Y, Park C, Pradel KC, et al. Nanogenerator based on zinc blende CdTe micro/nanowires. Nano Energy 2012;2(3):387−93.

[127] Wu JM, Xu C, Zhang Y, Yang Y, Zhou Y, Wang ZL. Flexible and transparent nanogenerators based on a composite of lead-free ZnSnO₃ triangular belts. Adv Mater 2012;24(45):6094−9.

[128] Gu L, Cui N, Cheng L, Xu Q, Bai S, Yuan M, et al. Flexible fiber nanogenerator with 209 V output voltage directly powers a light-emitting diode. Nano Lett 2012;13(1):91−4.

[129] Lee S, Bae SH, Lin L, Yang Y, Park C, Kim SW, et al. Super-flexible nanogenerator for energy harvesting from gentle wind and as an active deformation sensor. Adv Funct Mater 2012;23(19):2445−9.

[130] Shakouri A. Recent developments in semiconductor thermoelectric physics and materials. Annu Rev Mater Res 2011;41(1):399−431.

[131] Yang Y, Guo W, Pradel KC, Zhu G, Zhou Y, Zhang Y, et al. Pyroelectric nanogenerators for harvesting thermoelectric energy. Nano Lett 2012;12(6):2833−8.
[132] Snyder GJ, Toberer ES. Complex thermoelectric materials. Nat Mater 2008;7 (2):105−14.
[133] Lim JR, Whitacre JF, Fleurial JP, Huang CK, Ryan MA, Myung NV. Fabrication method for thermoelectric nanodevices. Adv Mater 2005;17(12):1488−92.
[134] Chen G, Dresselhaus M, Dresselhaus G, Fleurial J-P, Caillat T. Recent developments in thermoelectric materials. Int Mater Rev 2003;48(1):45−66.
[135] Lin Y-M, Dresselhaus M. Thermoelectric properties of superlattice nanowires. Phys Rev B 2003;68(7):075304.
[136] Liang W, Rabin O, Hochbaum AI, Fardy M, Zhang M, Yang P. Thermoelectric properties of p-type PbSe nanowires. Nano Res 2009;2(5):394−9.
[137] Yang R, Chen G, Dresselhaus MS. Thermal conductivity of simple and tubular nanowire composites in the longitudinal direction. Phys Rev B 2005;72(12):125418.
[138] Seol JH, Moore AL, Saha SK, Zhou F, Shi L, Ye QL, et al. Measurement and analysis of thermopower and electrical conductivity of an indium antimonide nanowire from a vapor−liquid−solid method. J Appl Phys 2007;101(2):023706.
[139] Chen X, Wang Y, Ma Y. High thermoelectric performance of Ge/Si core−shell nanowires: first-principles prediction. J Phys Chem C 2010;114(19):9096−100.
[140] Liang G, Huang W, Koong CS, Wang J-S, Lan J. Geometry effects on thermoelectric properties of silicon nanowires based on electronic band structures. J Appl Phys 2010;107(1):014317.
[141] Lewis NS. Toward cost-effective solar energy use. Science 2007;315(5813):798−801.
[142] Ibrahim H, Ilinca A, Perron J. Energy storage systems—characteristics and comparisons. Renewable Sustainable Energy Rev 2008;12(5):1221−50.
[143] Debe MK. Electrocatalyst approaches and challenges for automotive fuel cells. Nature 2012;486(7401):43−51.
[144] Tang JM, Jensen K, Waje M, Li W, Larsen P, Pauley K, et al. High performance hydrogen fuel cells with ultralow Pt loading carbon nanotube thin film catalysts. J Phys Chem C 2007;111(48):17901−4.
[145] Bashyam R, Zelenay P. A class of non-precious metal composite catalysts for fuel cells. Nature 2006;443(7107):63−6.
[146] Bing Y, Liu H, Zhang L, Ghosh D, Zhang J. Nanostructured Pt-alloy electrocatalysts for PEM fuel cell oxygen reduction reaction. Chem Soc Rev 2010;39 (6):2184−202.
[147] Chen Z, Waje M, Li W, Yan Y. Supportless Pt and PtPd nanotubes as electrocatalysts for oxygen-reduction reactions. Angew Chem Int 2007;46(22):4060−3.
[148] Liang HW, Cao X, Zhou F, Cui CH, Zhang WJ, Yu SH. A free-standing Pt-nanowire membrane as a highly stable electrocatalyst for the oxygen reduction reaction. Adv Mater 2011;23(12):1467−71.
[149] Etacheri V, Marom R, Elazari R, Salitra G, Aurbach D. Challenges in the development of advanced Li-ion batteries: a review. Energy Environ Sci 2011;4(9):3243−62.
[150] Chan CK, Peng H, Liu G, McIlwrath K, Zhang XF, Huggins RA, et al. High-performance lithium battery anodes using silicon nanowires. Nat Nanotechnol 2007; 3(1):31−5.
[151] Zhang W-J. A review of the electrochemical performance of alloy anodes for lithium-ion batteries. J Power Sources 2011;196(1):13−24.
[152] Chan CK, Patel RN, O'Connell MJ, Korgel BA, Cui Y. Solution-grown silicon nanowires for lithium-ion battery anodes. ACS Nano 2010;4(3):1443−50.

[153] Li Y, Tan B, Wu Y. Mesoporous Co_3O_4 nanowire arrays for lithium ion batteries with high capacity and rate capability. Nano Lett 2008;8(1):265−70.

[154] Chan CK, Zhang XF, Cui Y. High capacity Li ion battery anodes using Ge nanowires. Nano Lett 2008;8(1):307−9.

[155] Park M-H, Kim MG, Joo J, Kim K, Kim J, Ahn S, et al. Silicon nanotube battery anodes. Nano Lett 2009;9(11):3844−7.

[156] Wang G, Zhang L, Zhang J. A review of electrode materials for electrochemical supercapacitors. Chem Soc Rev 2012;41(2):797−828.

[157] Shaijumon MM, Ou FS, Ci L, Ajayan PM. Synthesis of hybrid nanowire arrays and their application as high power supercapacitor electrodes. Chem Commun 2008;20:2373−5.

[158] Yu G, Xie X, Pan L, Bao Z, Cui Y. Hybrid nanostructured materials for high-performance electrochemical capacitors. Nano Energy 2013;2(2):213−34.

4 Lanthanide-Doped Nanoparticles: Synthesis, Property, and Application

Hongli Wen and Feng Wang

Department of Physics and Materials Science, City University of Hong Kong, Hong Kong SAR, PR China

4.1 Introduction

Lanthanides represent a group of 15 chemically similar elements from lanthanum (La) to lutetium (Lu) that are associated with the filling of the 4f shell. These 15 lanthanide elements, along with the chemically similar elements scandium and yttrium, are often collectively known as the rare earth (RE) elements (Figure 4.1). The stunning story of lanthanides starts more than two centuries ago, but practical applications of these elements were not realized until early 1900s when their physical and chemical properties are properly understood [1−3]. In recent decades, lanthanides have contributed essentially to a range of modern day materials and generated increasing interests in fundamental research and technological applications in various fields ranging from environment and energy to biological sciences [4−7].

Lanthanides essentially exist in their most stable oxidation state as trivalent ions (Ln^{3+}). The trivalent lanthanide ions feature an electron configuration of $[Xe]4f^n$ ($0 < n < 14$) and the arrangements of electrons within this configuration are substantially diverse. As a result of the Coulomb interaction and the spin−orbit interaction between f electrons, different distributions of 4f electrons typically result in a fairly large number of energy levels that can be marked as $^{2S+1}L_J$ according to the Russell−Saunders notation [8,9]. When Ln^{3+} ions are incorporated into the host lattice of a crystalline material, these energy levels are hardly affected because of the shielding effect by 5s and 5p orbitals. However, the surrounding crystal field could lift the parity selection rule, thereby promoting intraconfigurational f−f electron transitions and giving rise to optical absorption and emission with reasonable intensities. The optical property associated with intraconfigurational f−f transitions has several unique features, including large Stokes shifts, long excited-state lifetimes, sharp emission bandwidths, and high resistance to optical blinking. Importantly, lanthanide-doped luminescent materials show extremely high resistance to photobleaching and photochemical degradation because 4f electrons do not

Nanocrystalline Materials. DOI: http://dx.doi.org/10.1016/B978-0-12-407796-6.00004-X

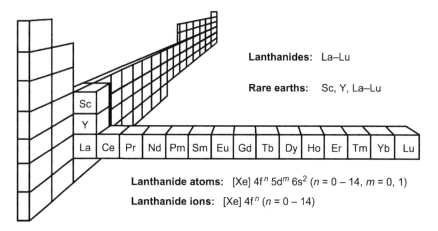

Figure 4.1 Position of the rare earth elements in the periodic table. The electron configurations of lanthanide atoms and ions are depicted to highlight the partially filled 4f orbitals.

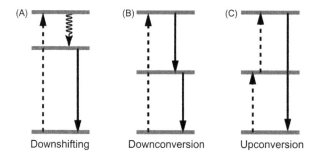

Figure 4.2 Simplified energy level diagrams describing the basic principle of DF, DC, and UC processes through use of 4f energy levels of lanthanide ions. The dashed, curly, and full arrows represent photon excitation, multiphonon relaxation, and emission processes, respectively. Note that only two basic photon processes are involved in B and C. For comprehensive illustrations of DC and UC, see Refs. [10,11].

play a role in chemical bonding and therefore no chemical bonds can be broken in the photocycle [1].

Electronic transitions within enormously complex energy levels of the lanthanide ions can result in several distinct optical processes, including downshifting (DF), downconversion (DC), and upconversion (UC) (Figure 4.2). DF is a common photoluminescence process that converts a high-energy excitation photon to a lower energy emission photon [1]. The quantum yield defined as the ratio of emitted photons to absorbed photons in a DF process is never beyond unity. DC is similar to a DF process except that one high-energy excitation photon results in several lower energy emission photons, with an overall quantum yield typically larger than unity [10]. UC is a reverse process of DC and features the combination of several low-energy excitation photons to generate one higher energy emission photon [11]. Through these well-established optical processes, lanthanide ions virtually can be used for spectral conversion in the whole spectral region spanning from ultraviolet (UV) to near infrared (NIR).

In recent years, lanthanide-doped nanoparticles have emerged as a new class of luminescent materials with expanded range of applications [12−20]. Nanoscale manipulation of lanthanide-doped nanoparticles can lead to important modification of their optical properties in excited-state dynamics, emission profiles, and efficiency. For example, the incompatible lanthanide dopant ions can be confined in different regions of a nanoparticle with well-defined energy exchange interactions, thereby resulting in unprecedented optical emissions that are typically inaccessible by conventional bulk materials. The study of lanthanide-doped nanoparticles is now a highly interdisciplinary field that has been rapidly expanding at the frontiers of photochemistry, biophysics, solid state physics, and materials science.

In this chapter, we discuss several substantial topics of current interests in the synthesis, property tuning, and technological application of lanthanide-doped nanoparticles. The general synthetic protocols to lanthanide-doped nanoparticles are described in Section 4.2. Specially, we focus on the wet-chemistry methods for monodisperse fluorides nanoparticles that represent the majority of nanoparticles reported in the literatures. In Section 4.3, we attempt to provide general strategies for optical tuning and surface modification of lanthanide-doped nanoparticles. In Section 4.4, we highlight the emerging applications of these lanthanide-doped nanoparticles in various fields.

4.2   Wet-Chemical Synthesis of Nanoparticles

Efforts were made to prepare lanthanide-doped nanoparticles in the late 1990s. However, the resulting products from these attempts generally exhibited poor dispersibility and had dimensions in the submicrometer range due to high-temperate treatment in the fabrication process [21]. Therefore, research is mainly directed to the wet-chemical synthesis at relatively low temperatures. Advances in the synthesis of nanoparticles have now enabled rational fabrication of lanthanide-doped nanoparticles with various compositions and complex core−shell structures. It is worth noting that the quality of the as-synthesized nanoparticles not only is affected by a specific synthetic method but also is strongly dependent on the nature of the materials, such as chemical composition and crystal structure. Some nanoparticles, especially those with narrow size distribution and high dispersity, require a special method or a combination of several methods, but others can be prepared by many complementary approaches. In this section, we focus on the state-of-the-art wet-chemistry approaches for the synthesis of lanthanide-doped nanoparticles that display intense optical emission in solutions.

4.2.1 Synthesis of Core Nanoparticles

Wet-chemistry synthetic methods typically involve a coprecipitation or a thermal decomposition reaction that results in the growth of solid nanoparticles in liquid reaction media. Except for only a few examples [22−25], heat treatment at above 100°C is required to prompt the reaction and to improve crystallinity of the resulting nanoparticles. To date, most nanoparticles are synthesized through hydro(solvo)thermal processing, coprecipitation in high boiling point solvent, and thermal decomposition.

The hydro(solvo)thermal synthesis utilizes a thick-walled steel cylinders with a hermetic seal for carrying out chemical reactions under pressures and at temperatures above the critical point of the solvent [26,27]. Possible advantages of this method include the ability to create highly crystalline phases at relatively low temperatures (100−200°C). Importantly, the reaction media that typically comprise water is compatible with a wide variety of inorganic metal salt precursors, thereby allowing cost-effective and environmentally benign synthesis of a large diversity of nanoparticles. Notably, Li et al. [28] have recently devised a water/ethanol/linoleic acid system to synthesize lanthanide-doped fluoride nanoparticles with different chemistries and properties and with high dispersity. The strategy is based on a general phase transfer and separation occurring at the interfaces of the liquid, solid, and solution phases during the synthesis. The liquid−solid−solution (LSS) strategy has been demonstrated to be a convenient route toward high quality fluoride nanoparticles, including $NaYF_4$, LaF_3, and CaF_2 by the same group and by several other groups (Figure 4.3A and B) [29,33−40].

Although hydro(solvo)thermal processing is effective for the synthesis of lanthanide-doped nanoparticles, the method typically suffers from drawbacks, such as the need for specialized reaction vessels known as autoclaves and the difficulty of observing the nanoparticle as it grows. To prepare nanoparticles in a common glass flask, people have attempted to carry out the reaction in high boiling point organic solvents. This process usually utilizes metal−oleate precursors that can be prepared by incubating the metal salts with sodium oleate at 70°C [41] or by heating inorganic metal salts in oleic acid at elevated temperatures ($\sim 150°C$) [30]. By mixing the metal−oleate precursor solution with a polar solution comprising fluorine anions at around 50°C, a phase transfer of fluorine across the liquid−liquid interface will occur and result in the formation of crystal nucleus in the organic phase. Subsequent heat treatment at high temperatures ($\sim 260−300°C$) will prompt the growth of monodisperse nanoparticles of various compositions, such as $NaYF_4$, $LiYF_4$, and $KMnF_3$, depending on the metal precursor involved (Figure 4.3C) [42−47].

Thermal decomposition is an alternative approach for the synthesis of lanthanide-doped fluoride nanoparticles in high boiling point organic solvents. The approach typically involves a single precursor of metal trifluoroacetate (a mixture of several metal trifluoroacetates in case of doped nanoparticles) that contains all the elements constituting the nanoparticles. Metal trifluoroacetates are typically obtained by incubating corresponding metal oxide with trifluoroacetic acid at $\sim 80°C$ overnight [48]. When heating at high temperatures ($\sim 300°C$), the metal

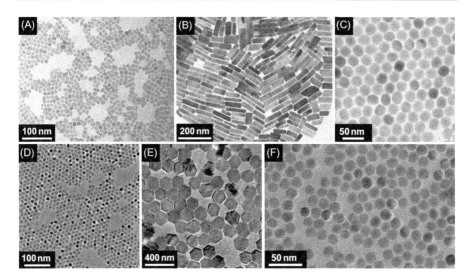

Figure 4.3 TEM images of typical lanthanide-doped nanoparticles. (A and B) α-NaYF$_4$:Yb/Er nanocubes and β-NaYF$_4$:Yb/Er nanorods prepared by Wang and Li through solvothermal methods. (C) β-NaYF$_4$:Yb/Er nanoparticles prepared by Li and Zhang through coprecipitation in high boiling point organic solvent. (D and E) α-NaYF$_4$:Yb/Er nanoparticles and β-NaYF$_4$:Yb/Er nanoplates prepared by Mai et al. through thermal decomposition. (F) β-NaGdF$_4$:Yb/Ho nanoplate nanoparticles prepared by Naccache et al. through thermal decomposition.
Source: (A and B) From Wang and Li [29], (C) from Li and Zhang [30], (D and E) from Mai et al. [49], and (F) from Naccache et al. [32].

trifluoroacetates will be quickly decomposed that triggers a burst nucleation process followed by the growth of high quality fluoride nanoparticles. The thermal decomposition method was developed by Yan's group [31] to synthesize highly monodisperse LaF$_3$ nanoparticles. The approach was later extended as a common route to the synthesis of various fluoride nanoparticles, including NaYF$_4$, NaGdF$_4$, LiYF$_4$, BaYF$_4$, and SrF$_2$ (Figure 4.3D−F) [32,49−60]. It is worth noting that the strategy typically requires accurate temperature control for temporal separation of nucleation and crystal growth for narrow nanoparticle size distributions. In addition, the synthesis must be handled with care as the decomposition of the metal trifluoroacetates produces various fluorinated and oxyfluorinated carbon species, such as trifluoroacetic anhydride (CF$_3$CO)$_2$O and trifluoroacetyl fluoride CF$_3$CF$_2$COF that are considered toxic.

4.2.2   Synthesis of Core−Shell Nanoparticles

A lanthanide-doped nanoparticle typically contains a large number of dopant ions spreading throughout host lattice. Notably, a high portion of dopant ions are exposed at or close to the nanoparticle surface and readily lose their excitation energy to the surrounding high-energy oscillators, resulting in poor optical emission of the nanoparticles. A common strategy for overcoming this limitation is to spatially confine the dopant ions in the interior of a core—shell structure.

Core—shell nanoparticles are usually synthesized by an epitaxial growth process, which involves the use of preformed core nanoparticles as seeds to mediate the growth of the shell layer following a similar procedure to that for preparing the core nanoparticles. To date, most core—shell nanoparticles were synthesized in high boiling point organic solvent, perhaps due to easy manipulation and monitor of the nanoparticle growth process, such as injection of reagents and extraction of intermediary products. For example, Yi and Chow [61] have demonstrated a one pot synthesis for $NaYF_4$:Yb/Er@$NaYF_4$ through hot injection. In their thermal decomposition synthesis, CF_3COONa, $(CF_3COO)_3Y$, $(CF_3COO)_3Yb$, and $(CF_3COO)_3Er$ were first heated in oleylamine at 340°C for 30 min to incubate the doped core nanoparticles. Then a mixture comprising CF_3COONa and $(CF_3COO)_3Y$ in oleylamine was injected into the reaction for epitaxial growth of an undoped shell layer.

Recently, van Veggel et al. [62] have envisaged an epitaxial growth technique by using sacrificial nanoparticles as solid shell precursors. They demonstrated that the injection of small sacrificial nanoparticles into a solution of core nanoparticles will result in a size-focusing process owing to Ostwald ripening, which leads to the dissolution of the small nanoparticles and subsequent deposition on the core nanoparticles. Following this strategy, a serial of core—shell nanoparticles including $NaYF_4$:Yb/Er@$NaYF_4$, $NaYF_4$:Yb/Er@$NaGdF_4$, and $NaGdF_4$:Yb/Er@$NaYF_4$ were obtained and the shell thickness was readily controlled by manipulating the number of sacrificial nanoparticles injected. Significantly, the protocol allows quick growth of multiple shell layers by successive self-focusing cycle without the need for tedious multiple reactions or stringent control over the introduction of shell components.

Instead of hot injection, several groups choose to extract the core nanoparticles from the reaction media and transfer these presynthesized nanoparticles to a fresh reaction for epitaxial shell coating [63—68]. Albeit laborious to some extent, this strategy avoids dangerous injection procedures at high temperatures. Another advantage might be the flexibility for controlling the formation of core—shell nanoparticles. For example, the core nanoparticles can be synthesized through solvothermal processing followed by a dissimilar coating process carried out in high boiling point organic solvent, yielding large core particle with a very thin shell layer [69]. Note that during the course of shell coating, phase separation often occurs as a result of kinetically favored nucleation of shell precursors, leading to the formation of a binary mixture of nanoparticles. Therefore, the experimental variables should be carefully controlled for the synthesis of uniform core—shell nanoparticles [69,70].

Other than growing an additional coating layer around the core nanoparticle, van Veggel et al. [71] reported an interesting demonstration to form core—shell nanostructures through partial cation exchange reaction in preformed nanoparticles. They found that a thin surface layer (~ 1 nm) of 20-nm $NaYF_4$:Yb/Tm nanoparticles can be converted to an $NaGdF_4$ shell by exposing the nanoparticles in a water solution comprising concentrated Gd^{3+} ($GdCl_3$) salts. They also showed that the shell thickness of the $NaYF_4$:Yb/Tm@$NaGdF_4$ core−shell nanoparticles can be tuned by varying the conditions of the cation exchange process, including incubation temperature and Gd^{3+} concentration. The challenges for this method perhaps include the formation of thick shells (> 2 nm) as well as the synthesis of complex core−shell−shell nanostructures.

It is worth noting that the verification of a core−shell structure is still relatively challenging currently [72]. A regular transmission electron microscopy (TEM) image typically cannot distinguish the shell from the core due to the low-contrast difference between the lanthanides. Core−shell structure is usually inferred from a change in nanoparticle size based on TEM analysis coupled with enhanced photoluminescence and increased lifetime based on spectroscopic investigations. Recently, high-resolution TEM techniques including high-angle annular dark-field imaging, energy-dispersive spectroscopy or energy-dispersive X-ray spectroscopy, and electron energy loss spectroscopy in line scan mode or in two-dimensional elemental mapping mode were frequently used to provide useful information for confirming a core−shell structure. However, fluoride nanoparticles are fairly vulnerable to the high-energy electron beams, making characterization rather difficult. An advanced technique with atomic resolution is highly desired for imaging lanthanide-doped core−shell nanoparticles and for studying element diffusion between different layers.

4.3 Tuning Nanoparticle Properties

Nanoparticle characteristics, such as size, shape, optical emission, and surface properties, are fundamental issues that ultimately influence the application of these nanoparticles. The rapid development of nanofabricating techniques coupled with improved understanding of energy exchange interactions at the nanoscale has now enabled exquisite control over the properties of lanthanide-doped nanoparticles.

4.3.1 Tuning Shape and Size

Shape and size of lanthanide-doped nanoparticles are perhaps the most significant parameters that affect their utilizations in technological applications. For example, nanoparticles for biomedical application should have a small particle size and good dispersity for integration with biological molecules and macromolecules.

The shape and size of lanthanide-doped nanoparticles can be readily tuned by controlling the thermodynamic or kinetic growth of the nanoparticles through manipulation of experimental variables, such as temperature, reaction time, and relative concentrations of the precursors [73−76]. For example, Zhang et al. [77] utilized the LSS strategy for the synthesis of $NaYF_4$ nanorods, nanotubes, and flower-patterned nanodisks by adjusting the oleic acid to NaOH ratio and NaF concentration (Figure 4.4A−C). Notably, it was found that the thickness of the

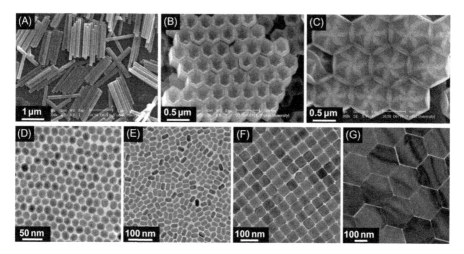

Figure 4.4 Tuning shape and size of lanthanide-doped nanoparticles. (A−C) Scanning electron microscopy (SEM) images of arrays of β-NaYF₄ by Zhang et al. The nanorods, nanotubes, and flower-patterned hexagonal nanodisks were synthesized through control precursor compositions in a solvothermal process. (D−G) TEM images of the β-NaYF₄:Yb/Er (20/2 mol%) nanoparticles by Ye et al. The nanoparticle morphology was tuned from spherical, to nanorods, to hexagonal nanoprisms, and finally to hexagonal nanoplates by adjusting the reaction time and/or the precursor composition in a thermal decomposition process.
Source: (A−C) From Zhang et al. [77] and (D−G) from Ye et al. [78].

hexagonal NaYF₄ prisms can be continuously tuned by altering the NaF-to-Y^{3+} molar ratio in the reaction system. In another outstanding demonstration, Ye et al. [78] have shown that monodisperse NaYF₄ nanoparticles with controlled shapes can be prepared by using the thermal decomposition method. They showed that under appropriate conditions the as-synthesized nanoparticles can be assembled into large-area superlattices (Figure 4.4D−G).

Recently, lanthanide doping has been envisaged as a promising alternative for facile control over the shape and size of NaYF₄ nanoparticles. In stark contrast to conventional synthetic techniques that require stringent control over several experimental variables for controlling crystallite size and shape, the lanthanide doping approach requires modification of only a single variable (dopant concentration), while enabling tremendous improvements in our ability to control the formation of nanoparticles. For example, we have demonstrated that hexagonal phase NaYF₄ with a tunable size down to 10 nm (Figure 4.5) can be readily synthesized through doping of Gd^{3+} ions at precisely defined concentrations [80]. By comparison, conventional synthesis usually requires air-sensitive, toxic organometallic precursors and hazardous coordinating solvents for accessing sub-20-nm hexagonal-phase NaYF₄ nanoparticles [51]. The versatility of the doping approach was also confirmed and used to control the formation of various NaREF₄ nanoparticles by

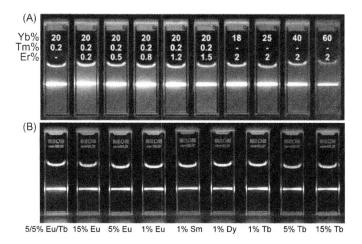

Figure 4.5 Multicolor fine-tuning of lanthanide-doped nanoparticles. (A) Multicolor tuning in $NaYF_4$ nanoparticles through control of dopant concentrations. (B) Multicolor tuning in $NaGdF_4$:Yb/Tm@$NaGdF_4$:Ln@$NaYF_4$ core−shell−shell nanoparticles comprising different lanthanide activator ions at defined dopant concentrations.
Source: (A) From Wang and Liu [79] and (B) from Su et al. [70].

several other groups [81−85]. More recently, Wang et al. [86,87] have extended the doping strategy to control the shape and size of MF_2 (M = Ca, Sr, and Ba) and LnF_3 (Ln = La, Ce, and Pr) nanoparticles.

4.3.2   Tuning Optical Emission

The ability to manipulate the color output of lanthanide-doped nanoparticles is particularly important for their applications. For example, as a liquid crystal display backlight, the nanoparticles should generate white-color emission. However for multiplexed labeling, a set of nanoparticles that show well-resolved emission spectra are required.

The strategies for tuning the color output of lanthanide-doped nanoparticles typically involve manipulating dopant/host combinations and dopant concentrations. Each lanthanide ion has a unique set of energy levels and therefore could offer a set of distinct emission peaks with distinguishable spectroscopic fingerprints. Importantly, different emission peaks of a given lanthanide ion could be selectively enhanced or suppressed through control of its surrounding environment in the host lattice, thereby leading to enormously tunable optical emissions.

The emission color of a nanoparticle is commonly tuned by selecting different emitter (activator) ions. An antenna (sensitizer) is usually codoped to collect the single wavelength excitation and to subsequently promote the emitter ions to the excited states through nonradiative energy transfer. Haase et al. [88,89] have pioneered multicolor tuning of lanthanide-doped nanoparticles in colloidal solutions.

In one of their representative work [88], ligand-capped lanthanum phosphate nano-particles were synthesized and codoped with Yb/Er or Yb/Tm. Here Yb^{3+} dopant servers as the antenna responsible for harvesting NIR irradiation light at 980 nm. As a result of successive energy transfer steps, these nanoparticles show tunable UC emissions spanning from the visible-to-NIR spectral regions, corresponding to electronic transitions within enormously complex energy levels of Er^{3+} and Tm^{3+} ions. By using Ce^{3+} ion or $[VO_4]^{3-}$ group as antenna, multicolor tuning by single wavelength UV excitation has also been demonstrated with other lanthanide ions, including Tb^{3+}, Eu^{3+}, Dy^{3+}, Sm^{3+}, and Nd^{3+}[90−93].

Notably, a lanthanide ion may display dramatically different emission profiles in different nanoparticles as a result of dissimilar dopant−host interactions. For exam-ple, Er^{3+} ion typically displays two main emission peaks in the green and red spec-tral regions, respectively. When doped in β-$NaYF_4$ featuring cubic crystal structure, the green emission peak is prominent [43]. By contrast, the red emission peak usually dominates in hexagonal phase α-$NaYF_4$[79]. However, when $KMnF_3$ is employed as the host material, a single peak red emission will appear due to a dopant−host energy transfer process [94]. Mn^{2+} ions in the host lattice can capture the energy of Er^{3+} ions in the $^2H_{9/2}$ and $^4S_{3/2}$ excited states, followed by back energy transfer to a ground state Er^{3+} that enhances the population in red emitting state ($^4F_{9/2}$). The energy exchange interactions are very efficient even at low tem-peratures of 10 K.

The emission color of lanthanide-doped nanoparticles has been fine-tuned by manipulation of relative intensity of the multipeak emission through control of dop-ant concentrations [43,79]. Dopant concentration dictates average distance between dopant ions in the host lattice, which affects energy exchange interactions among dopant ions and consequently the optical transitions. For example, the introduction of an elevated amount of Yb^{3+} dopants in the $NaYF_4$:Yb/Er host lattice can lead to emission color tuning from yellow to red (Figure 4.5A) [79]. This effect is attrib-uted to facilitation of back energy transfer from Er^{3+} to Yb^{3+} caused by decreased Yb...Er interionic distance. As a result, the population in excited states of $^2H_{9/2}$, $^2H_{11/2}$, and $^4S_{3/2}$ was suppressed by the energy transfer, resulting in the decrease of blue ($^2H_{9/2}\rightarrow{}^4I_{15/2}$) and green ($^2H_{11/2}$, $^4S_{3/2}\rightarrow{}^4I_{15/2}$) light emissions with respect to the red ($^4F_{9/2}\rightarrow{}^4I_{15/2}$). Similarly, the NIR emission of Tm^{3+} can be considerably enhanced with respect to the blue emission by increasing the concentration of Tm^{3+} ions in $NaYF_4$:Yb/Tm nanoparticles [79]. The phenomenon is primarily attributed to enhanced population of the $^3H_4$ level generated by the energy resonant between $^1G_4\rightarrow{}^3H_4$ and $^3F_2\leftarrow{}^3F_4$ at elevated dopant concentration of Tm^{3+}. The control of dopant concentration has also proven a versatile approach to fine-tuning the optical emission in other lanthanide-doped luminescent nanoparticles [95−100].

It is worth noting that the tunable optical emissions demonstrated in the afore-mentioned nanoparticles in principle stem from similar optical processes occurring in their corresponding bulk counterparts. By contrast, nanoscale manipulation of lanthanide-doped nanoparticles has opened up new possibilities toward optical tun-ing. For example, Yan et al. [101] have demonstrated that the green-to-red emission ratio of Er^{3+} in $NaYF_4$ nanoparticles can be largely tuned by increasing the nanoparticle size from 5 to 8 nm. It is important to note that the size-dependent optical property results from surface effects rather than quantum confinement effects [102]. The emission peaks of lanthanide-doped nanoparticles typically correspond to a sum of optical emissions contributed from dopant ions at the surface and in the interior of the nanoparticles. When compared to the interior dopant ions, the surface dopant ions should display a dissimilar emission behavior owing to quenching of the excitation energy by surface defects, impurities, ligands, and solvents. As the nanoparticle size decreases, the concentration of the surface dopant ions increases, leading to a variation of the relative emission intensity.

Intriguingly, core−shell nanostructures have also been utilized for optical tuning. Core−shell design allows the integration of incompatible lanthanide ions into a single nanoparticle with well-defined energy exchange interactions and provides additional flexibilities for emission modulation [63,66,103,104]. For example, we previously showed that by combining Gd-mediated energy migration and core − shell structure engineering, unprecedented optical emission is possible for lanthanide activators (Tb^{3+}, Eu^{3+}, Dy^{3+}, and Sm^{3+}) by using NIR excitation (Figure 4.5B) [69,70]. The $NaGdF_4$:Yb/Tm@$NaGdF_4$:Ln (Ln = Tb^{3+}, Eu^{3+}, Dy^{3+}, and Sm^{3+}) core − shell structure separates the Yb/Tm pair from the activators and eliminates deleterious cross-relaxation. Upon excitation, the low-energy NIR photons promote Tm^{3+} in the core layer to a high-lying excited state. The excitation energy then migrates over the Gd sublattice for a substantial distance to the activators, which are confined in the shell layer. Recently, we have improved our design by coating an inert $NaYF_4$ shell around the $NaGdF_4$@$NaGdF_4$ nanoparticles. We showed that the $NaYF_4$ coating impedes the migrating energy in Gd sublattice from trapping by surface quenchers, thereby promoting energy trapping by the activators [70]. The effect can lead to efficient activator emission without stringent control of activator concentration and should represent a key step toward the rational design of novel lanthanide-based luminescent nanomaterials.

4.3.3 Tuning Surface Properties and Functionality

Lanthanide-doped nanoparticles are typically prepared in the presence of organic ligands, such as oleic acid or oleylamine, that bind to the nanoparticle surface to control the particle growth and stabilize the particles against aggregation. Nanoparticles capped with hydrocarbon chains either have no intrinsic aqueous dispersibility or lack reactive organic moieties for further conjugation with functional molecules. In this regard, a collection of surface modification methods has been developed to improve dispersibility of the nanoparticles in various solvents and to provide a platform that integrates small molecules for various applications (Figure 4.6).

A common surface modification approach is based on ligand exchange that involves displacement of original ligands by hydrophilic polymeric molecules. A wide variety of bifunctional organic molecules, including 6-aminohexanoic acid, poly(ethylene glycol) (PEG) diacid, hexanedioic acid, citrate, polyacrylic acid (PAA), polyvinylpyrrolidone, and phosphate-derived molecules, have been used to replace the hydrophobic ligands on the surface of the as-prepared nanoparticles.

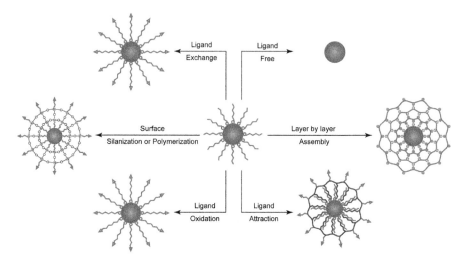

Figure 4.6 Schematic presentations illustrating the typical strategies that are commonly used by the community for surface modification of lanthanide-doped nanoparticles. *Source*: Reproduced from Wang and Liu [13] with permission of the Royal Society of Chemistry (http://pubs.rsc.org/en/Content/ArticleLanding/2009/CS/b809132n).

For example, Yi and Chow [51] have reported the ligand exchange preparation of hydrophilic $NaYF_4$:Yb/Er nanoparticles that were previously stabilized with oleylamine ligands. The technique utilized PEG 600 diacid molecules that replace the amine ligand to provide a water-soluble carboxyl-functionalized surface.

Ligand exchange with inorganic chemicals has also been used for the synthesis of ligand-free hydrophilic nanoparticles, which can be sequentially functionalized by reaction with bifunctional organic molecules. For example, Capobianco et al. [105] showed that oleate ligand at the surface of $NaYF_4$:Yb/Er nanoparticles can be removed through treatment with inorganic acid, resulting in ligand-free nanoparticles that are dispersible in water solutions of varying pH (2.6−10). Owing to strong coordination capability of Ln^{3+} ions, the ligand-free nanoparticles would facilitate direct conjugation with molecules comprising groups, such as −COOH and −NH_2. In parallel demonstration, Murray et al. [106] recently reported a general ligand exchange strategy by employing nitrosonium tetrafluoroborate ($NOBF_4$) to replace the original ligand. Notably, $NOBF_4$ can stabilize nanoparticles in various polar and hydrophilic media, including *N,N*-dimethylformamide, dimethylsulfoxide, and acetonitrile for years. Importantly, the $NOBF_4$-modified nanoparticles are amenable to a secondary ligand exchange process with a variety of organic molecules, thereby imparting different surface functionalities to the nanoparticles.

Alternatively, surface functionalization can also be achieved through chemical treatment of the original ligand molecules. For example, Li et al. [107] devised a ligand oxidation method to modify oleate-stabilized hydrophobic $NaYF_4$ nanoparticle. The oxidation process utilized Lemieux−von Rudloff reagent to transform monounsaturated carbon–carbon double bonds into carboxylic acid groups that impart the nanoparticles with water dispersibility and provide reactive functional moieties. Oxidation with Lemieux–von Rudloff reagent typically suffers from limitations, such as a relatively long reaction time and low yields. To this end, the group further developed an epoxidation approach by using 3-chloroperoxybenzoic acid as a peroxide reagent [108]. The method yields highly reactive epoxide intermediates that can be further conjugated with a broad range of molecules containing –SH, –NH$_2$, or –COOH groups. In a parallel development, Yan et al. [109] demonstrated oxidation of oleate ligand with ozone to selectively generate carboxyl or aldehyde groups with no noticeable influence on the morphology and optical emission of the nanoparticles. Interestingly, the functionalized surface not only allows easy conjugation with small molecules but also provides a platform for attachment of small gold nanoparticles.

Surface functionalization of lanthanide-doped nanoparticles has also been achieved through subsequent surface coating without discarding the original ligand molecules. Hydrocarbon chains on the nanoparticle surface can be utilized to immobilize amphiphilic block polymers via hydrophobic interactions. The hydrophilic outer block of the polymer then permits aqueous dispersion and further surface derivatization. For example, Yi and Chow [61] have modified commercially available PAA with 25% octylamine and 40% isopropylamine for use as the coating material. They showed that the amphiphilic polymer facilitates dispersion of the nanoparticles in various polar solvents, including ethanol, water, phosphate-buffered saline, and tris–borate–EDTA, to form stable transparent solutions. In addition to amphiphilic polymers, polyion molecules were also assembled on nanoparticle surface through layer-by-layer deposition based on electrostatic interactions. By sequential adsorption of positively charged poly(allylamine hydrochloride) and negatively charged poly (sodium 4-styrenesulfonate), Li et al. [110] have successfully modified NaYF$_4$:Yb/Er nanoparticles with stable amino-rich surfaces.

A more general method of surface coating is based on polymerization or condensation of various monomer molecules at the nanoparticle surface [111–119]. Important for these coating methods is the possibility to encapsulate additional functional nanoparticles or molecules in the coating shell. At the same time, surface coverage with molecules consisting of additional functional groups would allow further reactions with biological entities. Among various surface coating methods, silica coating enjoys common usage largely owing to the sophisticated Stöber method for coating various nanoparticles and well-established surface chemistry of silica coating for facile bioconjugation. Another reason perhaps lies in that the silica coating features porous structures that can be precisely controlled by several methods, thereby permitting rational trap and release of small molecules for a variety of biological applications. Impressively, Lin et al. [120–122] have prepared a diversity of composite nanoparticles comprising a porous silica shell. Their method generally utilizes cetyltrimethyl ammonium bromide (CTAB) as a cofactor to control the condensation of tetraethyl orthosilicate on the nanoparticle surface. Follow-up calcination at elevated temperatures ($\sim 500°C$) is typically involved to remove CTAB and thereby to promote the formation of mesoporous silica coatings.

4.4 Technological Applications

Owing to their small feature size and amenability to surface modification, lanthanide-doped nanoparticles can be readily conjugated with biomolecules for biological studies or used as building blocks for integrated optoelectronic devices, thereby providing promising applications ranging from biological diagnostic and therapy to photoelectric conversion.

4.4.1 Bioapplication

The advantages of lanthanide-doped nanoparticles for bioapplication, include low toxicity, characteristic optical emission, and high resistance to photoblinking and photochemical degradation. Especially, lanthanide-doped UC nanoparticles can display UV and visible emission by NIR irradiation that is essentially not absorbed by biological samples, thereby allowing maximal light penetration through tissues and minimal photodamage to cells. These nanoparticles are now widely accepted as attractive alternatives to organic dyes and quantum dots for sensitive biodetection, high-contrast bioimaging, as well as effective treatment of diseases like cancer.

4.4.1.1 Biodetection

In terms of the signal detection formats, the currently established biological assays can be generally divided into two major categories: heterogeneous bioassay and homogeneous bioassay [17]. Both detection formats involves the attachment of biomolecules on the nanoparticle surface for probing target molecules through highly specific binding affinity between biological molecules, such as an antigen and an antibody.

A heterogeneous assay platform makes use of the particular interaction between target analyte present in a liquid phase and the capture molecule immobilized on a solid substrate. Depending on whether the analytes are binding to the lanthanide-doped nanoparticles (the reporters), heterogeneous assay can be further divided into sandwich assay and competitive assay as illustrated in Figure 4.7. In a sandwich assay, target analyte is firstly conjugated with the capture molecule incubated on the solid substrate and then marked by the capture molecule labeled by lanthanide-doped nanoparticles. After completely washing off excess amount of unbound lanthanide-doped nanoparticles labeled capture molecules, the concentration of the analyte can be measured by detection of the optical signal of the lanthanide-doped nanoparticle labels. Differently, in a competitive assay, lanthanide-doped nanoparticles labeled target analyte instead of nanoparticle-labeled capture molecule is added to bind the capture molecule anchored on the substrate competing with the preadded free analyte, thus concentration of the free analyte will be inversely proportional to the detected optical signal [17].

Lanthanide-doped nanoparticles feature substantially long excited states (up to several milliseconds), which make the time-resolved luminescence detection technique readily feasible to minimize the interference of short-lived background

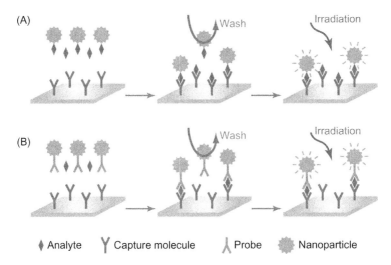

(A)

(B)

♦ Analyte   Y Capture molecule   人 Probe   🟢 Nanoparticle

Figure 4.7 Schematic illustration of two types of heterogeneous assays based on lanthanide-doped nanoparticles: (A) sandwich assay and (B) competitive assay.
Source: Reproduced from Wang et al. [123] with permission of the Royal Society of Chemistry (http://pubs.rsc.org/en/Content/ArticleLanding/2010/AN/c0an00144a).

autofluorescence. A DF-based heterogeneous assay has recently been combined with the time-resolved photoluminescence (TRPL) detection technique to minimize the unexpected interference of scattered lights and autofluorescence background from biological species due to the UV excitation. Since the PL lifetime of trivalent lanthanide ion typically in the order of a few to tens of milliseconds is much longer than that of scattered light and autofluorescence from biological samples generally in the nanosecond scale, the TRPL signal in bioassays can be measured with remarkably higher detection sensitivity and better signal-to-noise ratio by setting an appropriate delay time and gate time to minimize the interference of short-lived background noise [17]. Chen et al. [124] demonstrated the effectiveness of such TRPL detection technique when utilizing water-soluble and carboxyl group functionalized lanthanide-doped GdF_3 nanoparticles as the TRPL probes to detect trace amount of avidin with the delay time and gate time setting as 200 μs and 1 ms, respectively.

Especially, UC is a unique process associated with lanthanide-doped nanoparticles and visible background fluorescence is typically unexpected when an NIR diode laser (980 nm) is employed to excite lanthanide-doped UC nanoparticle reporters [125]. As a result, extremely high signal-to-noise ratio can be obtained without the use of expensive time-resolved spectrometers, thereby allowing for the development of high-performance compact and portable devices. For example, Niedbala et al. [126] reported a lateral flow-based strip assay for coincident detection of amphetamine, methamphetamine, phencyclidine, and opiates in saliva by use of multicolor $NaYF_4$:Yb,Er UC nanoparticles. Within this study, green-emitting

(550 nm) particles were coupled to antibodies for phencyclidine and amphetamine, and blue-emitting (475 nm) particles to antibodies for methamphetamine and morphine. The drug molecules were then successfully detected based on phosphor color and position after examination of the test strip for each colored phosphor.

Though heterogeneous assay can provide highly sensitive biodetection, it is labor intensive and time-consuming because of tedious separation and washing prior to optical measurements. In stark contrast, the homogeneous bioassay can be performed in one step and is particularly convenient to handle. Homogeneous assay is generally based on sensitive and reliable fluorescence resonance energy transfer (FRET) technique. Various homogeneous bioassay methods, such as conventional FRET, TR-FRET, and UC-FRET, have been developed for the most convenient biodetection by using lanthanide-doped nanoparticles as luminescent labels [17].

The conventional FRET assays achieve detection of the analyte through the distance-dependent energy transfer between a donor and an acceptor via long-range dipole—dipole interactions. In this process, excitation of the donor causes energy transfer from donor to acceptor under the circumstances of a closer distance between each other, spectral overlap between the acceptor absorption and the donor emission, and then the acceptor emitting light at its given wavelength [17]. Recently, Li and coworkers [127] developed a conventional FRET pair for glucose concentration determination by using glucose-modified LaF_3:Ce/Tb nanoparticles as energy donor and 3-aminophenyl boronic acid-modified rhodamine B isothiocyanate (RhBITC) as energy acceptor.

TR-FRET-based homogeneous assays, combining the merits of near-zero background noise from the TR technique and elimination of sample separation for homogeneous assay from FRET, accomplish quick detection without suffering from interference of background autofluorescence [128,129]. A TR-FRET assay shares the same main principle as that for conventional FRET, except that the fluorescence signal in TR-FRET is collected after a suitable delay time, typically $50-100\ \mu s$ after excitation [17]. Recently, Chen et al. [129] synthesized amine-functionalized and biocompatible lanthanide-doped $KGdF_4$ nanocrystals via a facile one-step solvothermal route by employing polyethylenimine (PEI) as capping ligand. The resultant nanoparticles have been demonstrated to be sensitive TR-FRET bioprobes to detect avidin protein at a concentration down to 5.5 nM.

UC-FRET biodetection utilizes UC nanoparticles as energy donors and the interference of background noise due to biological substances can be thoroughly removed in that the NIR irradiation can only excite the UC nanoparticles. False detection signals due to direct excitation into the acceptor as usually occurs in conventional FRET process can be largely exempted [17,123]. Morgan and Mitchell [130] first proposed the concept of UC-FRET bioassay and Kuningas et al. [131] made the afterward experimental validation and successfully demonstrated the sensitive detection of biotin by using streptavidin-conjugated La_2O_2S:Yb/Er or Y_2O_2S:Yb/Er UC phosphor as energy donors and biotinylated phycobiliprotein (bio-BPE) as acceptor. Rantanen et al. [132] later made an intriguing advance by combination of UC-FRET with conventional FRET for a fluorescence-quenching-based enzyme activity assay. The design incorporates $NaYF_4$:Yb/Er UC nanoparticles to donate energy to an

FRET pair comprising Alexa Fluor 680-Black and Berry Quencher 650. Compared to direct light excitation into Alexa Fluor 680 fluorophore at 655 nm, such UC-based system achieves an eightfold increase in signal-to-noise ratio.

4.4.1.2 Bioimaging

A luminescent material, when introduced into cells, tissues, and living organisms, can be readily visualized with a luminescence microscope, thereby providing useful information on what is happening in the biological samples under investigation. Although in principle DF and DC nanoparticles are also feasible for bioimaging [111,133−137], current studies almost invariably employ UC nanoparticles as contrast stains, largely owing to minimized background autofluorescence, reduced cytotoxicity, and deep tissue reaching by NIR excitation [13,20,123,138].

High-performance tissue imaging using lanthanide-doped UC nanoparticles was first explored in 1999 by Zijlmans et al. [139]. Within this study, they used submicron-sized Y_2O_2S:Yb/Tm particles to study the distribution of prostate-specific antigen in paraffin-embedded sections of human prostate tissue by standard immunohistological techniques. The nonspecific background autofluorescence related with the high-energy excitation from the tissue itself was thoroughly removed under NIR excitation, which enables high-resolution imaging. Also it was demonstrated that UC particle reporters do not bleach after continuous irradiation of high excitation energy, which therefore allows convenient keeping of tissue samples labeled with UC particles for permanent records.

With the recent development of readily obtainable high quality UC nanoparticles, the UC-based nanoparticles have been widely used for high-resolution cellular imaging. Vetrone et al. [140] showed that nonfunctionalized $NaYF_4$:Er/Yb nanoparticles can be endocytosed by HeLa cells and subsequently provide high-contrast luminescence micrographs of the cells with total absence of autofluorescence. Water-soluble and citrate-coated $NaYF_4$:Yb/Er UC nanoparticles were investigated by Cao et al. [141] to achieve nonspecific imaging of HeLa cells. Significantly, Park et al. [142] and Wu et al. [143] have demonstrated reliable single-molecule imaging by using $NaGdF_4$:Yb/Er and $NaYF_4$:Yb/Er UC nanoparticles (Figure 4.8), respectively. Equally important, Yu et al. [144] revealed that the UC-based visualization technique exhibits few fading effect over time, implying excellent ability of UC nanoparticles for long-term observation of cells.

Tumor-targeted cell imaging has also been widely investigated by using UC nanoparticles functionalized with biomolecular recognition moieties [118,145,146]. According to Wang et al. [118], highly specific staining and imaging of HeLa cells with antigen expressed on the cell membrane have been demonstrated by using of $NaYF_4$:Yb/Er UC nanoparticles conjugated with antibody. In another representative work, Zako et al. [146] demonstrated that Y_2O_3:Er nanoparticles modified with cyclic arginine−glycine−asparatic acid (RGD) peptide can specifically bind to cancer cells with elevated integrin $\alpha_v\beta_3$ expression.

UC nanoparticles have become an exceptional tool for *in vivo* organism and animal imagings. As a pioneering work carried out on live organism imaging through

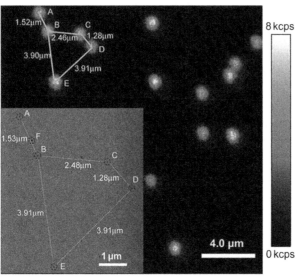

Figure 4.8 Confocal luminescence image of individual UC nanoparticles. The transmission mode-scanning electron microscope (TM-SEM) image (inset) taken at the upper left corner region of the optical image shows that the individual diffraction-limited luminescent spots are emitted from individual nanoparticles. *Source*: From Wu et al. [143].

the use of UC particles, Lim et al. [147] demonstrated that Y_2O_3:Yb/Er nanoparticles in the size range of 50–150 nm can be inoculated into live nematode *Caenorhabditis elegans* worms and then the digestive system of the worms can be imaged with IR excitation at 980 nm, clearly displaying the statistical distribution of the nanoparticles in the intestines. Especially, the nanoparticles have shown good biocompatibility as the worms behave normally in feeding. The facile preparation of small-sized UC nanoparticles with strong emission intensity and high aqueous dispersity has facilitated the *in vivo* small animal imaging by fast intravenous or intradermal injection of nanoparticles. *In vivo* animal imaging was first demonstrated by Chatterjee et al. [148], where PEI-coated $NaYF_4$:Yb/Er nanoparticles were injected underneath abdominal and back skin of anesthetized rats, and then clearly showed visible fluorescence when exposed to a 980-nm NIR laser. In a parallel demonstration by Nyk et al. [125], Maestro whole-body *in vivo* imaging of a Balb-c mouse has been achieved through use of NIR-to-NIR UC nanoparticles. The advantage of this technique is that both the excitation and emission are in the NIR range, allowing substantially imaging depth.

Great interests have been raised recently to investigate *in vivo* target imaging for tumor detection and drug delivery using lanthanide-doped UC nanoparticles conjugated with biomolecular recognition moieties. Xiong et al. [149,150] achieved detection of HeLa and U87MG tumors inside athymic nude mice by using folic acid-modified UC nanoparticles and RGD peptide-conjugated UC nanoparticles. UC nanoparticles have also been used for tracking transplanted cells *in vivo*. Recently, Idris et al. [151] have injected UC nanoparticle-loaded live myoblast cells into a living mouse model of cryoinjured hind limb and the distribution and activity of the delivered cells were studied by *in vivo* confocal imaging. The tumor-targeted molecular imaging described in these works is vitally important in tumor diagnosis and prognosis [20].

UC nanoparticles are also useful in diffuse optical tomography (DOT) that utilizes scattered light to probe structural variations in tissue. Typically, a DOT experiment is carried out by illuminating a highly scattering tissue medium using a narrow collimated beam and collecting the light which propagates through the medium using a series of detectors attached to the tissue surface. As tumorous tissue has different absorption and scattering properties, the recorded optical data will determine whether a tumor or other anomaly exists within the tissue [123]. Intrinsic or extrinsic fluorophores are usually utilized to enhance the imaging contrast [152−154]. However, traditional Stokes-shifting fluorophores raise much background tissue autofluorescence in the DOT scanning. In contrast, the anti-Stokes-shifted luminescence from UC nanoparticles upon NIR excitation can provide an autofluorescence-free environment for detection of signals [155]. Xu et al. [156] have recently developed autofluorescence insensitive DOT scanning with the use of $NaYF_4$:Yb/Tm nanoparticles. Liu et al. [157] further demonstrated multiple tomographic images obtained in DOT scanning through simultaneous excitation of the UC nanoparticles by two or more excitation beams, which can provide additional information and improved reconstruction of the optical data.

Multimodal imaging that combines several different imaging modalities within the setting of a single examination has also been widely studied based on UC nanoparticles. As a paramagnetic relaxation agent, gadolinium (Gd^{3+}) is extensively used in magnetic resonance (MR) imaging. UC nanoparticles comprising Gd^{3+} ions (e.g., $NaGdF_4$) therefore can simultaneously act as optical and magnetic contrast agents [123]. Park et al. [142], Zhou et al. [158], Kumar et al. [159], and Das et al. [160] have developed Gd^{3+}-based UC phosphors as multimodal imaging agents. A representative work, made by Hyeon and coworkers [142], demonstrated the use of $NaGdF_4$:Yb/Er nanoparticles for optical and MR imaging in breast cancer cells (SK-BR-3). Recently, Li and coworkers [158,161−163] extended the multimodal imaging with $NaGdF_4$-based UC nanoparticles to live small animals. They explored dual-modal optical and MR *in vivo* imaging using hydrophilic and carboxylic acid-functionalized $NaGdF_4$:Tm/Er/Yb UC nanoparticles in mice [158]. After that they used ^{18}F-labeled rare earth nanoparticles as a triple-model imaging probe for combined optical, MR and positron emission tomography (PET) imaging in animals [161−163]. In addition, several other related approaches have also demonstrated to develop UC-based nanoparticles as multimodal imaging probes by Zhu et al. [164], He et al. [165], and Xing et al. [166].

UC nanoparticles also provide a convenient platform for integration of additional imaging agents. Several groups have developed iron oxide (IO) nanoparticles (T_2-imaging)−UC nanoparticles nanocomposites for multimodal T_2-weighted MR and optical imaging, as super-paramagnetic IO nanoparticles have found wide application in MR imaging [167,168]. Gai et al. [120] demonstrated a facile strategy to synthesize multifunctional core−shell nanostructured material Fe_3O_4@$nSiO_2$@$mSiO_2$@UC nanoparticle using the encapsulation of silica-encapsulated magnetite (Fe_3O_4) nanoparticles in a mesoporous silica followed by functionalization on the outer surface of the mesoporous silica by UC nanoparticles. Chen et al. [169] have demonstrated both T_2-weighted MR and optical imaging using combination of IO nanoparticles and UC nanoparticles into heteronanoparticles by silica shielding through a "neck-formation" strategy. Zhu et al. [164] developed a stepwise synthetic method for the fabrication of core–shell Fe_3O_4@UC nanoparticle nanostructures with multifunctional properties for multimodal imaging. In addition, Cheng et al. [170] developed a novel class of multifunctional nanoparticles based on UC nanoparticle@IO nanoparticle@Au used for multimodal imaging. Recently, uniform core–shell structured fluorescent magnetic colloidal nanoparticles with Fe_3O_4/poly(St-co-GMA) nanoparticles as the seed synthesized at the first step and $Eu(AA)_3Phen$ copolymerized with the remaining St and GMA to form the fluorescent polymer shell at the second step, exhibiting a mean diameter of 120 nm and saturation magnetization of 1.92 emu/g, were developed by Zhu et al. [171] to demonstrate cytocompatibility, superparamagnetism, and excellent fluorescent properties, which make them an ideal candidate for using as biological imaging probes in MR and optical imaging.

4.4.1.3   Therapeutics

Lanthanide-doped nanoparticles alone are not effective therapeutic materials. They frequently act as a cofactor to promote the effect of other therapeutic reagents, for example, by functioning as a carrier for controlled delivery of anticancer drugs and as a light transducer for improving the optical response of a photosensitizer.

Lanthanide-doped nanoparticles can be made porous or surface-coated with a porous layer to store and deliver drugs. Through proper surface modification and functionalization, the nanoparticle platform allows targeted drug delivery to specific diseased cells or tissues and controlled drug release, leading to minimized side effects. For example, Lin et al. [121] developed a bifunctional core–shell nanocomposite comprising a Gd_2O_3:Er core and a mesoporous silica shell. They showed that ibuprofen molecules can be adsorbed onto the surface of mesoporous silica in hexane solution and released to media mimicking body fluids by a diffusion-controlled mechanism. Intriguingly, Branda and coworkers [172,173] reported that the UC emission of lanthanide-doped nanoparticles can be harnessed to trigger the release of cage compounds covalently bound to the nanoparticle surface through use of NIR light, providing an effective route to spatial and temporal control of drug delivery.

Lanthanide-doped nanoparticles are particularly useful to promote the effect of photodynamic therapy (PDT). PDT is an emerging medical method to treat a variety of diseases, such as cancers and cardiovascular [174,175], by using light to excite a photosensitizer to produce photochemical effects for the purpose of killing diseased cells or tissues. As one of the most promising modern medical technologies after chemotherapy, radiotherapy, and operative therapy [176,177], PDT approach offers a great deal of advantages, including simple operation, minimal damage, reduced side effects, repeatability, low cost, and increasing effectiveness [174,178]. However, current PDT techniques are limited to shallow lesions or small tumors, mainly attributing to excitation of photosensitizers with visible light that is strongly absorbed by biological samples and thus unlikely to penetrate deep into tissues [178].

NIR radiation that displays high penetration depth in biological tissues can be used for PDT when conventional photosensitizers are integrated with lanthanide-doped UC nanoparticles. UC nanoparticles can convert NIR radiation into visible light that can subsequently activate photosensitizers through localized energy transfer process. Y_2O_3:Yb/Er nanoparticles coated with a porous, thin layer of silica doped with the photosensitizer merocyanine 540 were used by Zhang et al. [179] in 2007 to first demonstrate the application of lanthanide-doped UC nanoparticles based PDT in MCF-7/AZ breast cancer cells. The obtained photosensitizer-impregnated nanoparticles were further functionalized with mouse monoclonal antibodies for killing targeted MCF-7/AZ breast cancer cells *in vitro* under NIR radiation, observed with several times increase in penetration depth compared with the visible irradiation excited PDT. Later, $NaYF_4$:Yb/Er UC nanoparticles with PEI coating bounded with a zinc phthalocyanine (ZnPC) photosensitizer were developed by Chatterjee and Zhang [180] for NIR light induced PDT. On NIR irradiation, UC nanoparticles can emit red light and activate photosensitizer ZnPC to produce singlet oxygen to kill cancer cells *in vitro*. Other studies using UC nanoparticles with mesoporous silica shell incorporated with photosensitizer molecules for NIR light induced PDT have also been developed and show promising results of *in vitro* cancer killing studies [181−183]. Researchers have also attempted to use UC nanoparticles for *in vivo* cancer PDT [184,185]. PEGylated $NaYF_4$:Yb/Er UC nanoparticles adsorbed with photosensitizer chlorine 6 (Ce6) to form an UC nanoparticle−Ce6 nanocomplex was used by Wang et al. [184] to demonstrate highly efficient cancer PDT treatment in a mouse model upon intratumoral injection of UCNP−Ce6, followed by NIR light exposure. It was further uncovered that UC nanoparticles were gradually eliminated from mouse organs after PDT treatment, showing nontoxicity to the treated animals. Moreover, a remarkably increased tissue penetration depth was demonstrated for NIR light-triggered PDT using UCNP−Ce6 compared to the visible light induced PDT.

4.4.2 Photovoltaic Application

Photovoltaic (PV) cells can convert sunlight directly into electricity, providing promising strategy for green and renewable energy generation. The current existing major problem in the efficiency improvement of PV energy conversion comes from the spectral mismatch between the incident solar spectrum and the bandgap of the semiconductor used for PV device fabrication [186,187]. Photons with energy lower than the bandgap are not absorbed by the solar cell for carrier generation (known as sub-bandgap transmission loss). Photons with energy higher than the bandgap are absorbed with excess energy lost as heat (known as lattice thermalization loss). The above-mentioned two energy loss mechanisms in solar cells account for the fundamental limitation of PV energy conversion efficiency [188]. The upper theoretical limit for the PV conversion efficiency, called the detailed balance limit of efficiency, known as the Shockley−Queisser limit, is found to be around 30% for crystalline silicon (c-Si) solar cell [189].

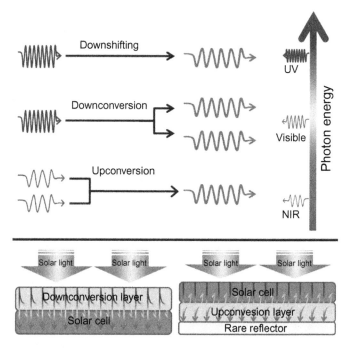

Figure 4.9 Spectral conversion PV design applied with DF, DC, and UC luminescent materials.
Source: From Wang et al. [191].

Solar cell performance can be optimized by reducing the absorption mismatch losses through spectral modification with lanthanide-doped nanoparticles. DS and DC can shift short-wavelength photon to longer wavelength one that is more sensitive to PV spectral response, thereby reducing thermalization losses. UC is useful for harvesting lower energy photons to overcome the sub-bandgap losses in solar cells. Typically, a photon conversion layer is coupled with the existing solar cell to enhance the energy conversion efficiency, by placing DS or QC layer at the front surface of a solar cell as the downconverted low-energy photons may be absorbed by the solar cell, and putting UC layer on the rear side of a solar cell to capture transmitted sub-bandgap light [190]. Figure 4.9 illustrates the photon conversion behaviors (DS, QC, and UC) and its application in designing of PV device [191].

4.4.2.1   DF Nanoparticles for PV Application

Chemical stability and energy conversion efficiency are two critical issues for practical use of dye-sensitized solar cells (DSSCs). UV irradiation was found to consume iodine irreversibly, degrade dye, and induce photooxidation of electrolyte solvents in DSSCs, and therefore strongly influence the chemical stability of DSSCs. Recently, a transparent luminescent film consisting of uniform Dy^{3+}-doped

$LaVO_4$ nanoparticles with capability of absorbing UV light and converting it to visible light was used by Liu et al. [192] to investigate the application in DSSCs and achieve a higher energy conversion efficiency relative to the referenced solar cell coated with an undoped $LaVO_4$ film, and greatly enhanced lifetime of the solar cell. In a parallel development, Hafez et al. [193] fabricated a TiO_2:Eu nanorod/TiO_2 nanoparticle bilayer film electrode in a DSSC and demonstrate the improved energy conversion efficiency, compared with that of the cell with undoped bilayer electrode, which is attributed to improvement in harvesting the UV light through downshifting via Eu^{3+} ion.

4.4.2.2 DC Nanoparticles for PV Application

DC typically occurs in special lanthanide ions, such as $Ce^{3+}-Yb^{3+}$ couple and the emissions mainly lie in the NIR spectral region suitable for improving the performance of c-Si solar cells. Notably, homogeneous $NaYF_4$:Er^{3+} and $NaYF_4$:Yb/Er nanoparticles with different sizes (10 and 200 nm) and various doping concentrations (Yb^{3+}: 0−20 mol%, Er^{3+}: 2 mol%) were used by Xu et al. [194] to obtain a wealth of strong IR emission bands through multiwavelength excitation (443, 488, and 520 nm). Several novel channels of visible-to-IR DC emission bands ranging from 1100 to 2000 nm were demonstrated for enhancing the efficiency of IR solar cells. In another example, crisscrossed Eu^{3+}-doped $Y(OH)_3$ nanotubes were developed by Cheng et al. as DC materials to demonstrate efficiency enhancement of more than 2% from 15.2 to 17.2% in screen-printed monocrystalline silicon solar cells (SPMSSCs) by spinning cast of the Eu^{3+}-doped $Y(OH)_3$ nanotubes on the front surface of the SPMSSCs [195].

4.4.2.3 UC Nanoparticles for PV Application

Despite the need of special lanthanide ions to facilitate the multiple excitation steps involved in UC, UC emission can be rationally tuned by a set of complimentary approaches to cover a wide spectral range from UV to NIR. Therefore UC nanoparticles can virtually be used to burst the efficiency of diverse solar cells. For example, $NaYF_4$:Yb/Er/Gd (18/2/30 mol%) nanorods coated with Au nanoparticles or Au shell nanostructures were developed by Li et al. [196] to demonstrate 16- to 72-fold improvement of the photocurrent under 980 nm light for the a-Si:H solar cell applied with the UC nanorods as the upconverters on the front side, compared with the cell without an upconverter. However, UC nanoparticles typically suffer from major drawbacks, including low photoconversion efficiency (<3%) and weak absorption of sunlight, which currently limit the practical application of upconverters for solar cells.

4.4.3 Other Applications

Apart from the biological and PV applications, lanthanide-doped nanoparticles are also proved useful in some other applications. UC nanoparticles of the $NaYF_4$:Yb/Er type were discussed by Mader and Wolfbeis [197] and Ali et al. [198] for applications to optical ammonium sensor and carbon dioxide sensor. Recent reports revealed the applications of the combination of UC nanoparticles and dithienylethene for optical memory [199,200]. Photopatterns based on Er/Yb or Tm/Yb codoped α-NaYF$_4$ UC nanoparticles with surface modification by incorporation of a photopatternable ligand like t-butoxycarbony were used for potential applications in security [201−203]. Lanthanide-doped nanoparticles can also be easily incorporated into various matrixes, such as transparent polymers and thin films, yielding composite bulk materials with promising applications in optical telecommunication and three-dimensional volumetric display [80,204−208]. Recently, lanthanide metal oxide nanoparticles with spherical shape and average particle size in range of 11−30 nm were used by Singh et al. [209] to demonstrate good catalytic effect on the thermal decomposition of ammonium perchlorate. Other reports include nanoparticles of rare earth-doped TiO$_2$ type useful for photocatalytic activities [210,211].

4.5 Conclusion

Lanthanide-doped nanoparticles as an emerging class of luminescent nanomaterials have generated diverse research interests covering fundamental research and technological applications. Recent advances in nanoparticle research have enabled exquisite control over crystal composition, phase, size, and shape. Rational structural engineering of lanthanide-doped nanoparticles has led to the construction of new optical materials with exciting photoconversion properties. Benefitting from the sophisticated surface modification and nanofabrication techniques, these nanoparticles also have provided valuable tools to promote biotechnology, information technology, energy and environmental sciences. Despite the achievements, our understanding of nanoparticle structures and luminescence mechanism is still limited partly owing to the lack of effective facilitates for characterization of nanoparticles with atomic resolution. Other challenges for the future include the development of small nanoparticles with higher quantum efficiency and more sensitive response to excitation light.

Acknowledgment

F. Wang acknowledges City University of Hong Kong for supporting this work.

References

[1]   Blasse G, Grabmaier BC. Luminescent materials. Berlin: Springer-Verlag; 1994.
[2]   Werts MHV. Making sense of lanthanide luminescence. Sci Prog 2005;88(2):101−31. Available from: http://dx.doi.org/10.3184/003685005783238435 [accessed May 2005].

 [3] Eliseeva SV, Bünzli J-CG. Rare earths: jewels for functional materials of the future. New J Chem 2011;35(6):1165−76. Available from: http://dx.doi.org/10.1039/C0NJ00969E [accessed 02.02.11].
 [4] Bünzli J-CG, Piguet C. Taking advantage of luminescent lanthanide ions. Chem Soc Rev 2005;34(12):1048−77. Available from: http://dx.doi.org/10.1039/B406082M [accessed 20.09.05].
 [5] Binnemans K. Lanthanide-based luminescent hybrid materials. Chem Rev 2009;109 (9):4283−374. Available from: http://dx.doi.org/10.1021/cr8003983 [accessed 04.08.09].
 [6] Kobayashi H, Ogawa M, Alford R, Choyke PL, Urano Y. New strategies for fluorescent probe design in medical diagnostic imaging. Chem Rev 2010;110(5):2620−40. Available from: http://dx.doi.org/10.1021/cr900263j [accessed 15.12.09].
 [7] Borisov SM, Wolfbeis OS. Optical biosensors. Chem Rev 2008;108(2):423−61. Available from: http://dx.doi.org/10.1021/cr068105t [accessed 30.01.08].
 [8] Carnall WT, Goodman GL, Rajnak K, Rana RS. A systematic analysis of the spectra of the lanthanides doped into single crystal LaF_3. J Chem Phys 1989;90(7):3443. Available from: http://dx.doi.org/10.1063/1.455853 [accessed 01.04.89].
 [9] Webb CE, Jones JDC. Handbook of laser technology and applications: laser design and laser systems. London: Taylor & Francis; 2004.
[10] Wegh RT, Donker H, Oskam KD, Meijerink A. Visible quantum cutting in $LiGdF_4$: Eu^{3+} through downconversion. Science 1999;283(5402):663−6. Available from: http://dx.doi.org/10.1126/science.283.5402.663 [accessed 29.01.99].
[11] Auzel F. Upconversion and anti-stokes processes with f and d ions in solids. Chem Rev 2004;104(1):139−74. Available from: http://dx.doi.org/10.1021/cr020357g [accessed 18.11.03].
[12] Shen J, Sun L-D, Yan C-H. Luminescent rare earth nanomaterials for bioprobe applications. Dalton Trans 2008;(42):5687−97. Available from: http://dx.doi.org/10.1039/B805306E [accessed 06.08.08].
[13] Wang F, Liu X. Recent advances in the chemistry of lanthanide-doped upconversion nanocrystals. Chem Soc Rev 2009;38(4):976−89. Available from: http://dx.doi.org/10.1039/B809132N [accessed 12.02.09].
[14] Zhou J, Liu Z, Li F. Upconversion nanophosphors for small-animal imaging. Chem Soc Rev 2012;41(3):1323−49. Available from: http://dx.doi.org/10.1039/C1CS15187H [accessed 19.10.11].
[15] Wang G, Peng Q, Li Y. Lanthanide-doped nanocrystals: synthesis, optical-magnetic properties, and applications. Acc Chem Res 2011;44(5):322−32. Available from: http://dx.doi.org/10.1021/ar100129p [accessed 11.03.11].
[16] Zhang C, Lin J. Defect-related luminescent materials: synthesis, emission properties and applications. Chem Soc Rev 2012;41(23):7938−61. Available from: http://dx.doi.org/10.1039/C2CS35215J [accessed 27.09.12].
[17] Liu Y, Tu D, Zhu H, Ma E, Chen X. Lanthanide-doped luminescent nano-bioprobes: from fundamentals to biodetection. Nanoscale 2013;5(4):1369−84. Available from: http://dx.doi.org/10.1039/C2NR33239F [accessed 12.11.12].
[18] Chatterjee DK, Gnanasammandhan MK, Zhang Y. Small upconverting fluorescent nanoparticles for biomedical applications. Small 2010;6(24):2781−95. Available from: http://dx.doi.org/10.1002/smll.201000418 [accessed 09.11.10].
[19] Wang M, Abbineni G, Clevenger A, Mao C, Xu S. Upconversion nanoparticles: synthesis, surface modification and biological applications. Nanomedicine 2011;7 (6):710−29. Available from: http://dx.doi.org/10.1016/j.nano.2011.02.013 [accessed December 2011].

[20] Cheng L, Wang C, Liu Z. Upconversion nanoparticles and their composite nanostructures for biomedical imaging and cancer therapy. Nanoscale 2013;5(1):23−37. Available from: http://dx.doi.org/10.1039/C2NR32311G [accessed 10.10.12].

[21] Tissue BM. Synthesis and luminescence of lanthanide ions in nanoscale insulating hosts. Chem Mater 1998;10(10):2837−45. Available from: http://dx.doi.org/10.1021/cm9802245 [accessed 16.09.98].

[22] Stouwdam JW, van Veggel FCJM. Near-infrared emission of redispersible Er^{3+}, Nd^{3+}, and Ho^{3+} doped LaF_3 nanoparticles. Nano Lett 2002;2(7):733−7. Available from: http://dx.doi.org/10.1021/nl025562q [accessed 15.05.02].

[23] Huignard A, Gacoin T, Boilot J-P. Synthesis and luminescence properties of colloidal YVO_4:Eu phosphors. Chem Mater 2000;12(4):1090−4. Available from: http://dx.doi.org/10.1021/cm990722t [accessed 29.03.00].

[24] Wang F, Zhang Y, Fan X, Wang M. Facile synthesis of water-soluble LaF_3:Ln^{3+} nanocrystals. J Mater Chem 2006;16(11):1031−4. Available from: http://dx.doi.org/10.1039/B518262J [accessed 10.02.06].

[25] Schäfer H, Ptacek P, Eickmeier H, Haase M. Synthesis of hexagonal Yb^{3+}, Er^{3+}-doped $NaYF_4$ nanocrystals at low temperature. Adv Funct Mater 2009;19(19):3091−7. Available from: http://dx.doi.org/10.1002/adfm.200900642 [accessed 31.08.09].

[26] Wang F, Chatterjee DK, Li Z, Zhang Y, Fan X, Wang M. Synthesis of polyethylenimine/$NaYF_4$ nanoparticles with upconversion fluorescence. Nanotechnology 2006;17 (23):5786−91. Available from: http://dx.doi.org/10.1088/0957-4484/17/23/013 [accessed 10.11.06].

[27] Liu C, Chen D. Controlled synthesis of hexagon shaped lanthanide-doped LaF_3 nanoplates with multicolor upconversion fluorescence. J Mater Chem 2007;17(37):3875−80. Available from: http://dx.doi.org/10.1039/B707927C [accessed 19.07.07].

[28] Wang X, Zhuang J, Peng Q, Li Y. A general strategy for nanocrystal synthesis. Nature 2005;437(7055):121−4. Available from: http://dx.doi.org/10.1038/nature03968 [accessed 01.09.05].

[29] Wang L, Li Y. Controlled synthesis and luminescence of lanthanide doped $NaYF_4$ nanocrystals. Chem Mater 2007;19(4):727−34. Available from: http://dx.doi.org/10.1021/cm061887m [accessed 20.01.07].

[30] Li Z, Zhang Y. An efficient and user-friendly method for the synthesis of hexagonal-phase $NaYF_4$:Yb, Er/Tm nanocrystals with controllable shape and upconversion fluorescence. Nanotechnology 2008;19(34):345606. Available from: http://dx.doi.org/10.1088/0957-4484/19/34/345606 [accessed 16.07.08].

[31] Zhang Y-W, Sun X, Si R, You L-P, Yan C-H. Single-crystalline and monodisperse LaF_3 triangular nanoplates from a single-source precursor. J Am Chem Soc 2005;127 (10):3260−1. Available from: http://dx.doi.org/10.1021/ja042801y [accessed 17.02.05].

[32] Naccache R, Vetrone F, Mahalingam V, Cuccia LA, Capobianco JA. Controlled synthesis and water dispersibility of hexagonal phase $NaGdF_4$:Ho^{3+}/Yb^{3+} nanoparticles. Chem Mater 2009;21(4):717−23. Available from: http://dx.doi.org/10.1021/cm803151y [accessed 22.01.09].

[33] Xie T, Li S, Peng Q, Li Y. Monodisperse BaF_2 nanocrystals: phases, size transitions, and self-assembly. Angew Chem Int Ed 2008;48(1):196−200. Available from: http://dx.doi.org/10.1002/anie.200804528 [accessed 03.12.08].

[34] Liu J, Li YD. Synthesis and self-assembly of luminescent Ln^{3+}-doped $LaVO_4$ uniform nanocrystals. Adv Mater 2007;19(8):1118−22. Available from: http://dx.doi.org/10.1002/adma.200600336 [accessed 16.04.07].

[35] Wang X, Zhuang J, Peng Q, Li Y. Hydrothermal synthesis of rare-earth fluoride nanocrystals. Inorg Chem 2006;45(17):6661−5. Available from: http://dx.doi.org/10.1021/ic051683s [accessed 19.07.06].

[36] Zeng JH, Xie T, Li ZH, Li Y. Monodispersed nanocrystalline fluoroperovskite up-conversion phosphors. Cryst Growth Des 2007;7(12):2774−7. Available from: http://dx.doi.org/10.1021/cg070477n [accessed 05.12.07].

[37] Wang G, Peng Q, Li Y. Upconversion luminescence of monodisperse $CaF_2:Yb^{3+}/Er^{3+}$ nanocrystals. J Am Chem Soc 2009;131(40):14200−1. Available from: http://dx.doi.org/10.1021/ja906732y [accessed 22.09.09].

[38] Hu H, Chen Z, Cao T, Zhang Q, Yu M, Li F, et al. Hydrothermal synthesis of hexagonal lanthanide-doped LaF_3 nanoplates with bright upconversion luminescence. Nanotechnology 2008;19(37):375702. Available from: http://dx.doi.org/10.1088/0957-4484/19/37/375702 [accessed 01.08.08].

[39] Zhang X, Quan Z, Yang J, Yang P, Lian H, Lin J. Solvothermal synthesis of well-dispersed MF_2 (M = Ca, Sr, Ba) nanocrystals and their optical properties. Nanotechnology 2008;19(7):075603. Available from: http://dx.doi.org/10.1088/0957-4484/19/7/075603 [accessed 29.01.08].

[40] Zhang F, Li J, Shan J, Xu L, Zhao D. Shape, size, and phase-controlled rare-earth fluoride nanocrystals with optical up-conversion properties. Chem Eur J 2009;15(41):11010−9. Available from: http://dx.doi.org/10.1002/chem.200900861 [accessed 08.09.09].

[41] Park J, An K, Hwang Y, Park J-G, Noh H-J, Kim J-Y, et al. Ultra-large-scale syntheses of monodisperse nanocrystals. Nat Mater 2004;3(12):891−5. Available from: http://dx.doi.org/10.1038/nmat1251 [accessed December 2004].

[42] Wang J, Wang F, Xu J, Wang Y, Liu Y, Chen X, et al. Lanthanide-doped $LiYF_4$ nanoparticles: synthesis and multicolor upconversion tuning. C R Chim 2010;13(6−7):731−6. Available from: http://dx.doi.org/10.1016/j.crci.2010.03.021 [accessed 07.05.10].

[43] Wang F, Wang J, Xu J, Xue X, Chen H, Liu X. Tunable upconversion emissions from lanthanide-doped monodisperse β-$NaYF_4$ nanoparticles. Spectrosc Lett 2010;43(5):400−5. Available from: http://dx.doi.org/10.1080/00387010.2010.487018 [accessed 01.07.10].

[44] Johnson NJJ, Oakden W, Stanisz GJ, Prosser RS, van Veggel FCJM. Size-tunable, ultrasmall $NaGdF_4$ nanoparticles: insights into their T_1 MRI contrast enhancement. Chem Mater 2011;23(16):3714−22. Available from: http://dx.doi.org/10.1021/cm201297x [accessed 22.07.11].

[45] Teng X, Zhu Y, Wei W, Wang S, Huang J, Naccache R, et al. Lanthanide-doped Na_xScF_{3+x} nanocrystals: crystal structure evolution and multicolor tuning. J Am Chem Soc 2012;134(20):8340−3. Available from: http://dx.doi.org/10.1021/ja3016236 [accessed 16.04.12].

[46] Chen G, Ohulchanskyy TY, Law WC, Ågren H, Prasad PN. Monodisperse $NaYbF_4$: $Tm^{3+}/NaGdF_4$ core/shell nanocrystals with near-infrared to near-infrared upconversion photoluminescence and magnetic resonance properties. Nanoscale 2011;3(5):2003−8. Available from: http://dx.doi.org/10.1039/C0NR01018A [accessed 03.03.11].

[47] Bao L, Li Z, Tao Q, Xie J, Mei Y, Xiong Y. Controlled synthesis of uniform LaF_3 polyhedrons, nanorods and nanoplates using NaOH and ligands. Nanotechnology 2013;24(14):145604. Available from: http://dx.doi.org/10.1088/0957-4484/24/14/145604 [accessed 18.03.13].

[48] Roberts JE. Lanthanum and neodymium salts of trifluoroacetic acid. J Am Chem Soc 1961;83(5):1087−8. Available from: http://dx.doi.org/10.1021/ja01466a020.

[49] Mai H-X, Zhang Y-W, Si R, Yan Z-G, Sun L-D, You L-P, et al. High-quality sodium rare-earth fluoride nanocrystals: controlled synthesis and optical properties. J Am Chem Soc 2006;128(19):6426−36. Available from: http://dx.doi.org/10.1021/ja060212h [accessed 20.04.06].

[50] Boyer J-C, Vetrone F, Cuccia LA, Capobianco JA. Synthesis of colloidal upconverting NaYF$_4$ nanocrystals doped with Er^{3+}, Yb^{3+} and Tm^{3+}, Yb^{3+} via thermal decomposition of lanthanide trifluoroacetate precursors. J Am Chem Soc 2006;128(23):7444−5. Available from: http://dx.doi.org/10.1021/ja061848b [accessed 23.05.06].

[51] Yi GS, Chow GM. Synthesis of hexagonal-phase NaYF$_4$:Yb, Er and NaYF$_4$:Yb,Tm nanocrystals with efficient up-conversion fluorescence. Adv Funct Mater 2006;6(18):2324−9. Available from: http://dx.doi.org/10.1002/adfm.200600053 [accessed 27.10.06].

[52] Shan J, Ju Y. Controlled synthesis of lanthanide-doped NaYF$_4$ upconversion nanocrystals via ligand induced crystal phase transition and silica coating. Appl Phys Lett 2007;91(12):123103. Available from: http://dx.doi.org/10.1063/1.2783476 [accessed 18.09.07].

[53] Du Y-P, Zhang Y-W, Yan Z-G, Sun L-D, Gao S, Yan C-H. Single-crystalline and near-monodispersed NaMF$_3$ (M = Mn, Co, Ni, Mg) and LiMAlF$_6$ (M = Ca, Sr) nanocrystals from cothermolysis of multiple trifluoroacetates in solution. Chem Asian J 2007;2(8):965−74. Available from: http://dx.doi.org/10.1002/asia.200700054 [accessed 30.05.07].

[54] Du Y-P, Zhang Y-W, Sun L-D, Yan C-H. Optically active uniform potassium and lithium rare earth fluoride nanocrystals derived from metal trifluoroacetate precursors. Dalton Trans 2009;(40):8574−81. Available from: http://dx.doi.org/10.1039/B909145A [accessed 02.09.09].

[55] Du Y-P, Sun X, Zhang Y-W, Yan Z-G, Sun L-D, Yan C-H. Uniform alkaline earth fluoride nanocrystals with diverse shapes grown from thermolysis of metal trifluoroacetates in hot surfactant solutions. Cryst Growth Des 2009;9(4):2013−9. Available from: http://dx.doi.org/10.1021/cg801371r [accessed 02.03.09].

[56] Sun X, Zhang Y-W, Du Y-P, Yan Z-G, Si R, You L-P, et al. From trifluoroacetate complex precursors to monodisperse rare-earth fluoride and oxyfluoride nanocrystals with diverse shapes through controlled fluorination in solution phase. Chem Eur J 2007;13(8):2320−32. Available from: http://dx.doi.org/10.1002/chem.200601072 [accessed 13.12.06].

[57] Mahalingam V, Vetrone F, Naccache R, Speghini A, Capobianco JA. Colloidal Tm^{3+}/Yb^{3+}-doped LiYF$_4$ nanocrystals: multiple luminescence spanning the UV to NIR regions via low-energy excitation. Adv Mater 2009;21(40):4025−8. Available from: http://dx.doi.org/10.1002/adma.200901174 [accessed 29.06.09].

[58] Vetrone F, Mahalingam V, Capobianco JA. Near-infrared-to-blue upconversion in colloidal BaYF$_5$:Tm^{3+}, Yb^{3+} nanocrystals. Chem Mater 2009;21(9):1847−51. Available from: http://dx.doi.org/10.1021/cm900313s [accessed 14.04.09].

[59] Mahalingam V, Vetrone F, Naccache R, Speghini A, Capobianco JA. Structural and optical investigation of colloidal Ln^{3+}/Yb^{3+} co-doped KY$_3$F$_{10}$ nanocrystals. J Mater Chem 2009;19(20):3149−52. Available from: http://dx.doi.org/10.1039/B900300B [accessed 24.03.09].

[60] Quan Z, Yang D, Yang P, Zhang X, Lian H, Liu X, et al. Uniform colloidal alkaline earth metal fluoride nanocrystals: nonhydrolytic synthesis and luminescence properties. Inorg Chem 2008;47(20):9509−17. Available from: http://dx.doi.org/10.1021/ic8014207 [accessed 26.09.08].

[61] Yi G-S, Chow G-M. Water-soluble NaYF$_4$:Yb, Er(Tm)/NaYF$_4$/Polymer core/shell/shell nanoparticles with significant enhancement of upconversion fluorescence. Chem Mater

2007;19(3):341−3. Available from: http://dx.doi.org/10.1021/cm062447y [accessed 30.12.06].

[62] Johnson NJJ, Korinek A, Dong C, van Veggel FCJM. Self-focusing by Ostwald ripening: a strategy for layer-by-layer epitaxial growth on upconverting nanocrystals. J Am Chem Soc 2012;134(27):11068−71. Available from: http://dx.doi.org/10.1021/ja302717u [accessed 26.06.12].

[63] Qian H-S, Zhang Y. Synthesis of hexagonal-phase core−shell NaYF$_4$ nanocrystals with tunable upconversion fluorescence. Langmuir 2008;24(21):12123−5. Available from: http://dx.doi.org/10.1021/la802343f [accessed 08.10.08].

[64] Chen F, Bu W, Zhang S, Liu X, Liu J, Xing H, et al. Positive and negative lattice shielding effects co-existing in Gd(III) ion doped bifunctional upconversion nanoprobes. Adv Funct Mater 2011;21(22):4285−94. Available from: http://dx.doi.org/10.1002/adfm.201101663 [accessed 28.09.11].

[65] Liu X, Kong X, Zhang Y, Tu L, Wang Y, Zeng Q, et al. Breakthrough in concentration quenching threshold of upconversion luminescence via spatial separation of the emitter doping area for bioapplications. Chem Commun 2011;47(43):11957−9. Available from: http://dx.doi.org/10.1039/C1CC14774A [accessed 04.10.11].

[66] Chen D, Lei L, Yang A, Wang Z, Wang Y. Ultra-broadband near-infrared excitable upconversion core/shell nanocrystals. Chem Commun 2012;48(47):5898−900. Available from: http://dx.doi.org/10.1039/C2CC32102E [accessed 23.04.12].

[67] Zhang F, Che R, Li X, Yao C, Yang J, Shen D, et al. Direct imaging the upconversion nanocrystal core/shell structure at the subnanometer level: shell thickness dependence in upconverting optical properties. Nano Lett 2012;12(6):2852−8. Available from: http://dx.doi.org/10.1021/nl300421n [accessed 30.04.12].

[68] Schäfer H, Ptacek P, Zerzouf O, Haase M. Synthesis and optical properties of KYF$_4$/Yb, Er nanocrystals, and their surface modification with undoped KYF$_4$. Adv Funct Mater 2008;18 (19):2913−8. Available from: http://dx.doi.org/10.1002/adfm.200800368 [accessed 22.09.08].

[69] Wang F, Deng R, Wang J, Wang Q, Han Y, Zhu H, et al. Tuning upconversion through energy migration in core−shell nanoparticles. Nat Mater 2011;10 (12):968−73. Available from: http://dx.doi.org/10.1038/nmat3149 [accessed December 2011].

[70] Su Q, Han S, Xie X, Zhu H, Chen H, Chen C-K, et al. The effect of surface coating on energy migration-mediated upconversion. J Am Chem Soc 2012;134(51):20849−57. Available from: http://dx.doi.org/10.1021/ja3111048 [accessed 04.12.12].

[71] Dong C, Korinek A, Blasiak B, Tomanek B, van Veggel FCJM. Cation exchange: a facile method to make NaYF$_4$:Yb, Tm-NaGdF$_4$ core−shell nanoparticles with a thin, tunable, and uniform shell. Chem Mater 2012;24(7):1297−305. Available from: http://dx.doi.org/10.1021/cm2036844 [accessed 21.03.12].

[72] van Veggel FCJM, Dong C, Johnson NJJ, Pichaandi J. Ln^{3+}-doped nanoparticles for upconversion and magnetic resonance imaging: some critical notes on recent progress and some aspects to be considered. Nanoscale 2012;4(23):7309−21. Available from: http://dx.doi.org/10.1039/C2NR32124F [accessed 25.09.12].

[73] Liang X, Wang X, Zhuang J, Peng Q, Li Y. Synthesis of NaYF$_4$ nanocrystals with predictable phase and shape. Adv Funct Mater 2007;17(15):2757−65. Available from: http://dx.doi.org/10.1002/adfm.200600807 [accessed 23.08.07].

[74] Li Z, Zhang Y, Jiang S. Multicolor core/shell-structured upconversion fluorescent nanoparticles. Adv Mater 2008;20(24):4765−9. Available from: http://dx.doi.org/10.1002/adma.200801056 [accessed 20.10.08].

[75] Shan J, Ju Y. A single-step synthesis and the kinetic mechanism for monodisperse and hexagonal-phase NaYF$_4$:Yb, Er upconversion nanophosphors. Nanotechnology 2009;20(27):275603. Available from: http://dx.doi.org/10.1088/0957-4484/20/27/275603 [accessed 17.06.09].

[76] Liu C, Wang H, Li X, Chen D. Monodisperse, size-tunable and highly efficient β-NaYF$_4$:Yb,Er(Tm) up-conversion luminescent nanospheres: controllable synthesis and their surface modifications. J Mater Chem 2009;19(21):3546−53. Available from: http://dx.doi.org/10.1039/B820254K [accessed 17.04.09].

[77] Zhang F, Wan Y, Yu T, Zhang F, Shi Y, Xie S, et al. Uniform nanostructured arrays of sodium rare-earth fluorides for highly efficient multicolor upconversion luminescence. Angew Chem Int Ed 2007;46(42):7976−9. Available from: http://dx.doi.org/10.1002/anie.200702519 [accessed 11.09.07].

[78] Ye X, Collins JE, Kang Y, Chen J, Chen DTN, Yodh AG, et al. Morphologically controlled synthesis of colloidal upconversion nanophosphors and their shape-directed self-assembly. Proc Natl Acad Sci USA 2010;107(52):22430−5. Available from: http://dx.doi.org/10.1073/pnas.1008958107 [accessed 10.12.10].

[79] Wang F, Liu X. Upconversion multicolor fine-tuning: visible to near-infrared emission from lanthanide-doped NaYF$_4$ nanoparticles. J Am Chem Soc 2008;130(17):5642−3. Available from: http://dx.doi.org/10.1021/ja800868a [accessed 05.04.08].

[80] Wang F, Han Y, Lim CS, Lu Y, Wang J, Xu J, et al. Simultaneous phase and size control of upconversion nanocrystals through lanthanide doping. Nature 2010;463(7284):1061−5. Available from: http://dx.doi.org/10.1038/nature08777 [accessed 25.02.10].

[81] Yu X, Li M, Xie M, Chen L, Li Y, Wang Q. Dopant-controlled synthesis of water-soluble hexagonal NaYF$_4$ nanorods with efficient upconversion fluorescence for multicolor bioimaging. Nano Res 2010;3(1):51−60. Available from: http://dx.doi.org/10.1007/s12274-010-1008-2 [accessed January 2010].

[82] Zeng S, Ren G, Xu C, Yang Q. Modifying crystal phase, shape, size, optical and magnetic properties of monodispersed multifunctional NaYbF$_4$ nanocrystals through lanthanide doping. CrystEngComm 2011;13(12):4276−81. Available from: http://dx.doi.org/10.1039/C0CE00833H [accessed 24.02.11].

[83] Liu Q, Sun Y, Yang T, Feng W, Li C, Li F. Sub-10 nm hexagonal lanthanide-doped NaLuF$_4$ upconversion nanocrystals for sensitive bioimaging *in vivo*. J Am Chem Soc 2011;33(43):17122−5. Available from: http://dx.doi.org/10.1021/ja207078s [accessed 29.09.11].

[84] Chen D, Yu Y, Huang F, Yang A, Wang Y. Lanthanide activator doped NaYb$_{1-x}$Gd$_x$F$_4$ nanocrystals with tunable down-, up-conversion luminescence and paramagnetic properties. J Mater Chem 2011;21(17):6186−92. Available from: http://dx.doi.org/10.1039/C1JM10266D [accessed 18.03.11].

[85] Chen D, Huang P, Yu Y, Huang F, Yang A, Wang Y. Dopant-induced phase transition: a new strategy of synthesizing hexagonal upconversion NaYF$_4$ at low temperature. Chem Commun 2011;47(20):5801−3. Available from: http://dx.doi.org/10.1039/C0CC05722C [accessed 14.04.11].

[86] Chen D, Yu Y, Huang F, Huang P, Yang A, Wang Y. Modifying the size and shape of monodisperse bifunctional alkaline-earth fluoride nanocrystals through lanthanide doping. J Am Chem Soc 2010;132(29):9976−8. Available from: http://dx.doi.org/10.1021/ja1036429 [accessed 01.07.10].

[87] Chen D, Yu Y, Huang F, Wang Y. Phase transition from hexagonal LnF$_3$ (Ln = La, Ce, Pr) to cubic Ln$_{0.8}$M$_{0.2}$F$_{2.8}$ (M = Ca, Sr, Ba) nanocrystals with enhanced upconversion induced by alkaline-earth doping. Chem Commun 2011;47(9):2601−3. Available from: http://dx.doi.org/10.1039/C0CC04846A [accessed 13.01.11].

[88] Heer S, Lehmann O, Haase M, Güdel H-U. Blue, green, and red upconversion emis-
      sion from lanthanide-doped LuPO$_4$ and YbPO$_4$ nanocrystals in a transparent colloidal
      solution. Angew Chem Int Ed 2003;42(27):3179−82. Available from: http://dx.doi.
      org/10.1002/anie.200351091 [accessed 10.07.03].

[89] Heer S, Kömpe K, Güdel H-U, Haase M. Highly efficient multicolour upconversion
      emission in transparent colloids of lanthanide-doped NaYF$_4$ nanocrystals. Adv Mater
      2004;16(23−24):2102−5. Available from: http://dx.doi.org/10.1002/adma.200400772
      [accessed 16.12.04].

[90] Wang F, Fan X, Wang M, Zhang Y. Multicolour PEI/NaGdF$_4$:Ce^{3+}, Ln^{3+} nanocrys-
      tals by single-wavelength excitation. Nanotechnology 2007;18(2):025701. Available
      from: http://dx.doi.org/10.1088/0957-4484/18/2/025701 [accessed 15.12.06].

[91] Stouwdam JW, Raudsepp M, van Veggel FCJM. Colloidal nanoparticles of Ln^{3+}-
      doped LaVO$_4$: energy transfer to visible- and near-infrared-emitting lanthanide ions.
      Langmuir 2005;21(15):7003−8. Available from: http://dx.doi.org/10.1021/la0505162
      [accessed 23.06.05].

[92] Li F, Li N, Wang M, Xu S, Zhang H. Intense visible multicolored luminescence from
      lanthanide ion-pair codoped NaGdF$_4$ nanocrystals. Luminescence 2010;25(5):394−8.
      Available from: http://dx.doi.org/10.1002/bio.1166 [accessed 27.08.09].

[93] Liu C, Wang H, Zhang X, Chen D. Morphology- and phase-controlled synthesis of
      monodisperse lanthanide-doped NaGdF$_4$ nanocrystals with multicolor photolumines-
      cence. J Mater Chem 2009;19(4):489−96. Available from: http://dx.doi.org/10.1039/
      B815682D [accessed 09.12.08].

[94] Wang J, Wang F, Wang C, Liu Z, Liu X. Single-band upconversion emission in lan-
      thanide-doped KMnF$_3$ nanocrystals. Angew Chem Int Ed 2011;50(44):10369−72.
      Available from: http://dx.doi.org/10.1002/anie.201104192 [accessed 14.09.11].

[95] Wang F, Xue X, Liu X. Multicolor tuning of (Ln, P)-doped YVO$_4$ nanoparticles by
      single-wavelength excitation. Angew Chem Int Ed 2008;47(5):906−9. Available
      from: http://dx.doi.org/10.1002/anie.200704520 [accessed 04.12.07].

[96] Hou Z, Yang P, Li C, Wang L, Lian H, Quan Z, et al. Preparation and luminescence
      properties of YVO$_4$:Ln and Y(V, P)O$_4$:Ln (Ln = Eu^{3+}, Sm^{3+}, Dy^{3+}) nanofibers and
      microbelts by sol−gel/electrospinning process. Chem Mater 2008;20(21):6686−96.
      Available from: http://dx.doi.org/10.1021/cm801538t [accessed 16.10.08].

[97] Mahalingam V, Mangiarini F, Vetrone F, Venkatramu V, Bettinelli M, Speghini A, et
      al. Bright white upconversion emission from Tm^{3+}/Yb^{3+}/Er^{3+}-doped Lu$_3$Ga$_5$O$_{12}$
      nanocrystals. J Phys Chem C 2008;112(46):17745−9. Available from: http://dx.doi.
      org/10.1021/jp8076479 [accessed 23.10.08].

[98] Dong B, Song H, Qin R, Bai X, Lu S, Ren X, et al. Upconversion white light devices:
      Ln^{3+}-tridoped NaYF$_4$ nanoparticles and PVP modified films. J Nanosci Nanotechnol
      2008;8(8):3921−5. Available from: http://dx.doi.org/10.1166/jnn.2008.182 [accessed
      01.08.08].

[99] Chen GY, Liu Y, Zhang YG, Somesfalean G, Zhang ZG, Sun Q, et al. Bright white upcon-
      version luminescence in rare-earth-ion-doped Y$_2$O$_3$ nanocrystals. Appl Phys Lett 2007;91
      (13):133103. Available from: http://dx.doi.org/10.1063/1.2787893 [accessed 25.09.07].

[100] Yang J, Zhang C, Peng C, Li C, Wang L, Chai R, et al. Controllable red, green, blue
      (RGB) and bright white upconversion luminescence of Lu$_2$O$_3$:Yb^{3+}/Er^{3+}/Tm^{3+} nano-
      crystals through single laser excitation at 980 nm. Chem Eur J 2009;15(18):4649−55.
      Available from: http://dx.doi.org/10.1002/chem.200802106 [accessed 18.03.09].

[101] Mai H-X, Zhang Y-W, Sun L-D, Yan C-H. Highly efficient multicolor up-conversion
      emissions and their mechanisms of monodisperse NaYF$_4$:Yb,Er core and core/shell-

structured nanocrystals. J Phys Chem C 2007;111(37):13721−9. Available from: http://dx.doi.org/10.1021/jp073920d [accessed 24.08.07].

[102] Wang F, Wang J, Liu X. Direct evidence of a surface quenching effect on size-dependent luminescence of upconversion nanoparticles. Angew Chem Int Ed 2010;49(41):7456−60. Available from: http://dx.doi.org/10.1002/anie.201003959 [accessed 31.08.10].

[103] Liu Y, Tu D, Zhu H, Li R, Luo W, Chen XA. Materials 2010;22(30):3266−71. Available from: http://dx.doi.org/10.1002/adma.201000128 [accessed 08.06.10].

[104] Vetrone F, Naccache R, Mahalingam V, Morgan CG, Capobianco JA. The active-core/active-shell approach: a strategy to enhance the upconversion luminescence in lanthanide-doped nanoparticles. Adv Funct Mater 2009;19(18):2924−9. Available from: http://dx.doi.org/10.1002/adfm.200900234 [accessed 16.07.09].

[105] Bogdan N, Vetrone F, Ozin GA, Capobianco JA. Synthesis of ligand-free colloidally stable water dispersible brightly luminescent lanthanide-doped upconverting nanoparticles. Nano Lett 2011;11(2):835−40. Available from: http://dx.doi.org/10.1021/nl1041929 [accessed 18.01.11].

[106] Dong A, Ye X, Chen J, Kang Y, Gordon T, Kikkawa JM, et al. A generalized ligand-exchange strategy enabling sequential surface functionalization of colloidal nanocrystals. J Am Chem Soc 2011;133(4):998−1006. Available from: http://dx.doi.org/10.1021/ja108948z [accessed 22.12.10].

[107] Chen Z, Chen H, Hu H, Yu M, Li F, Zhang Q, et al. Versatile synthesis strategy for carboxylic acid-functionalized upconverting nanophosphors as biological labels. J Am Chem Soc 2008;130(10):3023−9. Available from: http://dx.doi.org/10.1021/ja076151k [accessed 16.02.08].

[108] Hu H, Yu M, Li F, Chen Z, Gao X, Xiong L, et al. Facile epoxidation strategy for producing amphiphilic up-converting rare-earth nanophosphors as biological labels. Chem Mater 2008;20(22):7003−9. Available from: http://dx.doi.org/10.1021/cm801215t [accessed 28.10.08].

[109] Zhou H-P, Xu C-H, Sun W, Yan C-H. Lean and flexible modification strategy for carboxyl/aldehyde-functionalized upconversion nanoparticles and their optical applications. Adv Funct Mater 2009;19(24):3892−900. Available from: http://dx.doi.org/10.1002/adfm.200901458 [accessed 20.11.09].

[110] Wang L, Yan R, Huo Z, Wang L, Zeng J, Bao J, et al. Fluorescence resonant energy transfer biosensor based on upconversion-luminescent nanoparticles. Angew Chem Int Ed 2005;44(37):6054−7. Available from: http://dx.doi.org/10.1002/anie.200501907 [accessed 24.08.05].

[111] Beaurepaire E, Buissette V, Sauviat M-P, Giaume D, Lahlil K, Mercuri A, et al. Functionalized fluorescent oxide nanoparticles: artificial toxins for sodium channel targeting and imaging at the single-molecule level. Nano Lett 2004;4(11):2079−83. Available from: http://dx.doi.org/10.1021/nl049105g [accessed 26.10.04].

[112] Lu H, Yi G, Zhao S, Chen D, Guo L-H, Cheng J. Synthesis and characterization of multi-functional nanoparticles possessing magnetic, up-conversion fluorescence and bio-affinity properties. J Mater Chem 2004;14(8):1336−41. Available from: http://dx.doi.org/10.1039/B315103D [accessed 04.03.04].

[113] Sivakumar S, Diamente PR, van Veggel FCJM. Silica-coated Ln^{3+}-doped LaF3 nanoparticles as robust down- and upconverting biolabels. Chem Eur J 2006;12 (22):5878−84. Available from: http://dx.doi.org/10.1002/chem.200600224 [accessed 01.06.06].

[114] Li Z, Zhang Y. Monodisperse silica-coated polyvinylpyrrolidone/$NaYF_4$ nanocrystals with multicolor upconversion fluorescence emission. Angew Chem Int Ed 2006;45

(46):7732—5. Available from: http://dx.doi.org/10.1002/anie.200602975 [accessed 07.11.06].

[115] Qian H, Li Z, Zhang Y. Multicolor polystyrene nanospheres tagged with up-conversion fluorescent nanocrystals. Nanotechnology 2008;19(25):255601. Available from: http://dx.doi.org/10.1088/0957-4484/19/25/255601 [accessed 14.05.08].

[116] Liu Z, Yi G, Zhang H, Ding J, Zhang Y, Xue J. Monodisperse silica nanoparticles encapsulating upconversion fluorescent and superparamagnetic nanocrystals. Chem Commun 2008;6:694—6. Available from: http://dx.doi.org/10.1039/B715402J [accessed 04.12.07].

[117] Ehlert O, Thomann R, Darbandi M, Nann T. A four-color colloidal multiplexing nanoparticle system. ACS Nano 2008;2(1):120—4. Available from: http://dx.doi.org/10.1021/nn7002458 [accessed 04.01.08].

[118] Wang M, Mi C, Wang W, Liu C, Wu Y, Xu Z, et al. Immunolabeling and NIR-excited fluorescent imaging of HeLa cells by using $NaYF_4$:Yb, Er upconversion nanoparticles. ACS Nano 2009;3(6):1580—6. Available from: http://dx.doi.org/10.1021/nn900491j [accessed 28.05.09].

[119] Johnson NJJ, Sangeetha NM, Boyer J-C, van Veggel FCJM. Facile ligand-exchange with polyvinylpyrrolidone and subsequent silica coating of hydrophobic upconverting β-$NaYF_4$:Yb^{3+}/Er^{3+} nanoparticles. Nanoscale 2010;2(5):771—7. Available from: http://dx.doi.org/10.1039/B9NR00379G [accessed 04.03.10].

[120] Gai S, Yang P, Li C, Wang W, Dai Y, Niu N, et al. Synthesis of magnetic up-conversion luminescent, and mesoporous core—shell-structured nanocomposites as drug carriers. Adv Funct Mater 2010;20(7):1166—72. Available from: http://dx.doi.org/10.1002/adfm.200902274 [accessed 04.03.10].

[121] Xu Z, Li C, Ma P, Hou Z, Yang D, Kang X, et al. Facile synthesis of an up-conversion luminescent and mesoporous Gd_2O_3:Er^{3+}@$nSiO_2$@$mSiO_2$ nanocomposite as a drug carrier. Nanoscale 2011;3(2):661—7. Available from: http://dx.doi.org/10.1039/C0NR00695E [accessed 22.11.10].

[122] Li C, Hou Z, Dai Y, Yang D, Cheng Z, Ma P, et al. A facile fabrication of upconversion luminescent and mesoporous core—shell structured β-$NaYF_4$:Yb^{3+}, Er^{3+}@$mSiO_2$ nanocomposite spheres for anti-cancer drug delivery and cell imaging. Biomater Sci 2013;1(2):213—23. Available from: http://dx.doi.org/10.1039/C2BM00087C [accessed 17.10.12].

[123] Wang F, Banerjee D, Liu Y, Chen X, Liu X. Upconversion nanoparticles in biological labeling, imaging, and therapy. Analyst 2010;135(8):1839—54. Available from: http://dx.doi.org/10.1039/C0AN00144A [accessed 18.05.10].

[124] Ju Q, Liu Y, Tu D, Zhu H, Li R, Chen X. Lanthanide-doped multicolor GdF3 nanocrystals for time-resolved photoluminescent biodetection. Chem Eur J 2011;17 (31):8549—54. Available from: http://dx.doi.org/10.1002/chem.201101170 [accessed 15.06.11].

[125] Nyk M, Kumar R, Ohulchanskyy TY, Bergey EJ, Prasad PN. High contrast *in vitro* and *in vivo* photoluminescence bioimaging using near infrared to near infrared upconversion in Tm^{3+} and Yb^{3+} doped fluoride nanophosphors. Nano Lett 2008;8 (11):3834—8. Available from: http://dx.doi.org/10.1021/nl802223f [accessed 17.10.08].

[126] Niedbala RS, Feindt H, Kardos K, Vail T, Burton J, Bielska B, et al. Detection of analytes by immunoassay using up-converting phosphor technology. Anal Biochem 2001;293(1):22—30. Available from: http://dx.doi.org/10.1006/abio.2001.5105 [accessed 04.05.01].

[127] Wang L, Li Y. Luminescent nanocrystals for nonenzymatic glucose concentration determination. Chem Eur J 2007;13(15):4203−7. Available from: http://dx.doi.org/10.1002/chem.200700005 [accessed 26.02.07].

[128] Liu Y, Zhou S, Tu D, Chen Z, Huang M, Zhu H, et al. Amine-functionalized lanthanide-doped zirconia nanoparticles: optical spectroscopy, time-resolved fluorescence resonance energy transfer biodetection, and targeted imaging. J Am Chem Soc 2012;134(36):15083−90. Available from: http://dx.doi.org/10.1021/ja306066a [accessed 22.08.12].

[129] Ju Q, Tu D, Liu Y, Li R, Zhu H, Chen J, et al. Amine-functionalized lanthanide-doped KGdF4 nanocrystals as potential optical/magnetic multimodal bioprobes. J Am Chem Soc 2012;134(2):1323−30. Available from: http://dx.doi.org/10.1021/ja2102604 [accessed 06.12.11].

[130] Morgan CG, Mitchell AC. Prospects for applications of lanthanide-based upconverting surfaces to bioassay and detection. Biosens Bioelectron 2007;22(8):1769−75. Available from: http://dx.doi.org/10.1016/j.bios.2006.08.024 [26.09.06].

[131] Kuningas K, Rantanen T, Ukonaho T, Lövgren T, Soukka T. Homogeneous assay technology based on upconverting phosphors. Anal Chem 2005;77(22):7348−55. Available from: http://dx.doi.org/10.1021/ac0510944 [accessed 06.10.05].

[132] Rantanen T, Järvenpää M-L, Vuojola J, Kuningas K, Soukka T. Fluorescence-quenching-based enzyme-activity assay by using photon upconversion. Angew Chem Int Ed 2008;47(20):3811−3. Available from: http://dx.doi.org/10.1002/anie.200705861 [accessed 10.04.08].

[133] Meiser F, Cortez C, Caruso F. Biofunctionalization of fluorescent rare-earth-doped lanthanum phosphate colloidal nanoparticles. Angew Chem Int Ed 2004;43 (44):5954−7. Available from: http://dx.doi.org/10.1002/anie.200460856 [accessed 10.11.04].

[134] Casanova D, Giaume D, Moreau M, Martin J-L, Gacoin T, Boilot J-P, et al. Counting the number of proteins coupled to single nanoparticles. J Am Chem Soc 2007;129 (42):12592−3. Available from: http://dx.doi.org/10.1021/ja0731975 [accessed 29.09.07].

[135] Gu J-Q, Shen J, Sun L-D, Yan C-H. Resonance energy transfer in steady-state and time-decay fluoro-immunoassays for lanthanide nanoparticles based on biotin and avidin affinity. J Phys Chem C 2008;112(17):6589−93. Available from: http://dx.doi.org/10.1021/jp801203w [accessed 04.04.08].

[136] Wong K-L, Law G-L, Murphy MB, Tanner PA, Wong W-T, Lam PK-S, et al. Functionalized europium nanorods for in vitro imaging. Inorg Chem 2008;47 (12):5190−6. Available from: http://dx.doi.org/10.1021/ic8000416 [accessed 21.05.08].

[137] Traina CA, Dennes TJ, Schwartz J. A modular monolayer coating enables cell targeting by luminescent yttria nanoparticles. Bioconjug Chem 2009;20(3):437−9. Available from: http://dx.doi.org/10.1021/bc800551x [accessed 18.02.09].

[138] Lin M, Zhao Y, Wang S, Liu M, Duan Z, Chen Y, et al. Recent advances in synthesis and surface modification of lanthanide-doped upconversion nanoparticles for biomedical applications. Biotechnol Adv 2012;30(6):1551−61. Available from: http://dx.doi.org/10.1016/j.biotechadv.2012.04.009 [accessed 27.04.12].

[139] Zijlmans HJMAA, Bonnet J, Burton J, Kardos K, Vail T, Niedbala RS, et al. Detection of cell and tissue surface antigens using up-converting phosphors: a new reporter technology. Anal Biochem 1999;267(1):30−6. Available from: http://dx.doi.org/10.1006/abio.1998.2965 [accessed 01.02.99].

[140] Vetrone F, Naccache R, de la Fuente AJ, Sanz-Rodriguez F, Blazquez-Castro A, Rodriguez EM, et al. Intracellular imaging of HeLa cells by non-functionalized $NaYF_4:Er^{3+}$, Yb^{3+} upconverting nanoparticles. Nanoscale 2010;2(4):495−8. Available from: http://dx.doi.org/10.1039/B9NR00236G [accessed 17.11.09].

[141] Cao T, Yang T, Gao Y, Yang Y, Hu H, Li F. Water-soluble $NaYF_4$:Yb/Er upconversion nanophosphors: synthesis, characteristics and application in bioimaging. Inorg Chem Commun 2010;13(3):392−4. Available from: http://dx.doi.org/10.1016/j.inoche.2009.12.031 [accessed 06.01.10].

[142] Park YI, Kim JH, Lee KT, Jeon K-S, Na HB, Yu JH, et al. Nonblinking and nonbleaching upconverting nanoparticles as an optical imaging nanoprobe and T1 magnetic resonance imaging contrast agent. Adv Mater 2009;21(44):4467−71. Available from: http://dx.doi.org/10.1002/adma.200901356 [accessed 29.06.09].

[143] Wu S, Han G, Milliron DJ, Aloni S, Altoe V, Talapin DV, et al. Non-blinking and photostable upconverted luminescence from single lanthanide-doped nanocrystals. Proc Natl Acad Sci USA 2009;106(27):10917−21. Available from: http://dx.doi.org/10.1073/pnas.0904792106 [accessed 18.06.09].

[144] Yu M, Li F, Chen Z, Hu H, Zhan C, Yang H, et al. Laser scanning up-conversion luminescence microscopy for imaging cells labeled with rare-earth nanophosphors. Anal Chem 2009;81(3):930−5. Available from: http://dx.doi.org/10.1021/ac802072d [accessed 06.01.09].

[145] Hu H, Xiong L, Zhou J, Li F, Cao T, Huang C. Multimodal-luminescence core−shell nanocomposites for targeted imaging of tumor cells. Chem Eur J 2009;15 (14):3577−84. Available from: http://dx.doi.org/10.1002/chem.200802261 [accessed 13.02.09].

[146] Zako T, Nagata H, Terada N, Utsumi A, Sakono M, Yohda M, et al. Cyclic RGD peptide-labeled upconversion nanophosphors for tumor cell-targeted imaging. Biochem Biophys Res Commun 2009;381(1):54−8. Available from: http://dx.doi.org/10.1016/j.bbrc.2009.02.004 [accessed 08.02.09].

[147] Lim SF, Riehn R, Ryu WS, Khanarian N, Tung C-K, Tank D, et al. *In vivo* and scanning electron microscopy imaging of upconverting nanophosphors in *Caenorhabditis elegans*. Nano Lett 2006;6(2):169−74. Available from: http://dx.doi.org/10.1021/nl0519175 [accessed 23.12.05].

[148] Chatteriee DK, Rufalhah AJ, Zhang Y. Upconversion fluorescence imaging of cells and small animals using lanthanide doped nanocrystals. Biomaterials 2008;29 (7):937−43. Available from: http://dx.doi.org/10.1016/j.biomaterials.2007.10.051 [accessed 03.12.07].

[149] Xiong L-Q, Chen Z-G, Yu M-X, Li F-Y, Liu C, Huang C-H. Synthesis, characterization, and *in vivo* targeted imaging of amine-functionalized rare-earth up-converting nanophosphors. Biomaterials 2009;30(29):5592−600. Available from: http://dx.doi.org/10.1016/j.biomaterials.2009.06.015 [accessed 28.06.09].

[150] Xiong L, Chen Z, Tian Q, Cao T, Xu C, Li F. High contrast upconversion luminescence targeted imaging *in vivo* using peptide-labeled nanophosphors. Anal Chem 2009;81 (21):8687−94. Available from: http://dx.doi.org/10.1021/ac901960d [accessed 09.10.09].

[151] Idris NM, Li Z, Ye L, Wei Sim EK, Mahendran R, Ho PC-L, et al. Tracking transplanted cells in live animal using upconversion fluorescent nanoparticles. Biomaterials 2009;30(28):5104−13. Available from: http://dx.doi.org/10.1016/j.biomaterials.2009.05.062 [accessed 17.06.09].

[152] Pogue BW, Poplack SP, McBride TO, Wells WA, Osterman KS, Osterberg UL, et al. Quantitative hemoglobin tomography with diffuse near-infrared spectroscopy: pilot

results in the breast. Radiology 2001;218(1):261−6. Available from: < http://radiol-ogy.rsna.org/content/218/1/261.html > [accessed January 2001].

[153] Ntziachristos V, Yodh AG, Schnall M, Chance B. Concurrent MRI and diffuse optical tomography of breast after indocyanine green enhancement. Proc Natl Acad Sci USA 2000;97(6):2767−72. Available from: http://dx.doi.org/10.1073/pnas.040570597 [accessed 07.03.00].

[154] Joshi A, Bangerth W, Hwang K, Rasmussen JC, Sevick-Muraca EM. Fully adaptive FEM based fluorescence optical tomography from time-dependent measurements with area illumination and detection. Med Phys 2006;33(5):1299−310. Available from: http://dx.doi.org/10.1118/1.2190330 [accessed 21.04.06].

[155] Xu CT, Svensson N, Axelsson J, Svenmarker P, Somesfalean G, Chen G, et al. Autofluorescence insensitive imaging using upconverting nanocrystals in scattering media. Appl Phys Lett 2008;93(17):171103. Available from: http://dx.doi.org/10.1063/1.3005588 [accessed 27.10.08].

[156] Xu CT, Axelsson J, Andersson-Engels S. Fluorescence diffuse optical tomography using upconverting nanoparticles. Appl Phys Lett 2009;94(25):251107. Available from: http://dx.doi.org/10.1063/1.3156857 [accessed 23.06.09].

[157] Liu H, Xu CT, Andersson-Engels S. Multibeam fluorescence diffuse optical tomography using upconverting nanoparticles. Opt Lett 2010;35(5):718−20. Available from: http://dx.doi.org/10.1364/OL.35.000718 [accessed 12.01.10].

[158] Zhou J, Sun Y, Du X, Xiong L, Hu H, Li F. Dual-modality in vivo imaging using rare-earth nanocrystals with near-infrared to near-infrared (NIR-to-NIR) upconversion luminescence and magnetic resonance properties. Biomaterials 2010;31(12):3287−95. Available from: http://dx.doi.org/10.1016/j.biomaterials.2010.01.040 [accessed 04.02.10].

[159] Kumar R, Nyk M, Ohulchanskyy TY, Flask CA, Prasad PN. Combined optical and MR bioimaging using rare earth ion doped $NaYF_4$ nanocrystals. Adv Funct Mater 2009;19(6):853−9. Available from: http://dx.doi.org/10.1002/adfm.200800765 [accessed 18.02.09].

[160] Das GK, Heng BC, Ng S-C, White T, Loo JSC, D'Silva L, et al. Gadolinium oxide ultranarrow nanorods as multimodal contrast agents for optical and magnetic reso-nance imaging. Langmuir 2010;26(11):8959−65. Available from: http://dx.doi.org/10.1021/la904751q [accessed 11.02.10].

[161] Sun Y, Yu M, Liang S, Zhang Y, Li C, Mou T, et al. Fluorine-18 labeled rare-earth nanoparticles for positron emission tomography (PET) imaging of sentinel lymph node. Biomaterials 2011;32(11):2999−3007. Available from: http://dx.doi.org/10.1016/j.biomaterials.2011.01.011 [accessed 03.02.11].

[162] Zhou J, Yu M, Sun Y, Zhang X, Zhu X, Wu Z, et al. Fluorine-18-labeled $Gd^{3+}/Yb^{3+}/Er^{3+}$ co-doped $NaYF_4$ nanophosphors for multimodality PET/MR/UCL imaging. Biomaterials 2011;32(4):1148−56. Available from: http://dx.doi.org/10.1016/j.bioma-terials.2010.09.071 [accessed 20.10.10].

[163] Liu Q, Sun Y, Li C, Zhou J, Li C, Yang T, et al. ^{18}F-labeled magnetic-upconversion nanophosphors via rare-earth cation-assisted ligand assembly. ACS Nano 2011;5 (4):3146−57. Available from: http://dx.doi.org/10.1021/nn200298y [accessed 08.03.11].

[164] Zhu X, Zhou J, Chen M, Shi M, Feng W, Li F. Core-shell Fe_3O_4@$NaLuF_4$:Yb,Er/Tm nanostructure for MRI, CT and upconversion luminescence tri-modality imaging. Biomaterials 2012;33(18):4618−27. Available from: http://dx.doi.org/10.1016/j.bio-materials.2012.03.007 [accessed 22.03.12].

[165] He M, Huang P, Zhang C, Hu H, Bao C, Gao G, et al. Dual phase-controlled synthesis of uniform lanthanide-doped NaGdF4 upconversion nanocrystals via an OA/ionic liquid two-phase system for *in vivo* dual-modality imaging. Adv Funct Mater 2011;21 (23):4470−7. Available from: http://dx.doi.org/10.1002/adfm.201101040 [accessed 22.09.11].

[166] Xing H, Bu W, Zhang S, Zheng X, Li M, Chen F, et al. Multifunctional nanoprobes for upconversion fluorescence, MR and CT trimodal imaging. Biomaterials 2012;33 (4):1079−89. Available from: http://dx.doi.org/10.1016/j.biomaterials.2011.10.039 [accessed 05.11.11].

[167] Veiseh O, Gunn JW, Zhang M. Design and fabrication of magnetic nanoparticles for targeted drug delivery and imaging. Adv Drug Deliv Rev 2010;62(3):284−304. Available from: http://dx.doi.org/10.1016/j.addr.2009.11.002 [accessed 10.11.09].

[168] Mikhaylov G, Mikac U, Magaeva AA, Itin VI, Naiden EP, Psakhye I, et al. Ferri-liposomes as an MRI-visible drug-delivery system for targeting tumours and their microenvironment. Nat Nanotechnol 2011;6(9):594−602. Available from: http://dx.doi.org/10.1038/nnano.2011.112 [accessed 07.08.11].

[169] Chen F, Zhang S, Bu W, Liu X, Chen Y, He Q, et al. A "neck-formation" strategy for an antiquenching magnetic/upconversion fluorescent bimodal cancer probe. Chem Eur J 2010;16(37):11254−60. Available from: http://dx.doi.org/10.1002/chem.201000525 [accessed 16.08.10].

[170] Cheng L, Yang K, Li Y, Zeng X, Shao M, Lee S-T, et al. Multifunctional nanoparticles for upconversion luminescence/MR multimodal imaging and magnetically targeted photothermal therapy. Biomaterials 2012;33(7):2215−22. Available from: http://dx.doi.org/10.1016/j.biomaterials.2011.11.069 [accessed 12.12.11].

[171] Zhu H, Shang Y, Wang W, Zhou Y, Li P, Yan K, et al. Fluorescent magnetic Fe_3O_4/ rare earth colloidal nanoparticles for dual-modality imaging. Small 2013. Available from: http://dx.doi.org/10.1002/smll.201300126 [accessed 06.03.13].

[172] Carling C-J, Nourmohammadian F, Boyer J-C, Branda NR. Remote-control photorelease of caged compounds using near-infrared light and upconverting nanoparticles. Angew Chem Int Ed 2010;49(22):3782−5. Available from: http://dx.doi.org/10.1002/anie.201000611 [accessed 14.04.10].

[173] Carling C-J, Boyer J-C, Branda NR. Remote-control photoswitching using NIR light. J Am Chem Soc 2009;131(31):10838−9. Available from: http://dx.doi.org/10.1021/ja904746s [accessed 17.07.09].

[174] Dolmans DEJGJ, Fukumura D, Jain RK. Photodynamic therapy for cancer. Nat Rev Cancer 2003;3(5):380−7. Available from: http://dx.doi.org/10.1038/nrc1071 [accessed May 2003].

[175] Konan YN, Gurny R, Allémann E. State of the art in the delivery of photosensitizers for photodynamic therapy. J Photochem Photobiol B 2002;66(2):89−106. Available from: http://dx.doi.org/10.1016/S1011-1344(01)00267-6 [accessed March 2002].

[176] Yano S, Hirohara S, Obata M, Hagiya Y, Ogura S-I, Ikeda A, et al. Current states and future views in photodynamic therapy. J Photochem Photobiol C Photochem Rev 2011;12(1):46−67. Available from: http://dx.doi.org/10.1016/j.jphotochemrev.2011.06.001 [accessed March 2011].

[177] Agostinis P, Berg K, Cengel KA, Foster TH, Girotti AW, Gollnick SO, et al. Photodynamic therapy of cancer: an update. CA Cancer J Clin 2011;61(4):250−81. Available from: http://dx.doi.org/10.3322/caac.20114 [accessed 26.05.11].

[178] Guo H, Sun S. Lanthanide-doped upconverting phosphors for bioassay and therapy. Nanoscale 2012;4(21):6692−706. Available from: http://dx.doi.org/10.1039/c2nr31967e [accessed 24.08.12].

[179] Zhang P, Steelant W, Kumar M, Scholfield M. Versatile photosensitizers for photodynamic therapy at infrared excitation. J Am Chem Soc 2007;129(15):4526−7. Available from: http://dx.doi.org/10.1021/ja0700707 [accessed 27.03.07].

[180] Chatterjee DK, Zhang Y. Upconverting nanoparticles as nanotransducers for photodynamic therapy in cancer cells. Nanomedicine 2008;3(1):73−82. Available from: http://dx.doi.org/10.2217/17435889.3.1.73 [accessed February 2008].

[181] Qian HS, Guo HC, Ho PC-L, Mahendran R, Zhang Y. Mesoporous-silica-coated upconversion fluorescent nanoparticles for photodynamic therapy. Small 2009;5 (20):2285−90. Available from: http://dx.doi.org/10.1002/smll.200900692 [accessed 13.07.13].

[182] Guo H, Qian H, Idris NM, Zhang Y. Singlet oxygen-induced apoptosis of cancer cells using upconversion fluorescent nanoparticles as a carrier of photosensitizer. Nanomedicine 2010;6(3):486−95. Available from: http://dx.doi.org/10.1016/j.nano.2009.11.004 [accessed June 2010].

[183] Qiao X-F, Zhou J-C, Xiao J-W, Wang Y-F, Sun L-D, Yan C-H. Triple-functional core−shell structured upconversion luminescent nanoparticles covalently grafted with photosensitizer for luminescent, magnetic resonance imaging and photodynamic therapy in vitro. Nanoscale 2012;4(15):4611−23. Available from: http://dx.doi.org/10.1039/C2NR30938F [accessed 23.05.12].

[184] Wang C, Tao H, Cheng L, Liu Z. Near-infrared light induced in vivo photodynamic therapy of cancer based on upconversion nanoparticles. Biomaterials 2011;32 (26):6145−54. Available from: http://dx.doi.org/10.1016/j.biomaterials.2011.05.007 [accessed September 2011].

[185] Cui S, Chen H, Zhu H, Tian J, Chi X, Qian Z, et al. Amphiphilic chitosan modified upconversion nanoparticles for in vivo photodynamic therapy induced by near-infrared light. J Mater Chem 2012;22(11):4861−73. Available from: http://dx.doi.org/10.1039/C2JM16112E [accessed 30.01.12].

[186] Huang X, Han S, Huang W, Liu X. Enhancing solar cell efficiency: the search for luminescent materials as spectral converters. Chem Soc Rev 2013;42(1):173−201. Available from: http://dx.doi.org/10.1039/C2CS35288E [accessed 16.10.12].

[187] Chen D, Wang Y, Hong M. Lanthanide nanomaterials with photon management characteristics for photovoltaic application. Nano Energy 2012;1(1):73−90. Available from: http://dx.doi.org/10.1016/j.nanoen.2011.10.004 [accessed January 2012].

[188] van Sark WGJHM, de Wild J, Rath JK, Meijerink A, Schropp REI. Upconversion in solar cells. Nanoscale Res Lett 2013;8:81. Available from: http://dx.doi.org/10.1186/1556-276X-8-81 [accessed 15.02.13].

[189] Shockley W, Queisser HJ. Detailed balance limit of efficiency of p−n junction solar cells. J Appl Phys 1961;32(3):510−9. Available from: http://dx.doi.org/10.1063/1.1736034.

[190] Richards BS. Enhancing the performance of silicon solar cells via the application of passive luminescence conversion layers. Sol Energ Mat Sol C 2006;90(15):2329−37. Available from: http://dx.doi.org/10.1016/j.solmat.2006.03.035 [accessed 22.09.06].

[191] Wang H-Q, Batentschuk M, Osvet A, Pinna L, Brabec CJ. Rare-earth ion doped upconversion materials for photovoltaic applications. Adv Mater 2011;23 (22−23):2675−80. Available from: http://dx.doi.org/10.1002/adma.201100511 [accessed 21.04.11].

[192] Liu J, Yao Q, Li Y. Effects of downconversion luminescent film in dye-sensitized solar cells. Appl Phys Lett 2006;88(17):173119. Available from: http://dx.doi.org/ 10.1063/1.2198825 [accessed 27.04.06].

[193] Hafez H, Wu J, Lan Z, Li Q, Xie G, Lin J, et al. Enhancing the photoelectrical performance of dye-sensitized solar cells using $TiO_2:Eu^{3+}$ nanorods. Nanotechnology 2010;21(41):415201. Available from: http://dx.doi.org/10.1088/0957-4484/21/41/ 415201 [accessed 16.09.10].

[194] Xu S, Xu W, Dong B, Bai X, Song H. Downconversion from visible to near infrared through multi-wavelength excitation in Er^{3+}/Yb^{3+} co-doped $NaYF_4$ nanocrystals. J Appl Phys 2011;110(11):113113. Available from: http://dx.doi.org/10.1063/ 1.3666068 [accessed 15.12.11].

[195] Cheng C-L, Yang J-Y. Hydrothermal synthesis of Eu^{3+}-doped $Y(OH)_3$ nanotubes as downconversion materials for efficiency enhancement of screen-printed monocrystalline silicon solar cells. Electron Devic Lett, IEEE 2012;33(5):697−9. Available from: http://dx.doi.org/10.1109/LED.2012.2187771 [accessed 19.03.12].

[196] Li ZQ, Li XD, Liu QQ, Chen XH, Sun Z, Liu C, et al. Core/shell structured $NaYF_4$: $Yb^{3+}/Er^{3+}/Gd^{3+}$ nanorods with Au nanoparticles or shells for flexible amorphous silicon solar cells. Nanotechnology 2012;23(2):025402. Available from: http://dx.doi. org/10.1088/0957-4484/23/2/025402 [accessed 14.12.12].

[197] Mader HS, Wolfbeis OS. Optical ammonia sensor based on upconverting luminescent nanoparticles. Anal Chem 2010;82(12):5002−4. Available from: http://dx.doi.org/ 10.1021/ac1007283 [accessed 19.05.10].

[198] Ali R, Saleh SM, Meier RJ, Azab HA, Abdelgawad II, Wolfbeis OS. Upconverting nanoparticle based optical sensor for carbon dioxide. Sens Actuators B Chem 2010;150(1):126−31. doi:10.1016/j.snb.2010.07.031 [accessed 21.09.10].

[199] Zhou Z, Hu H, Yang H, Yi T, Huang K, Yu M, et al. Up-conversion luminescent switch based on photochromic diarylethene and rare-earth nanophosphors. Chem Commun 2008;39:4786−8. Available from: http://dx.doi.org/10.1039/b809021a [accessed 12.08.08].

[200] Zhang C, Zhou H-P, Liao L-Y, Feng W, Sun W, Li Z-X, et al. Luminescence modulation of ordered upconversion nanopatterns by a photochromic diarylethene: rewritable optical storage with nondestructive readout. Adv Mater 2010;22(5):633−7. Available from: http://dx.doi.org/10.1002/adma.200901722 [accessed 28.10.09].

[201] Gupta BK, Haranath D, Saini S, Singh VN, Shanker V. Synthesis and characterization of ultra-fine $Y_2O_3:Eu^{3+}$ nanophosphors for luminescent security ink applications. Nanotechnology 2010;21(5):055607. Available from: http://dx.doi.org/10.1088/0957-4484/21/5/055607 [accessed 06.01.10].

[202] Kim WJ, Nyk M, Prasad PN. Color-coded multilayer photopatterned microstructures using lanthanide (III) ion co-doped $NaYF_4$ nanoparticles with upconversion luminescence for possible applications in security. Nanotechnology 2009;20(18):185301. Available from: http://dx.doi.org/10.1088/0957-4484/20/18/185301 [accessed 14.04.09].

[203] Blumenthal T, Meruga J, May PS, Kellar J, Cross W, Ankireddy K, et al. Patterned direct-write and screen-printing of NIR-to-visible upconverting inks for security applications. Nanotechnology 2012;23(18):185305. Available from: http://dx.doi.org/ 10.1088/0957-4484/23/18/185305 [accessed 13.04.12].

[204] Barber DB, Pollock CR, Beecroft LL, Ober CK. Amplification by optical composites. Opt Lett 1997;22(16):1247−9. Available from: http://dx.doi.org/10.1364/ OL.22.001247 [accessed 15.08.97].

[205] Downing E, Hesselink L, Ralston J, Macfarlane R. A three-color, solid-state, three-dimensional display. Science 1996;273(5279):1185−9. Available from: http://dx.doi.org/10.1126/science.273.5279.1185 [accessed 30.08.96].

[206] Hinklin TR, Rand SC, Laine RM. Transparent, polycrystalline upconverting nanoceramics: towards 3-D displays. Adv Mater 2008;20(7):1270−3. Available from: http://dx.doi.org/10.1002/adma.200701235 [accessed 20.03.08].

[207] Boyer JC, Johnson NJJ, van Veggel FCJM. Upconverting lanthanide-doped $NaYF_4$−PMMA polymer composites prepared by *in situ* polymerization. Chem Mater 2009;21(10):2010−2. Available from: http://dx.doi.org/10.1021/cm900756h [accessed 30.04.09].

[208] Lin CK, Berry MT, Anderson R, Smith S, May PS. Highly luminescent NIR-to-visible upconversion thin films and monoliths requiring no high-temperature treatment. Chem Mater 2009;21(14):3406−13. Available from: http://dx.doi.org/10.1021/cm901094m [accessed 19.06.09].

[209] Singh S, Srivastava P, Kapoor IPS, Singh G. Preparation, characterization, and catalytic activity of rare earth metal oxide nanoparticles. J Therm Anal Calorim 2013;111(2):1073−82. Available from: http://dx.doi.org/10.1007/s10973-012-2538-5 [accessed 01.02.13].

[210] Xu A-W, Gao Y, Liu H-Q. The preparation, characterization, and their photocatalytic activities of rare-earth-doped TiO_2 nanoparticles. J Catal 2002;207(2):151−7. Available from: http://dx.doi.org/10.1006/jcat.2002.3539 [accessed 25.04.02].

[211] Stengl V, Bakardjieva S, Murafa N. Preparation and photocatalytic activity of rare earth doped TiO_2 nanoparticles. Mater Chem Phys 2009;114(1):217−26. Available from: http://dx.doi.org/10.1016/j.matchemphys.2008.09.025 [accessed 15.03.09].

5 Self-Assembled Monolayer Covalently Fixed on Oxide-Free Silicon

Hiroyuki Sugimura

Department of Materials Science and Engineering, Kyoto University, Sakyo, Kyoto, Japan

5.1 Introduction

Much attention has been paid to the chemisorption of organic molecules onto semi-conductor surfaces, particularly, on silicon (Si), and the formation of monolayers from the viewpoint of functionalization of semiconductor surfaces [1,2]. The research in this field was driven at its beginning stage by fundamental scientific interests in the chemisorption behavior of organic molecules on Si surfaces in ultra-high vacuum conditions and the physical/chemical properties of these systems [3−5]. Actual chemical modification of Si surfaces in an environment around normal temperature and pressure has been conducted mainly by the use of organosilane self-assembled monolayers (SAMs) [6−8]. In this system, a thin oxide layer at least 1 nm or more in thickness is required to be prepared on an Si substrate prior to covering it with the organosilane SAM. However, in the last two decades, liquid phase processes, by which SAMs could be formed on oxide-free Si substrates, have been developed [9,10]. Namely, each organic molecule comprising such a SAM is covalently bonded to a surface Si atom of the substrate. Those direct-bonded SAMs have interesting potential as electronic materials [11−13], lubricating layers in micromechanical systems [14,15], anchoring layers in order to immobilize biomolecules, e.g., proteins and DNAs, on Si [16−18], and patternable resist films for micro−nano lithography [19−23].

5.2 Chemical Reactions of Si-H Groups with Organic Molecules

As schematically illustrated in Figure 5.1, the chemical reactions of hydrogen-terminated Si (Si-H) in the liquid phase processes are primarily employed in order to graft organic molecules onto Si. The research in this filed was initiated by the

Nanocrystalline Materials. DOI: http://dx.doi.org/10.1016/B978-0-12-407796-6.00005-1

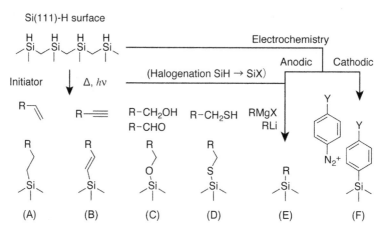

Figure 5.1 Molecular grafting on Si(111)-H through chemical reactions of Si-H with molecules having a terminal functional group of (A) vinyl, (B) ethynyl, (C) alcoholic hydroxyl or aldehyde, (D) mercapto, and molecules of (E) Grignard or alkyl lithium reagent and (F) diazonium compound.

first report in early 1990s [24]. Many techniques, in which unsaturated hydrocarbons, alcohols, and aldehydes reacted on Si-H surface, were subsequently developed [25–38]. An unsaturated hydrocarbon molecule, which has a vinyl or ethyl group, reacts with an Si-H group on an Si(111)-H surface so as to be grafted on the Si surface through a covalent bond of Si–C. This reaction is promoted by the use of a chemical initiator [24], thermal activation [25], photoexcitation [26,28,31,34], sonication [37], etc. The chemical reactions of Si-H groups with organic molecules have been reported to proceed even at room temperature in the dark [33], although SAMs grafted at a higher temperature or under photoillumination showed better qualities with a higher grafted molecular density than those of the SAM prepared at room temperature in the dark. The thermal grafting of alkenes on Si-H was assumed to be promoted through Si radical formation due to the homolysis of Si-H to Si + H [25]. However, this assumption was somewhat suspect, since hydrogen is difficult to desorb from an Si-H surface thermally at room temperature or even at the higher temperature of around 200°C frequently used for thermal grafting processes [39]. The effects of impurities, such as fluorine and oxygen, on the generation of surface Si radicals and other mechanisms for the molecular grafting have been considered [40–42].

Alcohol and aldehyde molecules are fixed on Si through Si–O–C bonds [29,32]. Organothiol molecules which have a similar molecular structure of R-SH to that of alcohol (R-OH) react with Si-H groups as well and graft probably through Si–S–C bonds resulting in the formation of SAMs [35,38]. It has been reported that Grignard reagents (RMgX) and alkyl lithium reagents (RLi) reacted on Si-H to form SAMs [43–50]. For example, thermal activation [44,45], photoexcitation [49,50], and anodic electrochemical reactions (electrochemical oxidation) [46,47] have been employed to promote the reactions of Si-H with reagents. In order to increase the chemical reactivities of the Si surface, surface-terminating hydrogen atoms of the Si-H surface were replaced with halogen atoms, such as iodine or chlorine, to form an Si-X surface [46,47]. When CH_3MgBr or CH_3Li is used as a reagent, methyl groups, i.e., the smallest alkyl group, can be grafted on the Si(111) surface [46,48,50]. When 1-alkene (R−CH=CH$_2$) molecules are grafted on an Si (111)-H surface, the substitution ratio of the Si-H groups with the grafted alkyl groups is reported to be around 50% due to the steric hindrance between the alkyl chains [27]. Even in the case of grafting 1-alkyne (R−C≡CH), of which steric hinderance is smaller, the ratio is near 60% at most [36]. However, methyl groups can substitute all of the Si-H groups on Si(111)-H, that is, the substitution ratio is 100% as reported previously [46]. An electrochemical grafting process of diazonium compounds has also been reported [51,52]: in this case, cathodic reactions were employed in order to generate phenyl radicals.

5.3 Photochemical Grafting of Alkyl SAMs on Si(111)-H

In the case of the photoexcitation process, it was considered that a photon, having energy high enough to cleave an Si−H bond and to generate an Si radical, is necessary [28]. However, successive studies have proved that visible light could promote the grafting process [31,34]. Light at a wavelength of around 447 nm was successfully applied to graft 1-alkenes on an Si-H surface [31]. It was also reported that the grafting of 1-alkene molecules onto Si-H was promoted by visible light even at 700 nm [34]. In this section, this visible-light activation process is described.

5.3.1 Molecular Grafting with Visible-Light Activation

An Si wafer [n-type Si(111) with a resistivity of $1−10$ Ωcm] was etched in 5% HF solution for 5 min at room temperature in the dark and, subsequently, in 40% NH_4F solution for 30 s at 80°C. Throughout these treatments, the native oxide layer on each sample was removed and, consequently, the underlying Si surface was exposed and terminated with hydrogen. The hydrogen-terminated Si(111) substrates [Si(111)-H] were modified with 1-hexadecene [HD, $H_3C(CH_2)_{13}CH=CH_2$] via three different methods as illustrated in Figure 5.2 Details of the experimental setup were described elsewhere [34]. Briefly, an Si(111)-H substrate was immersed in neat HD liquid and then heated or photoirradiated under an atmosphere purged with nitrogen. The experimental conditions are summarized in Table 5.1. For UV irradiation, a high pressure mercury lamp was used as the light source, while a xenon (Xe) lamp was employed for the visible-light excitation. The Xe lamp light was filtered with the use of a long pass filter (>420 nm) or one of three types of band-pass filters (400, 500, and 700 nm, Full width of half maximum (FWHM) 10 nm). Light longer than 850 nm was cut off by the VIS-mirror installed in the Xe lamp house. Band-pass light with a center wavelength at 300 nm was also prepared

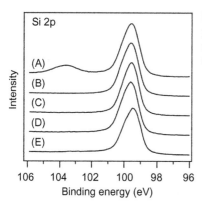

Figure 5.2 XPS-Si 2p spectra of (A) photochemically cleaned Si substrate and (B) Si-H substrates, and (C) Th-SAM, (D) UV-SAM, and (E) Vis-SAM coated samples.

Table 5.1 Experimental Conditions for SAM Formation

Activation Method	Temperature or Irradiance	Reaction Time (h)
Heat (thermal)	180°C	2
UV light	100 mW/cm^2	1
Visible light (>420 nm)	330 mW/cm^2	8
Visible light (400 nm)	8 mW/cm^2	8–32
Visible light (550 nm)	8 mW/cm^2	8–32
Visible light (700 nm)	8 mW/cm^2	8–32

using the UV lamp and a filter. The HD liquid was kept at room temperature during the photoillumination. By these surface modification processes, hexadecyl groups were grafted on the Si(111)-H substrates resulting in the formation of hexadecyl-SAMs covalently fixed on Si (HD-SAMs, $H_3C(CH_2)_{15}-Si\equiv$).

5.3.2 HD-SAMs on Si Prepared by the Three Different Methods

The HD-SAMs formed by the thermal, UV-light, and visible-light activation processes are named Th-SAM, UV-SAM, and Vis-SAM, respectively. Water contact angles of these HD-SAMs were all almost the same, regardless of the activation method. The values were 109, 109, and 108°, respectively, and close to that of closely packed methyl-terminated surfaces, 110° [53], and that of an octadecyl SAMs on Si fabricated by Linford et al. [25], who reported that the advancing and receding contact angles of the octadecyl SAMs prepared on an Si-H substrate by a thermal method were 113° and 110°, respectively. HD molecules were certainly attached to the Si(111)-H substrate surfaces so as to form HD-SAMs. X-ray photoelectron spectroscopy (XPS) was applied to the samples for characterization. XPS-Si 2p spectra of the samples are shown in Figure 5.2. While the spectrum of a cleaned Si substrate without H-termination has a small side peak centered at

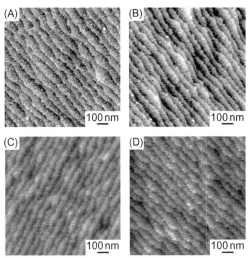

Figure 5.3 AMF topographic images of (A) Si-H substrate, (B) Th-SAM, (C) UV-SAM, and (D) Vis-SAM coated samples.

103.8 eV which corresponds to oxidized Si and indicates that the substrate is covered with a surface oxide, the other spectra consist of a single peak centered around 99.6 eV, which corresponds to oxide-free Si. These results indicate that there was no interfacial oxide layer on any of the samples and that the SAMs were directly attached to the Si substrates. From quantitative XPS analysis of the samples, the surface carbon amounts of the samples, i.e., Th-, UV- and Vis-SAMs, were confirmed to be 34−38 at.%, greatly increased from that of the Si(111)-H surface. The increase in the surface carbon concentration supports that HD molecules were chemisorbed on the samples. From the experimental results shown here, it can be concluded that the visible light activated the SAM formation reaction just as UV light did, although the visible light could not homolytically cleave an Si−H bond as mentioned above.

An atomic force microscope (AFM) image of the Si(111)-H sample is shown in Figure 5.3A. Its surface morphology shows that a surface consisting of atomically flat terraces and monoatomic steps has been successfully fabricated. The step height was estimated to be about 0.3 nm from a cross-sectional view of the AFM image. AFM images of the SAM-formed substrate surfaces (Figure 5.3B−D) show very similar surface morphologies to that of the Si(111)-H surface. The SAM surfaces have terraces with the same widths and monoatomic steps. The thicknesses of the SAMs were estimated to be 2.0−2.3 nm by ellipsometry. These values are reasonable for the thickness of monolayers, since the precursor molecule of HD has a length of about 2 nm. The thickness of a monolayer is determined by the density and the tilt angle of the adsorbed molecules, so there is little difference in film thickness, and consequently in molecular density and tilt angle, between Th-, UV-, and Vis-SAMs. It should be emphasized that the organic molecules are arranged uniformly with a high order molecular packing in these SAMs since the monoatomic steps of the H-Si(111) substrates remain without any distortions as demonstrated in the AFM images even though the substrates were coated with a SAM, the thickness of which was more than seven times greater than the step height.

5.3.3 Wavelength Dependence on the Photoactivation Process

From the discussion above, visible light certainly promoted the SAM formation reaction between Si-H and 1-alkene. The Vis-SAM so obtained proved to have the same features as Th- and UV-SAMs. For this Vis-SAM formation process, white light with a spectrum between 420 and 850 nm in wavelength was employed. However, it has not yet been elucidated whether light of any wavelength would result in the same SAMs on Si. Figure 5.4 demonstrates the wavelength dependence on Vis-SAM formation: (a) the water contact angle and (b) the ellipsometric thickness of the samples fabricated by irradiation at 300, 400, 550, and 700 nm in wavelength for 0, 8, 16, and 32 h. As the irradiation wavelength became shorter, the water contact angle increased faster. The thicknesses of the films grown on the samples estimated by means of ellipsometry also increased as irradiation time increased (Figure 5.4B). The growth rate became higher when the irradiation wavelength was shorter. XPS-Si 2p spectra of the samples fabricated by irradiation for 8 h at each wavelength are shown in Figure 5.5. The spectra of the irradiated samples consist of a single peak centered at around 99.6 eV, which corresponds to oxide-free Si, since there are no side peaks corresponding to oxidized Si.

These experimental results clearly demonstrate that the densities of HD molecules chemisorbed on the Si(111)-H substrates gradually increased with irradiation time and finally formed completed SAMs by visible-light irradiation even at a wavelength of 700 nm. AFM images of the sample surfaces as shown in Figure 5.6 indicate very similar surface morphologies to that of the Si(111)-H surface, demonstrating that the prepared HD-SAMs are all uniform and have ordered molecular

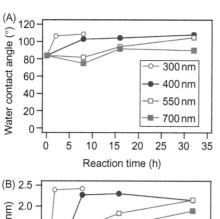

Figure 5.4 (A) Water contact angles and (B) ellipsometric thickness of the samples treated in HD liquid for 8–32 h with UV and visible-light irradiation at 300, 400, 550, and 700 nm in wavelength.

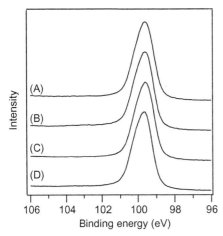

Figure 5.5 XPS-Si 2p spectra of SAM-coated samples prepared by visible-light irradiation at (A) 300, (B) 450, (C) 550, and (D) 700 nm in wavelength.

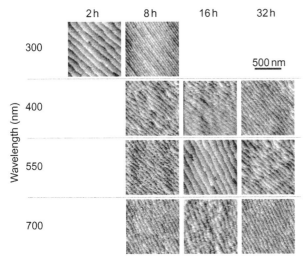

Figure 5.6 AMF topographic images of the SAM-coated samples prepared by UV and visible-light irradiation at 300, 400, 550, and 700 nm in wavelength.

packing. As seen in Figure 5.4, the growth rate became slower with the longer irradiation wavelengths. This is primarily dependent on the light absorption behavior of Si. The absorption coefficients and light penetration depths calculated from optical constants of Si [54] are summarized in Table 5.2. The light absorption became greater as its wavelength became shorter. The absorption coefficient at 300 nm is more than 700 times that at 700 nm. The penetration depth of light and its wavelength dependence should be noted. At 300 nm wavelength, almost all light is absorbed in the surface Si layer of 0.013 μm in depth, while at 700 nm, light penetrates to the inside of bulk Si by more than 10 μm. This means that the generation of holes and electrons due to excitation of Si becomes more concentrated in the surface layer as the irradiated light wavelength becomes shorter. We conclude that

Table 5.2 Optical Absorption of Si

Wavelength (nm)	Absorption Coefficient α (m^{-1})	Relative α Ratio	Penetration Depth[a] (μm)
300	1740×10^5	770	0.013
400	122×10^5	54	0.189
550	9.27×10^5	4	2.48
700	2.26×10^5	1	10.21

[a]The depth where the light intensity becomes 1/10.

the photoactivation of the reaction between 1-alkene and Si-H is governed by the generation of holes and electrons in the surface layer of the Si substrate.

A mechanism to promote the chemical reaction of 1-alkenes with an Si-H surface for SAM formation based on the excitation of the Si-H surface with visible light has been suggested in several papers: Stewart et al. suggested an exciton-based mechanism [52]. This mechanism sounds reasonable when SAMs are formed on H-terminated porous Si substrates, which have a direct band gap and, hence, are easily excited by visible light to provide hole/electron pairs. Sun et al. proposed that the same mechanism was also adaptable for SAM formation on Si(100)-H surfaces [31]. They have recently suggested another refined mechanism in which Si−Si bonds are cleaved in a concerted manner [55]. In these proposed mechanisms, the generation and separation of hole/electron pairs are hypothesized. First, hole/electron pairs are generated at the Si-H surface due to photoexcitation. The electrons temporarily transfer deep into the bulk Si. Finally, only the holes remaining at the surface participate the reaction. However, this hypothesis is not perfect, since photochemical hydrosilylation based on visible-light excitation can occur both on n- and p-type Si substrates [56]. On the p-type Si surface, in general, photogenerated holes migrate into the bulk Si and most electrons remain on the surface. In 2006, Coletti et al. [41] proposed a new concerted mechanism for thermal hydrosilylation as an alternative to the conventional radical-based mechanism. In this mechanism, the carbon−carbon double bond of a 1-alkene approaches the Si−H bond in a parallel fashion, and they then form a four-membered transition state, resulting in the evolution of an Si−C bond. They have explained using density functional calculations that the activation energy is much lower than that required for the radical-based mechanism. According to their calculations, the activation energy of this concerted mechanism is about 270 kJ/mol. Based on a simple arithmetic estimation, light with a wavelength shorter than 441 nm has a photon energy larger than 270 kJ/mol. However as described above, the reaction proceeded even under irradiation with lower photon energies at wavelengths of 550 and 700 nm. HD molecules do not absorb light with wavelengths between 300 and 800 nm. This indicates that the visible light was not absorbed by the precursor molecules but excited the Si substrate itself. In addition, the optical absorption coefficient of single crystal Si at 400 nm is about 50 times that at 700 nm as shown

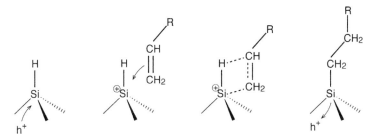

Figure 5.7 Grafting reaction of alkene to Si(111)-H at the hole-trapped site.

in Table 5.2. Thus, as the irradiation wavelength becomes shorter, the number of hole/electron pairs generated at the Si substrate surface increases and governs the SAM growth reaction. Considering that the reaction itself proceeds at room temperature in the dark, thus, the holes and electrons are not crucially necessary, but might accelerate the reaction. As schematically illustrated in Figure 5.7, at the first step of the reaction, holes and electrons are generated in the substrate Si due to photoexcitation. Next, some of the holes are trapped for a while at the surface Si atoms terminated with H. This step is more favorable for n-type Si than p-type Si, since the generated holes migrate to the surface due to the band bending of n-type Si. This Si-H site at which a hole is trapped might have an affinity to the electron-rich vinyl group of the 1-alkene so that it is more reactive than the normal Si-H site where a hole is not trapped. Consequently, the grafting of the alkyl group through the concerted mechanism proposed by Coletti et al. [27] is further promoted. The trapped hole might disappear due to migration or recombination with a photogenerated electron.

5.4 Photochemical Preparation of Methyl-Terminated Si(111)

Methyl ($-CH_3$) is the simplest of the alkyl groups ($-C_nH_{2n+1}$). It has the capability to completely cover the Si surface [46,57−59] because its van der Waals radius is small enough to fit between the spacing of Si(111). Furthermore, the bonding of carbon with the Si surface has superior stability compared to the bonding of other elements that may also fit in between the Si(111) spacing (e.g., hydrogen, halogens). To note: halogen-terminated Si surfaces have high chemical reactivity that make them unsuitable for stabilizing the Si surface. Conventional techniques of producing a CH_3-terminated Si(111) include (a) the chlorination−alkylation method [57,60−63] and (b) electrochemical method [46,47,64,65]. The chlorination−alkylation method uses a Cl-terminated Si(111) as an intermediate state between the H-terminate and CH_3-terminate surfaces. Chlorination is done either by using PCl_5 in chlorobenzene at an elevated temperature [60−62] or by using Cl_2 gas diluted in argon [57,63].

The electrochemical preparation uses anodic potential to trigger the CH_3-grafting process [46,64−66]. In this section, a photochemical grafting process for CH_3-termination of Si(111) surface [50] is described. Compared with the chlorination−alkylation process, photochemical preparation uses the Si-H surface directly as a grafting substrate, instead of using an Si−Cl surface. This avoids the use of either a hazardous thermal chlorination process or a gas-based chlorination process in an inert atmosphere. Compared with the electrochemical preparation, the photochemical preparation uses no external power supply. Thus, the photochemical grafting can avoid some complications, such as creating an ohmic contact to the Si substrate or an electrical connection from an external power supply in a hazardous liquid bath. The creation of an ohmic contact also introduces fixtures on Si that are permanent or difficult to remove. The photochemical technique can either offers a safer or a simpler procedure of producing methyl-terminated Si(111) compared with the conventional methods.

5.4.1 Methyl-Termination Process

An n-type Si(111) wafer with a resistivity of around $1−10\ \Omega$cm was used as the substrate. The photochemical preparation was performed by illuminating visible light from a Xe lamp onto an Si(111)-H substrate immersed in a 12% (ca. 1 mol/L) Grignard−tetrahydrofuran (CH_3MgBr−THF) solution at room temperature. The solution was placed inside an optical cell and was purged with a stream of nitrogen gas prior to and during the photochemical grafting process. Because the Grignard reagent is water sensitive, the grafting process was done inside a moisture-controlled chamber and all the materials used were confirmed to be dry.

5.4.2 Properties of CH_3-Terminated Si(111)

The sample photochemically treated in the Grignard−THF solution exhibits an oxidation-resistant property. Figure 5.8A shows that this sample has no distinct peak associated with the presence of silicon oxide, unlike the other samples that

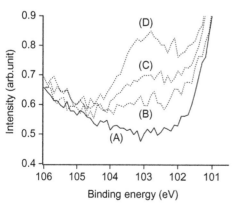

Figure 5.8 XPS-Si 2p spectra (focus on the region associated with silicon oxide) of the different samples. (A) Si(111)-H sample irradiated with visible light for 2 h in the CH_3MgBr−THF solution, (B) Si(111)-H sample, (C) Si(111)-H sample immersed in the Grignard−THF solution for 2 h in the dark, and (D) Si(111)-H sample irradiated with visible light for 2 h in THF without the Grignard reagent.

have been oxidized somewhat (Figure 5.8B–D). In contrast, the Si(111)-H sample (Figure 5.8B) shows the presence of silicon oxide. The oxide might have developed during the postcleaning treatment. Likewise, the sample that was immersed in the Grignard–THF solution for 2 h in the dark (Figure 5.8C) also showed the presence of Si oxide but its level was higher than the former. In this case, the absence of light would retard the methyl grafting that could protect the surface from oxidation. The immersion of the sample for 2 h in the Grignard–THF solution and the post-cleaning procedure promoted oxidation that destroyed the hydrogenated surface. The sample irradiated with visible light for 2 h in THF without the Grignard reagent (Figure 5.1D) showed the largest fraction of Si oxide. In this case, there was no source of methyl that could be grafted on the Si surface. Instead of grafting, the presence of light triggered extensive oxidation due to photogenerated holes on the Si(111)-H surface. The CH$_3$-termination certainly protected the Si surface from oxidation more completely than the H-termination. The resistivity to oxidation of the sample provided by the photochemical treatment is attributed to attachment of CH$_3$ groups on the Si surface via the formation of Si–C bonds. An XPS-C1s spectrum of this sample is shown in Figure 5.9. This spectrum has a peak at 284.1 eV, which is different from the peak energy of longer alkyl chains, $-(CH_2)_n-$, i.e., 285.2 eV. This peak shift is evident that the carbon species on the sample consists of Si-CH$_3$.

The holes generated on the Si surface during the illumination of n-type Si trigger the oxidative decomposition of CH$_3$MgBr that in turn liberates the methyl radicals that are needed in the successive steps of the grafting process. The following is our proposed photochemical reaction scheme patterned after the reaction scheme proposed by Fellah et al. [47]:

$$CH_3MgBr + h^+ \rightarrow {}^\bullet CH_3 + MgBr^+ \tag{5.1}$$

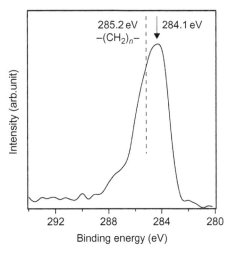

Figure 5.9 XPS-C1s spectra of the Si(111)-H sample irradiated with visible light for 2 h in the CH$_3$MgBr–THF solution.

$$\equiv Si\text{-}H + {}^\bullet CH_3 \rightarrow \ \equiv Si^\bullet + CH_4 \tag{5.2}$$

$$\equiv Si^\bullet + {}^\bullet CH_3 \rightarrow \ \equiv Si\text{-}CH_3 \tag{5.3}$$

Figure 5.10 shows an AFM topographical image of this sample. The image shows a stair-like structure with monoatomic steps, that is, a feature of the Si(111)-H that was accomplished using the HF and NH$_4$F treatments. Figure 5.10 demonstrates that the structure of the Si(111)-H surface has been preserved after the photochemical preparation. This preservation would be possible if the grafted molecules form a film with an atomically flat structure. The covalently bonded carbon is oriented perpendicularly to the Si(111) surface [62] with a bond length of 1.85 Å [67]. This orientation suggests that CH$_3$ groups can form a film on Si(111) surface with an atomically flat structure.

Figure 5.11A shows the water contact angles of the sample as a function of grafting time (at a light intensity of 35 mW/cm^2). The sample with zero grafting time was identical to that of Si(111)-H. Its water contact angle was 78.8°, lower than the water contact angle of Si(111)-H without the postcleaning treatment, i.e., 85.4°. This decrease is attributed to the destruction of the hydrogenated surface due to the postcleaning procedure; this is in agreement with the result in Figure 5.8B. In another aspect of the graph, the grafting process progresses with time. At a shorter grafting time (1 h), only a part of the hydrogen atoms on the Si substrate was substituted with CH$_3$ groups so that the substrate Si was oxidized either in the Grignard−THF solution or by the postcleaning treatment. Thus, the water contact angle was lower than that of the Si(111)-H samples. As grafting time progressed, the Si(111)-H surface became increasingly methylated, and thus the water contact angle increased. Figure 5.11B shows the water contact angles of the samples prepared by 2h-reaction as a function of light intensity. The increase in water contact angle with light intensity is attributed to a faster grafting process. The sample with zero light intensity was the same as that of the sample presented in Figure 5.8C, that is, the sample that was immersed in the Grignard−THF solution for 2 h in the dark. Its water contact angle was 49.2°. This low value means that the CH$_3$-passivation did not develop in the dark. The highest water contact angle

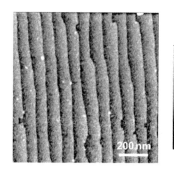

1.2 nm

200 nm

0

Figure 5.10 AFM topographical image of an Si (111)-CH$_3$ surface.

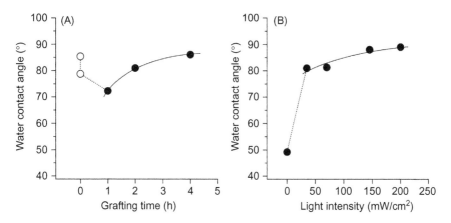

Figure 5.11 Graph of water contact angle with (A) grafting time at 35 mW/cm^2 light intensity and (B) light intensity for 2 h. The open circle in Figure 5.4A denotes the value of the water contact angle of an Si(111)-H.

that we achieved was $(89.0 \pm 0.80)°$. This value was obtained from the sample that was illuminated for 2 h at a light intensity of 200 mW/cm^2.

5.5 Chemical and Mechanical Properties of SAMs on Si

Fundamental knowledge about the chemical durability of SAMs is of primary importance in order to put these SAMs to practical use. For example, when the SAMs are used as resist films or when the SAM surfaces are further functionalized chemically, SAMs are required to resist etching solutions, such as acid, basic, and HF solutions. Linford et al. studied the chemical durabilities of 1-alkene SAM directly attached to Si by comparing it with organosilane SAM formed on bulk Si with an interfacial oxide layer [25]. In the research, they found that 1-alkene SAM has more durability to NH$_4$F solution than organosilane SAM, while both SAMs showed strong durability to H$_2$SO$_4$ solution. This different durability to NH$_4$F solution is due to the difference in the interfacial structure. Saito et al. have also compared the durability of SAMs directly attached to Si with those of organosilane SAMs with an interfacial oxide layer and concluded that, when immersed in HF and NH$_4$F solutions, the SAMs directly attached to Si are more resistive to the solutions than organosilane SAMs [68]. This strong resistance of SAMs directly attached to Si has been applied to photo, electron beam, and scanning probe lithography techniques [19−23]. Boukherroub et al. compared the n-aldehyde SAM and 1-alcohol SAM, both of which have Si−O−C linkages at the interface between the SAMs and Si, and they found that n-aldehyde SAM showed more durability to HCl and HF solutions due to the higher packing density of the molecules in the n-aldehyde SAM [29].

Although several studies have examined on the chemical durability as mentioned above, little is known about the difference in durability between the SAMs with Si—C linkages and those with Si—O—C linkages. The next section assesses the chemical durabilities and tribological performances dependent on properties of SAMs formed on Si(111)-H substrates by thermal activation from 1-alkene, 1-alcohol, and n-aldehyde; the former molecule gives the SAM with Si—C linkages, while the latter two give the SAMs with Si—O—C linkages [15,69].

5.5.1 SAMs Prepared for Chemical Durability Testing

An Si wafer [n-type Si(111) of $1-10\,\Omega$cm in resistivity] was hydrogen-terminated as described in Section 3.1. SAM formation was performed using a thermal activation method as described elsewhere [69]. 1-Hexadecene [HD, $CH_3(CH_2)_{13}CH = CH_2$], 1-hexadecanol [HDO, $CH_3(CH_2)_{15}OH$], 1-dodecanol [DDO, $CH_3(CH_2)_{11}OH$], and n-dodecanal [DDA, $CH_3(CH_2)_{10}CHO$] were used as precursor molecules. For the SAM preparation from HD molecules, the liquid was heated to 180°C and kept for 2 h. For the SAM preparation from HDO and DDO molecules, the liquid was heated to 150°C and kept for 16 h. For the SAM preparation from DDA molecules, the liquid was heated to 150°C and was for 2 h. The samples prepared from HD, HDO, DDO, and DDA are hereafter referred to as HD-SAM, HDO-SAM, DDO-SAM, and DDA-SAM, respectively. The chemical structures of these SAMs are shown in Figure 5.12A.

Water contact angles of these SAMs are shown in Table 5.3. The step and terrace structure of the Si(111)-H surface was retained after the formation of the SAMs in Figures 5.6 and 5.10. The water contact angles of the SAM surfaces are about 110°, indicating that each surface is terminated with close-packed alkyl groups, the top ends of which are terminated with methyl groups. Ellipsometric thicknesses are also listed in Table 5.3. Note that these thicknesses are relative volumes, not the absolutely true thicknesses, since we used an alternative refractive index data of SiO_2 to that of the organic layers for the estimation. However, the difference between the estimated and true thicknesses is considered to be not so large, since the values are comparable to the molecular lengths consisting the SAMs. HD-SAM and HDO-SAM gave almost the same thickness values, as is convincing from the length of the precursor molecule. The same is true for the combination of DDO-SAM and DDA-SAM.

The IR spectra of the SAM samples acquired by FTIR in attenuated total reflection (ATR) mode are shown in Figure 5.12B, where the Si(111)-H sample was used as the reference. Namely, a raw spectrum of each sample was divided by that of the Si(111)-H sample. All the spectra had a negative peak at $2083\,cm^{-1}$, which is attributed to Si-H stretching vibration, meaning that the number of Si—H bonds decreased through the SAM formation. All the spectra had positive peaks between 2800 and $3000\,cm^{-1}$, which are attributed to C—H stretching vibration, and the peaks at 1468 and $1380\,cm^{-1}$, which are attributed to CH_2 scissoring vibration and CH_3 symmetric bending vibration, respectively [70]. These results clearly

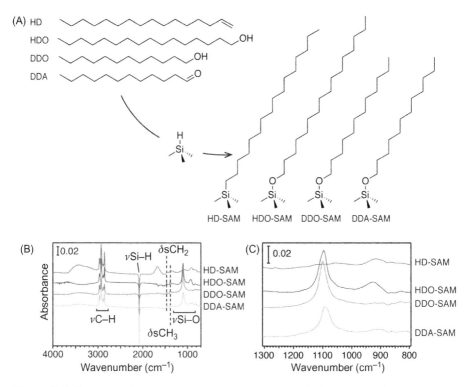

Figure 5.12 SAM samples and ATR-FTIR spectra: (A) chemical structures of HD-, HDO-, DDO-, and DDA-SAMs and (B and C) FTIR spectra.

Table 5.3 Water Contact Angles and Ellipsometric Thicknesses of the SAM Samples

SAM Sample	Water Contact Angle (°)	Thickness[a] (nm)
HD-SAM	109	2.2
HDO-SAM	109	2.1
DDO-SAM	110	1.5
DDA-SAM	110	1.9

[a]Measured by a spectroscopic ellipsometer.

demonstrate that a part of the Si-H groups was replaced with alkyl or alchoxy groups as illustrated in Figure 5.12A.

Figure 5.12C shows the enlarged ATR-FTIR spectra of the four samples in the region between 1300 and 800 cm^{-1}, where the absorption of Si$-$O stretching vibration appears. The spectra of HDO-SAM, DDO-SAM, and DDA-SAM samples have a sharp peak at around 1100 cm^{-1}, which is attributable to Si$-$O$-$C linkage at the interface between the monolayer and the substrate [32,66]. Since this Si$-$O$-$C linkage is not contained in 1-alcohol or n-aldehyde, that is, the precursor

Table 5.4 Densities, Thicknesses, and Substitution Ratios of the SAM Samples Estimated
from the GIXR Results

SAM Sample	Density (g/cm^3)	Thickness (nm)	Substitution Ratio (%)
HD-SAM	0.82	1.71	48
HDO-SAM	0.85	1.76	48
DDO-SAM	0.82	1.31	45
DDA-SAM	0.80	1.22	41

molecule of HDO-SAM, DDO-SAM, or DDA-SAM, we can conclude that the Si−O−C linkage was formed during the SAM formation where the H-Si surface reacted with 1-alcohol or n-aldehyde, which is also indicated by the disappearance of the absorption band due to the OH and CHO groups as mentioned above. The spectrum from the HD-SAM sample does not have an absorption band in the region for Si−O−C, indicating that HD-SAM has a different linkage at the interface between the monolayer and the substrate.

Based on the grazing incidence X-ray reflective (GIXR) technique, the densities and thicknesses of the SAM samples were measured [69]. The estimated results of the SAMs are shown in Table 5.4. Thicknesses of HD-SAM and HDO-SAM were measured to be 1.7 nm. Likewise, those of DDO-SAM and DDA-SAM were 1.3 and 1.2 nm, respectively. The densities of the monolayers are estimated to be 0.8−0.85 g/cm^3. Considering that the density of high-density polyethylene (HDPE) is 0.92−0.96 g/cm^3, the densities of the alkyl chains in the SAMs are 90% as large as that of HDPE. The thicknesses obtained from GIXR are smaller than those obtained from the ellipsometric measurements as shown in Table 5.4, since each measurement depends on different assumptions. However, the thicknesses are of the same order and HD and HDO-SAMs are certainly thicker than DDO- and DDA-SAMs depending on their alkyl chain lengths. The GIXR thicknesses are considered to be accurate enough to compare the four SAMs. A ratio substituting the surface Si-H sites with molecules immobilized through an Si−C bond or Si−O−C bond was roughly estimated from the fitted values of thickness and density of each SAM [69]. The ratio of 48% for HD-SAM is coincident with the maximum value of nearly 50% for grafting 1-alkane on Si(111) [27]. As shown in Table 5.4, the ratio for HDO-SAM is the same as that for HD-SAM, while those for DDO- and DDA-SAMs are smaller, indicating that these SAMs are more loosely packed. The alkyl chains in DDO- and DDA-SAMs are probably to interact with adjacent chains more weakly than those in HD- and HDO-SAMs, since the chain lengths are shorter.

5.5.2 Chemical Durability

The chemical durability of the SAMs was examined by immersing the samples in 5% HF aqueous solution for 60−3600 s. Durabilities of the four SAMs to 5% HF solution are shown in Figure 5.13A as changes in water contact angle due to the

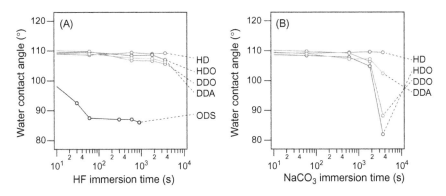

Figure 5.13 Changes in the water contact angle with immersion time in (A) 5% HF solution and (B) 300 mM Na$_2$CO$_3$ solution.

HF immersion. HD-SAM is the most durable to HF etching: the contact angle remains 109° even after immersion for 3600 s. Water contact angles of HDO-SAM, DDO-SAM, and DDA-SAM decreased as immersion time was prolonged. After HF immersion for 3600 s, the contact angles were around 107°, while those before immersion were around 109° for the three SAMs. This contact angle decrease of the three SAMs means that the grafted molecules were delaminated to some extent. However, HDO-, DDO-, and DDA-SAMs are still highly durable to HF compared to the alkylsilane SAM formed on an oxide-covered Si substrate. As a control experiment, the same HF immersion test was conducted using an alkylsilane SAM, which had been prepared from octadecyltrimethoxysilane (ODS) by the vapor phase method on an Si substrate covered with a surface oxide layer 2 nm in thickness [7]. The water contact angle of ODS-SAM/Si samples immersed in HF rapidly decreased to near 85° which is almost same with that of the Si(111)-H. Thus, ODS-SAM and the underlying oxide layer were completely delaminated at an immersion time of 100 s or less, most likely due to the dissociation of Si−O−Si linkages, by which ODS-SAM is immobilized, and etching of the oxide layer.

Figure 5.14 displays AFM topographic images of all the SAMs immersed in the HF solution for 600, 1800, and 3600 s. Before the immersion, HD-SAM had a similar morphology to Si(111)-H with flat terraces and steps. This feature is preserved throughout the HF immersion. The results mean that the molecules comprising SAMs are still as highly packed and ordered as before the immersion. There are, however, some pits near the down sides of steps on the SAM surfaces after the immersion. The pits are, as Gorostiza et al. supposed, due to the removal of molecules [71]. The sizes of the pits did not change with increasing immersion time. For the whole immersion time range, the depth of the pits was about 0.5 nm. At the down side of steps, a steric hindrance effect between absorbed molecules and the step is more significant, so that the molecular density becomes rather small there. Thus, such an area becomes weak toward etching. The pit depth observed by AFM

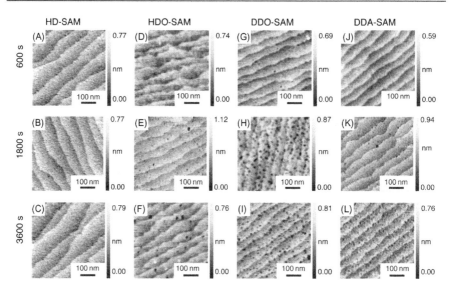

Figure 5.14 AFM topographic images of (A−C) HD-SAM, (D−F) HDO-SAM, (G−I) DDO-SAM, and (J−L) DDA-SAM after immersion in HF solution for 600, 1800, and 3600 s.

is about 0.5 nm, which is not consistent with the film thickness of about 2 nm, because the pit size was so small that the apex of the AFM tip could not reach to the bottom of the pits. For HDO-SAM, DDO-SAM, and DDA-SAM, the pits were also generated by the HF immersion, and a staircase-like structure was also observed. The size of the pits increased with immersion time, which is different from the case for HD-SAM. The difference comes from the interfacial bond differences: HD-SAM has interfacial Si−C linkages and the other three SAMs have interfacial Si−O−C linkages. The Si−O bond in the Si−O−C linkage is very likely to cleave by etching with HF. Once a pit is formed, HF molecules readily attack the Si−O−C linkages around the periphery of each pit and then the molecule is removed from the surface, resulting in the pit size enlargement on HDO-SAM, DDO-SAM, and DDA-SAM surfaces. Regarding the pit location, there was a difference between HD-SAM and the other SAMs. Many pits were present not only at the down side of the steps but also on the terraces. This is because the interfacial Si−O−C linkage is easier to cleave than the Si−C linkage of HD-SAM. The pit depths of 2, 1, and 1.5 nm at most for HDO-SAM, DDO-SAM, and DDA-SAM, respectively, are consistent with the film thicknesses obtained from the GIXR measurement: 1.76, 1.31, and 1.22 nm, respectively. For the images after immersion for 3600 s comparing the HDO-SAM with DDO-SAM, HDO-SAM has fewer pits than DDO-SAM, which means that HDO-SAM is more durable to HF etching than DDO-SAM. Likewise, comparing DDO-SAM with DDA-SAM, DDO-SAM has a similar number of pits to DDA-SAM, which means that they have similar durabilities to HF etching as each other.

The durabilities of the four SAMs to Na_2CO_3 solution were examined by immersion in 300 mM Na_2CO_3 solution. The results of water contact angle measurements are shown in Figure 5.13B. HD-SAM was the most durable to Na_2CO_3 etching: the contact angle remained 109° after etching for 3600 s. The contact angles of HDO-SAM, DDO-SAM, and DDA-SAM decreased drastically with increasing immersion time. After Na_2CO_3 immersion for 3600 s, the contact angles decreased to <103°, meaning a decrease in the number of methyl groups on the SAM surfaces. In particular, for HDO-SAM and DDO-SAM, the contact angles decreased greatly (to around 85°), indicating that many more molecules were removed from the surface for these SAMs. Figure 5.14 displays AFM topographic images of all the SAMs after immersion for 600, 1800, and 3600 s. Throughout the immersion, HD-SAM showed a similar terrace and step morphology to H-Si(111). For HDO-SAM, DDO-SAM, and DDA-SAM, destruction of the SAM structure was observed. For HDO-SAM, some pits were formed after immersion for 1800 s and the stair-like structure disappeared after immersion for 3600 s. For DDO-SAM and DDA-SAM, the size of the pits was larger and the number of the pits after immersion for 3600 s was greater than the case after immersion for 1800 s. More pits were observed on the down sides of steps than on the terraces, as with HF etching. Considering that Si oxide is dissolved with hydroxy ions, the pits were generated and enlarged by the reaction of hydroxy ions with interfacial Si—O bonds and underlying SiO_2.

In conclusion, the order of durability of the SAMs to the HF solution was: HD-SAM > HDO-SAM > DDO-SAM = DDA-SAM and that to the Na_2CO_3 solution was: HD-SAM > DDA-SAM $\cong$ DDO-SAM > HDO-SAM (Figure 5.15).

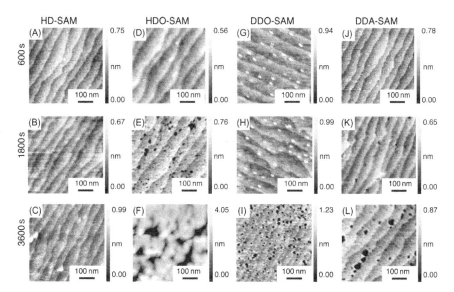

Figure 5.15 AFM topographic images of (A–C) HD-SAM, (D–F) HDO-SAM, (G–I) DDO-SAM, and (J–L) DDA-SAM after immersion in Na_2CO_3 solution for 600, 1800, and 3600 s.

5.5.3   Tribological Properties of Direct-Bonding SAMs on Si

The frictional properties of organic monolayers are of particular importance for their applications in tribology, for example, in order to attain low friction and anti-stick surfaces. The tribological properties of SAMs on minute scales have frequently been investigated, mainly by the use of scanning probe microscopy [72]. SAMs on Si substrates and their tribology have attracted attention, since Si has become a key material for microelectromechanical systems (MEMS). Tribological studies of organic monolayers formed on Si substrates have mainly been conducted on organosilane SAMs so far [73−78], for example, aiming for an application in MEMS [79]. However, although it has been reported that organosilane SAMs show reduced friction coefficients, stiction, and wear, the level of wear resistance was not sufficient for use as MEMS components [80]. The direct-bonded SAMs on Si, showing better chemical durabilities than the organosilane SAMs as discussed above, are expected to be more durable in tribology, since the direct-bonded SAMs are more firmly fixed on Si.

SAMs were prepared on Si(111)-H substrates by the thermal activation method. Each substrate was soaked in a 100 mM solution of HD, HDO, DDO, or 1-dodecene [DD, $CH_3(CH_2)_9CH = CH_2$] in mesitylene for 2 h at a temperature of 150°C. The frictional properties of the samples were examined using a pin-on-plate tribometer as illustrated in Figure 5.16A [15]. As shown in Figure 5.16B, the friction test for each sample was performed in a direction parallel to the steps to prevent any effect of morphologies on the tribological properties.

Figure 5.16C shows typical examples of the tribology test in which the friction coefficients on the SAMs and bare Si(111) surfaces are plotted as a function of sliding time. The normal load of 9.8 mN was applied and the values of the friction coefficients were average values for each 1 min of sliding. The bare Si(111) surface showed higher friction coefficients than the other surfaces with SAMs. The HD- and HDO-SAMs with longer alkyl chains (16 carbon atoms) exhibited lower friction coefficients than the DD- and DDO-SAMs (12 carbon atoms). In addition, HD-SAM had lower friction coefficients than HDO-SAM. DD- and DDO-SAMs had almost the same friction coefficients. The differences between the HD, HDO-SAMs and DD, DDO-SAMs are considered to be due to the effect of the packing density of the SAMs. It is known that SAMs formed by organosulfur [81] and organosilane [82] with longer alkyl chains have higher packing densities than those with shorter alkyl chains because the longer alkyl chains have stronger intermolecular interactions. Figure 5.16D shows the friction coefficients of the SAMs and the bare Si(111) surfaces as a function of the sliding time under a higher load of 98 mN. For the bare Si(111), DD- and DDO-SAMs, the friction coefficients increased immediately after the start of the friction tests, indicating that the sample surfaces were damaged due to this high-load friction. However, the friction coefficients of HDO-SAM were initially 0.04−0.06 and increased after 1000 s. Furthermore, HD-SAM showed low friction coefficients of 0.04−0.08 even after 4000 s of sliding, which indicated that HD-SAMs are the most durable to friction and promising for practical applications.

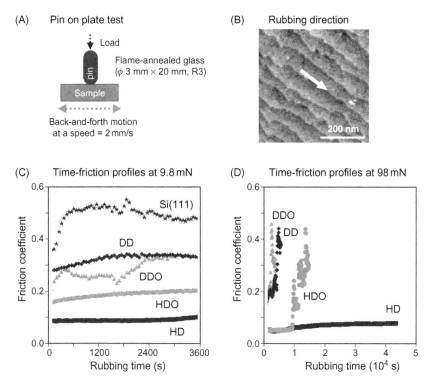

Figure 5.16 Tribological properties of HD-, HDO-, DD-, and DDO-SAMs. (A) Schematic illustration of the pin-on-plate test. (B) Sliding direction: The pin was moved along the direction approximately parallel to the steps. (C) Time-friction profiles acquired on the SAMs at a load of 9.8 mN. (D) Profiles at a load of 98 mN.

5.6   Electronic Functions of Direct-Bonding SAMs on Si

The electronic properties of the direct-bonding SAMs are of interest in applying them to microelectronics. Current—voltage characteristics have been measured by preparing a micro-fabricated device and by the use of an AFM tip or a mercury drop as a probing electrode. The properties of alkyl SAMs covalently bonded to Si as insulators have been fairly well elucidated, as have the band structure and electron transfer behavior at the alkyl SAM/Si interface [12,83—85]. Redox-active molecules attached to an electrode surface can receive and release a charge through the electrochemical reaction depending on the electrode potential. This charge storage/release function is repeatable. Consequently, SAMs containing redox-active molecules [86—91] are anticipated for use in sensors and solid-state memory devices. Some metal—organic complex molecules are known to show good repeatabilities in their redox reactions. In particular, ferrocene (Fc) and its derivatives have well-known electrochemical properties. Thus, Fc has attracted attention for immobilization on Si and for the basic research on charge storage SAMs [92].

5.6.1   SAMs as Dielectric Layers

In this section, the resistance of an alkyl SAM formed on Si(111) to the application of a dc bias voltage and resulting changes in surface and interfacial states are reported. On an Si(111) substrate (n-type, $1-10\,\Omega$cm), HD-SAM was formed by the thermal activation method as illustrated in Figure 5.17A. In this sample, the insulator thickness was estimated to be about 2.3 nm by ellipsometry. For the control experiment, an alkylsilane SAM, i.e., ODS-SAM, was prepared on the same Si substrate by the vapor phase method [7]. Prior to the silane coupling, an amorphous oxide layer with a hydroxylated surface was prepared on the substrate. The thickness of the oxide layer was about 2 nm. Thus, the total insulator thickness on this sample was 3.8 nm as estimated by ellipsometry. The surface of ODS-SAM showed a water contact angle of 103°, less hydrophobic than HD-SAM which showed a

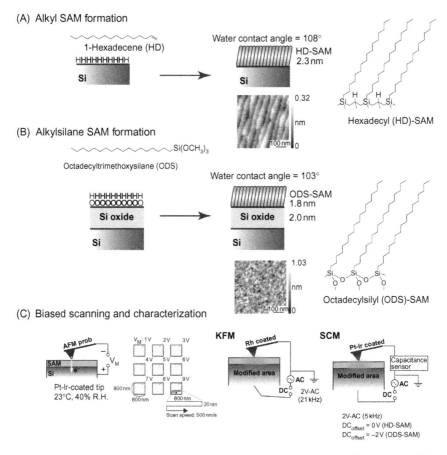

Figure 5.17 Sample preparation for the electrical characterization. (A) Alkyl SAM on Si (HD-SAM sample). (B) Alkylsilane SAM on oxide/Si (ODS-SAM sample). (C) Sample modification via scanning probe anodization and characterization by KFM and SCM.

water contact angle of 108°, indicating that HD-SAM is more closely packed than ODS-SAM. Nine square regions on both the samples were scanned by a metal-coated AFM tip. The scanning was conducted in air with conditions as indicated in Figure 5.17C. A dc bias voltage (V_M: tip-negative/substrate-positive) was applied during the scan of each square region.

After the scanning, topographic AFM and lateral force microscopy (LFM) images were acquired simultaneously as shown in Figure 5.18A. In the topographic and frictional images, there are certain changes in the square regions scanned at V_M >5 V. This is due to anodization of the organic molecules consisting of the SAMs and the substrate Si [93]. Conversely, the maximum voltage at which no damage was induced to the samples was found to be 5 V for each. Thus, the threshold field strengths for HD-SAM and ODS-SAM to remain damage-free can be estimated to be 2.1 and 1.3 V/nm, respectively, from the total insulating layer thicknesses. ODS-SAM itself might be damaged when a bias voltage of 2.4 V ($=1.3$ V/nm $\times$ 1.8 nm) is truly applied to the SAM. The dc bias durability of HD-SAM is concluded to be twice that of ODS-SAM.

In addition to the topographic and frictional imaging, the samples were further electrically characterized using Kelvin-probe force microscopy (KFM) and scanning capacitance microscopy (SCM) [94]. KFM and SCM images are shown in Figure 5.18C. Contrasts between the scanned and unscanned regions are summarized in Figure 5.18D. In the case of HD-SAM, no contrasts appear in the KFM and SCM images, when chemical changes were not induced in the scanned regions as demonstrated in Figure 5.18A and B, namely, in the bias range of $V_M \leqq 5$ V. In contrast, in the case of ODS-SAM, distinct image contrasts are present at $V_M \leqq 5$ V, even there are no changes in the topography and frictional images. This is most likely due to charge trapping at the interface (oxide−silicon and/or SAM/oxide) [93]. The interface of HD-SAM/Si(111) is demonstrated to be a trap-free interface, which might be useful for organic-Si electronics.

5.6.2 Redox-Active SAM on Si

In this section, the immobilization processes of two types of Fc derivatives, vinyl-ferrocene (VFc) and ferrocenecarboxaldehyde (FcA), on Si(111)-H and the electro-chemical responses of VFc-Si and FcA-Si [95,96] were examined. Since VFc readily polymerizes thermally, it was difficult to form uniform SAMs by the thermal process, and thus a photochemical process which could be conducted at room temperature was employed. Moreover, visible-light excitation was applied instead of the UV process in order to minimize photodecomposition of Fc molecules. Figure 5.19A and B shows the AFM surface images of VFc-SAM and FcA-SAM formed on Si(111) substrates (n-type, 0.001−0.004 Ωcm). An Si(111)-H substrate was immersed in a VFc or FcA solution in *n*-decane at a concentration of 10 mM and then irradiated with white light from the Xe lamp at an intensity of 200 mW/cm^2. Experimental details have been described in Ref. [96]. As shown in Figure 5.19A and B, uniform SAMs were successfully fabricated.

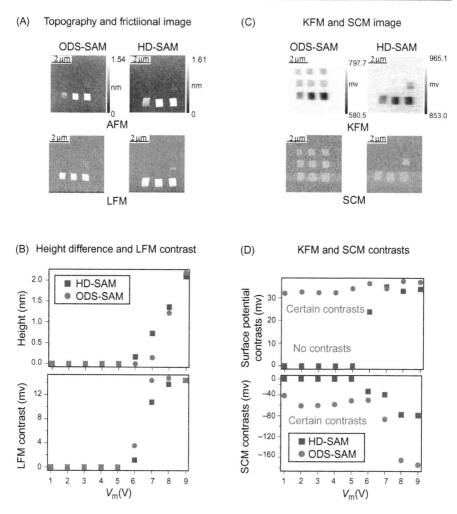

Figure 5.18 Changes in HD-SAM/Si and ODS-SAM/Si samples AFM probe scanning with applying a positive sample bias. (A) Topographic and frictional AFM images, (B) height and LFM contrasts, (C) KFM and SCM images, and (D) surface potential and SCM contrasts.

Cyclic voltammetry (CV) was carried out using 0.1 M $HClO_4$ aqueous solution as an electrolyte as shown in Figure 5.20. These samples were fabricated by the light irradiation for 4 h. The CV results imply:

1. electrochemical activities Fc parts in VFc and FcA are retained after immobilization on Si;
2. VFc and FcA-SAMs are linked electronically with each Si substrate;
3. the redox potential of VFc-SAM is 0.21 V versus Ag/AgCl, while that of FcA-SAM is 0.32 V versus Ag/AgCl.

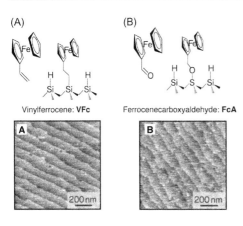

Figure 5.19 Attachment of Fc derivatives onto Si(111): (A) VFc-SAM and (B) FcA-SAM.

(A) Vinylferrocene: **VFc**

(B) Ferrocenecarboxyaldehyde: **FcA**

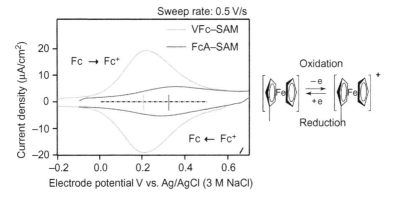

Figure 5.20 Electrochemical behaviors of VFc-SAM and FcA-SAM on high-doped n-Si (111).

The redox potential shift of FcA-SAM is most likely due to the electron negativity of the O atom linking Fc to Si. From the CV peak area shown in Figure 5.20, the molecular packing density of VFc-SAM was estimated to be 5.7×10^{13} molecules cm^{-2}, corresponding to a substitution ratio (Fc/Si-H ratio) of 7.3%. This low packing ratio was due to the bulkiness of Fc. Accordingly, a much larger steric hindrance was seen compared with the alkyl groups. The packing density of FcA-SAM was estimated similarly to be 2.6×10^{13} molecules cm^{-2} and 3.3% Fc/SiH ratio—less than half that of VFc-SAM, probably because FcA, namely $-CHO$, is less reactive than VFc, namely, $-CH = CH_2$.

The CV curves shown in Figure 5.20 were acquired using high-doped Si with low resistivity as a substrate material. Such a high-doped Si substrate acts similarly to a metal electrode. When the doping density is reduced, the semiconductor nature of Si appears more prominently on the electron transfer between Fc and Si. It becomes dependent on the band structure at the Fc/Si interface. VFc-SAMs were formed on p-Si $(1-30 \,\Omega cm)$ and n-Si $(1-10 \,\Omega cm)$ and their electrochemical responses were measured as shown in Figure 5.21A–C. All these CV curves were measured in 0.1 M $HClO_4$. Figure 5.21A shows the CV curves for the VFc-SAM/p-Si sample acquired in the dark. A series of anodic and cathodic waves with varied potential sweep rates were obtained, and the peak potentials were all around 0.29 V versus Ag/AgCl for all the sweep rates with small peak separations of <0.01 V, implying that the redox-active species are on the substrate surface. From the CV curve peak areas, the packing density and the substitution ratio of this FcA-SAM were estimated to be 4.0×10^{13} molecules cm^{-2} and 5.6% Fc/SiH. VFc-SAM on

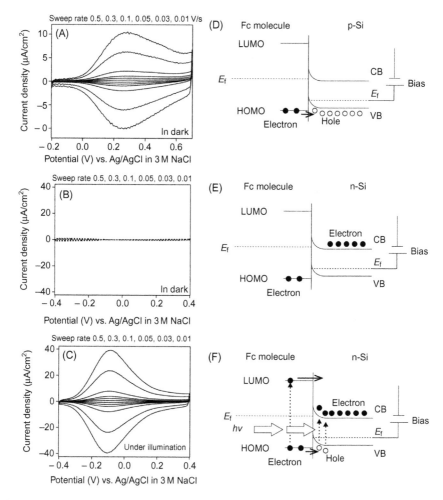

Figure 5.21 Electrochemical behaviors and band structures of VFc-SAM/Si. CV curves of (A) VFc-SAM/p-Si in the dark, (B) VFc-SAM/n-Si in dark, and (C) VFc-SAM/n-Si under illumination. Energy diagrams of (D) VFc-SAM/p-Si in dark, (E) VFc-SAM/n-Si in the dark, and (F) VFc-SAM/n-Si under illumination.

p-Si (middle-doped) was confirmed to show an electrochemical response similar to that of the high-doped Si. In contrast, the VFc-SAM/n-Si sample showed no electrochemical response in the dark as shown in Figure 5.21B. There was no electron transfer between Fc molecules and n-Si (middle-doped). However, when this sample was irradiated with visible light, the sample showed clear electrochemical responses. CV curves acquired at a light intensity of 20 mW/cm^2 are shown in Figure 5.21C. Potentials of anodic and cathodic wave peaks are around 0.09 V versus Ag/AgCl and did not change with the sweep rate. Their peak separation width is <0.01 V. These features suggest that the redox-active species are adsorbed on the sample surface and Fc moieties are tethered on the surface as well as those on p-Si. The packing density and the substitution ratio of this FcA-SAM on n-Si were estimated to be 14.0×10^{13} molecules cm^{-2} and 18.0% Fc/SiH, respectively.

The photoresponsivity of VFc-SAM is often explained by considering the energy diagram at the interface of the molecule and Si [97]. The energy diagrams are drawn in Figure 5.21D−F. The conduction band (CB) and valence band (VB) of Si are 4.05 and 5.17 eV from vacuum level and the energy gap is 1.12 eV [98]. The highest occupied molecular orbital (HOMO) level of the Fc molecule, which corresponds to the Fc/Fc + redox potential, is variously reported to be 4.5−5.4 eV from vacuum level [99−102]. These values are all around VB of Si. The gap between HOMO and the lowest unoccupied molecular orbital (LUMO) of Fc is experimentally obtained to be 2.8 eV from optical absorption spectra of VFC solution [96]. The LUMO level is thus considered to be 1.0−2.0 eV higher than CB of Si. For the energy diagram shown in Figure 5.21, the following three assumptions are considered:

1. Dipoles at the interface between the molecular film and Si substrate are neglected.
2. The molecular orbit of Fc moiety is not affected by the connections with the vinyl group and Si substrate.
3. The HOMO level of Fc moiety is as high as or a little lower than the VB of Si; namely, there is no electron transfer flow from HOMO of Fc to VB of Si, when no bias voltage is applied to the sample substrate.

When a sufficient negative bias is applied to the p-Si substrate, the bands of Si move downward and, consequently, the electrons in HOMO of Fc can flow into VB of Si, since there are sufficient holes even in the dark (Figure 5.21D). In contrast, for VFc-SAM/n-Si in the dark, there are few holes in VB. Thus, there is no electron transfer due to the hole−electron recombination (Figure 5.21E). On the other hand, when the sample is illuminated, electron−hole pairs are generated in Si so that electrons flow from HOMO of Fc to VB of Si, resulting in hole−electron recombination. Another possible mechanism is the excitation of Fc itself. The electrons of HOMO are excited to LUMO and then move to CB of Si.

5.7 Summary and Conclusion

The preparation and properties of SAMs covalently bonded on oxide-free Si were reviewed in this chapter. Such SAMs are formed through the chemical reaction of

Si-H with organic molecules having a chemically reactive moiety, such as alkenes, alkynes, aldehydes, alcohols, and thiols. The processes proceed at room temperature in the dark, but are activated thermally and photochemically. By these processes, the organic molecules are covalently attached to bulk Si through $Si-C$, $Si-O-C$, or $Si-S-C$ bonds. In particular, the process is promoted by light even at 700 nm and a hole-mediated grafting mechanism is proposed. Furthermore, the photochemical replacement of surface Si-H groups with CH_3 groups to form $Si-CH_3$ on Si(111)-H is demonstrated. This $Si-H/Si-CH_3$ replacement could be promoted by the photoelectrochemistry only by the visible-light irradiation of an Si-H substrate in CH_3MgBr solution. In addition to chemical and mechanical durabilities of the direct-bonding alkyl and alcohoxy SAMs, some of the electronic properties of several direct-bonded SAMs are described. One is the insulating dielectric properties of alkyl SAMs and others are the charge storage/release functions of redox-active SAMs consisting of Fc derivatives. The self-assembling approach to fabricate ultrathin films on Si using molecules as building blocks will increasingly become a crucial bottom-up nanotechnology in surface and interface engineering in future semiconductor electronics.

References

[1] Tao T, Bernasek SL, editors. Functionalization of semiconductor surfaces. New Jersey: John Wiley & Sons; 2012.

[2] Sugimura H. In: Tjong SC, editor. Nanocrystalline materials: their synthesis−structure−property relationships and applications. 1st ed. Oxford: Elsevier Science; 2006.

[3] Wolkow RA. Controlled molecular adsorption on silicon: laying a foundation for molecular devices. Annu Rev Phys Chem 1999;50:413−41.

[4] Hamers RJ, Coulter S, Ellison MD, Hovis JS, Padowitz DF, Schwartz MP, et al. Cycloaddition chemistry of organic molecules with semiconductor surfaces. Acc Chem Res 2000;33(9):617−24.

[5] Leftwich TR, Teplyakov AV. Chemical manipulation of multifunctional hydrocarbons on silicon surfaces. Surf Sci Rep 2008;63(1):1−71.

[6] Sagiv J. Organized monolayers by adsorption. 1. Formation and structure of oleophobic mixed monolayers on solid surfaces. J Am Chem Soc 1980;102(1):92−8.

[7] Sugimura H, Hozumi A, Kameyama T, Takai O. Organosilane self-assembled monolayers formed at the vapour/solid interface. Surf Interf Anal 2002;34(1):550−4.

[8] Onclin S, Ravoo BJ, Reinhoudt DN. Engineering silicon oxide surfaces using self-assembled monolayers. Angew Chem Int Ed. 2005;44(39):6282−304.

[9] Buriak JM. Organometallic chemistry on silicon and germanium surfaces. Chem Rev 2002;102(5):1271−308.

[10] Vilan A, Yaffe O, Biller A, Salomon A, Kahn A, Cahen D. Molecules on Si: electronics with chemistry. Adv Mater 2010;22(2):140−59.

[11] He T, He J, Lu M, Chen B, Pang H, Reus WF, et al. Controlled modulation of conductance in silicon devices by molecular monolayers. J Am Chem Soc 2006;128 (45):14537−41.

[12] Gergel-Hackett N, Zangmeister CD, Hacker CA, Richter LJ, Richter CA. Demonstration of molecular assembly on Si(100) for CMOS-compatible molecule-based electronic devices. J Am Chem Soc 2008;130(13):4259−61.

[13] Har-Lavan R, Ron I, Thieblemont F, Cahen D. Toward metal-organic insulator−semiconductor solar cells, based on molecular monolayer self-assembly on n-Si. Appl Phys Lett 2009;94(4):043308.

[14] Zhang L, Li L, Chen S, Jiang S. Measurements of friction and adhesion for alkyl monolayers on Si(111) by scanning force microscopy. Langmuir 2002;18 (14):5448−56.

[15] Nakano M, Ishida T, Sano H, Sugimura H, Miyake K, Ando Y, et al. Tribological properties of self-assembled monolayers covalently bonded to Si. Appl Surf Sci 2008;255(5):3040−5.

[16] Strother T, Cai W, Zhao X, Hamers RJ, Smith LM. Synthesis and characterization of DNA-modified silicon (111) surfaces. J Am Chem Soc 2002;122(6):1205−9.

[17] Lasseter TL, Clare BH, Abbott NL, Hamers RJ. Covalently modified silicon and diamond surfaces: resistance to nonspecific protein adsorption and optimization for biosensing. J Am Chem Soc 2004;126(33):10220−1.

[18] Ara M, Tsuji M, Tada H. Preparation of amino-terminated monolayers via hydrolysis of phthalimide anchored to Si(111). Surf Sci 2007;601(22):5012−98.

[19] Ara M, Graaf H, Tada H. Nanopatterning of alkyl monolayers covalently bound to Si (111) with an atomic force microscope. Appl Phys Lett 2002;80(14):2565−7.

[20] Balaur E, Djenizian T, Boukherroub R, Chazalviel JN, Ozanam F, Schmuki P. Electron beam-induced modification of organic monolayers on Si(111) surfaces used for selective electrodeposition. Electrochem Commun 2004;6(2):153−7.

[21] Sugimura H, Sano H, Lee KH, Murase K. Organic monolayers covalently bonded to Si as ultra thin photoresist films in vacuum ultra-violet lithography. Jpn J Appl Phys 2006;45(6B):5456−60.

[22] Takakusagi S, Uosaki K. Photopatterning of an organic monolayer formed on a Si single crystal surface via Si − C covalent bond by UV irradiation in an inert atmosphere. Jpn J Appl Phys 2006;45(11):8961−6.

[23] Sugimura H, Nanjo S, Sano H, Murase K. Gold nanoparticle arrays fabricated on a silicon substrate covered with a covalently bonded alkyl monolayer by electroless plating combined with scanning probe anodization lithography. J Phys Chem C 2009;113 (27):11643−6.

[24] Linford MR, Chidsey CDE. Alkyl monolayers covalently bonded to silicon surfaces. J Am Chem Soc 1993;115(26):12631−2.

[25] Linford MR, Fenter P, Eisenberger PM, Chidsey CDE. Alkyl monolayers on silicon prepared from 1-alkenes and hydrogen-terminated silicon. J Am Chem Soc 1995;117 (11):3145−55.

[26] Effenberger F, Götz G, Bidlingmaier B, Wezstein M. Photoactivated preparation and patterning of self-assembled monolayers with 1-alkenes and aldehydes on silicon hydride surfaces. Angew Chem Int Ed 1998;37(18):2462−4.

[27] Sieval AB, van den Hout B, Zuilhof H, Sudhöelter EJR. Molecular modeling of alkyl monolayers on the Si(111) surface. Langmuir 2000;16(7):2987−90.

[28] Cicero RL, Linford MR, Chidsey CDE. Photoreactivity of unsaturated compounds with hydrogen-terminated silicon(111). Langmuir 2000;16(13):5688−95.

[29] Boukherroub R, Morin S, Sharpe P, Wayner DDM, Allongue P. Insights into the formation mechanisms of Si-OR monolayers from the thermal reactions of alcohols and aldehydes with Si(111)-H. Langmuir 2000;16(19):7429−34.

[30] Saito N, Hayashi K, Sugimura H, Takai O. Microstructured π-conjugated organic monolayer covalently attached to silicon. Langmuir 2003;19(26):10632−4.

[31] Sun QY, de Smet LCPM, Lagen B, Wright A, Zuilhof H, Sudhölter EJR. Covalently attached monolayers on hydrogen terminated Si(100): extremely mild attachment by visible light. Angew Chem Int Ed 2004;43(11):1352−5.

[32] Hacker CA, Anderson KA, Richter LJ, Richter CA. Comparison of Si−O−C interfacial bonding of alcohols and aldehydes on Si(111) formed from dilute solution with ultraviolet irradiation. Langmuir 2005;21(3):882−9.

[33] Scheres L, Arafat A, Zuilhof H. Self-assembly of high-quality covalently bound organic monolayers onto silicon. Langmuir 2007;23(16):8343−6.

[34] Sano H, Maeda H, Matsuoka S, Lee KH, Murase K, Sugimura H. Self-assembled monolayers directly attached on silicon substrates formed from 1-hexadecene by thermal, ultra-violet and visible light activation methods. Jpn J Appl Phys 2008;47 (7):5659−64.

[35] Sano H, Ohno K, Ichii T, Murase K, Sugimura H. Alkanethiol self-assembled monolayers formed on silicon substrates. Jpn J Appl Phys 2010;49(1):01AE09.

[36] Scheres L, Rijksen B, Giesbers M, Zuilhof H. Molecular modeling of alkyl and alkenyl monolayers on hydrogen-terminated Si(111). Langmuir 2011;27(7):972−80.

[37] Zhong YL, Bernasek SL. Mild and efficient functionalization of hydrogen-terminated Si(111) via sonochemical activated hydrosilylation. J Am Chem Soc 2011;133 (21):8118−21.

[38] Lou JL, Shiu HW, Chang LY, Wu CP, Soo YL, Chen CH. Preparation and characterization of an ordered 1-dodecanethiol monolayer on bare Si(111) surface. Langmuir 2011;27(7):3436−41.

[39] Kosuri MR, Gerung H, Li Q, Han SM. Vapor-phase adsorption kinetics of 1-decene on H-terminated Si(100). Langmuir 2003;19(22):9315−20.

[40] Woods M, Carlsson S, Hong Q, Patole SN, Lie LH, Houlton H, et al. A Kinetic model of the formation of organic monolayers on hydrogen-terminated silicon by hydroxylation of alkenes. J Phys Chem B 2005;109(50):24035−45.

[41] Coletti C, Marrone A, Giorgi G, Sgamellotti A, Cerofolini G, Re N. Nonradical mechanisms for the uncatalyzed thermal functionalization of silicon surfaces by alkenes and alkynes: a density functional study. Langmuir 2006;22(24):9949−56.

[42] Scipioni MVR, Boero M, Silvestrelli PL, Ariga K. The initiation mechanisms for surface hydrosilylation with 1-alkenes. Phys Chem Chem Phys 2011;13(11):4862−7.

[43] Boukherroub R, Morin S, Bensebaa F, Wayner DDM. New synthetic routes to alkyl monolayers on the Si(111) surface. Langmuir 1999;15(11):3831−5.

[44] Sieval AB, Vleeming V, Zuilhof H, Sudhölter EJR. An improved method for the preparation of organic monolayers of 1-alkenes on hydrogen-terminated silicon surfaces. Langmuir 1999;15(23):8288−91.

[45] Royea WJ, Juang A, Lewis NS. Preparation of air-stable, low recombination velocity Si(111) surfaces through alkyl termination. Appl Phys Lett 2000;77(13):1988−90.

[46] Fidélis A, Ozanam F, Chazalviel JN. Fully methylated, atomically flat (111) silicon surface. Surf Sci 2000;444(1−3):L7−10.

[47] Fellah S, Teyssot A, Ozanam F, Chazalviel JN, Vigneron J, Etcheberry A. Kinetics of electrochemical derivatization of the silicon surface by Grignards. Langmuir 2002;18 (15):5851−60.

[48] Yamada Y, Inoue T, Yamada K, Takano N, Osaka T, Harada H, et al. Detection of C−Si covalent bond in CH$_3$ adsorbate formed by chemical reaction of CH$_3$MgBr and H:Si(111). J Am Chem Soc 2003;125(26):8039−42.

[49] Takakusagi S, Miyasaka T, Uosaki K. Photoanodic formation of an organic monolayer on a hydrogen-terminated Si(111) surface via Si−C covalent bond using a Grignard reagent and its application for one-step monolayer-patterning. J Electroanal Chem 2007;599(12):344−8.

[50] Herrera M, Ichii T, Murase K, Sugimura H. Photochemical preparation of methyl-terminated Si(111) using a Grignard reagent. Chem Lett 2012;41(9):902−4.

[51] de Villeneuve CH, Pinson J, Bernard MC, Allongue P. Electrochemical formation of close-packed phenyl layers on Si(111). J Phys Chem B 1997;101(14):2415−20.

[52] Stewart MP, Maya F, Kosynkin DV, Dirk SM, Stapleton JJ, McGuiness CL, et al. Direct covalent grafting of conjugated molecules onto Si, GaAs, and Pd surfaces from aryldiazonium salts. J Am Chem Soc 2004;126(1):370−8.

[53] Ulman A. An introduction to ultrathin organic films from Langmuir−Blogett to self-assembly. San Diego: Academic Press; 1991. p. 281

[54] Palik ED. Handbook of optical constants of solids. San Diego: Academic Press; 1998. p. 547

[55] Sun QY, de Smet LCPM, van Lagen B, Giesbers M, Thüne PC, van Engelenburg J, et al. Covalently attached monolayers on crystalline hydrogen-terminated silicon: extremely mild attachment by visible light. J Am Chem Soc 2005;127 (8):2514−23.

[56] Sano H, Yaku T, Ichii T, Murase K, Sugimura H. Soft processing for formation of SAM on hydrogen-terminated silicon surface based on visible-light excitation. J Vac Sci Technol B. 2009;27(2):858−62.

[57] Yamada T, Kawai M, Wawro A, Suto S, Kasuya A. HREELS, STM and STS study of CH3-terminated Si(111)-(1 × 1) surface. J Chem Phys 2004;121(21):10660−7.

[58] Nemanick EJ, Solares SD, Goddard III WA, Lewis NS. Quantum mechanics calculations of the thermodynamically controlled coverage and structure of alkyl monolayers on Si(111) surfaces. J Phys Chem B 2006;110(30):14842−8.

[59] Webb LJ, Nemanick EJ, Biteen JS, Knapp DW, Michalak DJ, Traub MC, et al. High-resolution X-ray photoelectron spectroscopic studies of alkylated silicon (111) surfaces. J Phys Chem B 2005;109(9):3930−7.

[60] Hunger R, Fritsche R, Jaeckel B, Jaegermann W, Webb LJ, Lewis NS. Chemical and electronic characterization of methyl-terminated Si(111) surfaces by high-resolution synchrotron photoelectron spectroscopy. Phys Rev B 2005;72(4):045317.

[61] Yu H, Webb LJ, Ries RS, Solares SD, Goddard WA, Heath JR, et al. Low temperature STM images of methyl-terminated Si(111) surfaces. J Phys Chem B 2005;109 (2):671−4.

[62] Webb LJ, Rivillon S, Michalak DJ, Chabal YJ, Lewis NS. Transmission infrared spectroscopy of methyl- and ethyl-terminated silicon (111) surfaces. J Phys Chem B 2006;110(14):7349−56.

[63] Waluyo I, Ogasawara H, Kawai M, Nilsson A, Yamada T. Direct interaction of water ice with hydrophobic methyl-terminated Si(111). J Phys Chem C 2010;114 (44):19004−8.

[64] Miyadera T, Koma A, Shimada T. Ultraviolet photoelectron spectroscopy of a methyl-terminated Si(111) surface. Surf Sci 2003;526(1−2):177−83.

[65] Nishiyama K, Tanaka Y, Harada H, Yamada T, Niwa D, Inoue T, et al. Adsorption of organic molecules by photochemical reaction on Cl:Si(111) and H:Si(111) evaluated by HREELS. Surf Sci 2006;600(100):1965−72.

[66] Kim NY, Laibinis PE. Thermal derivatization of porous silicon with alcohols. J Am Chem Soc 1997;119(9):2297−8.

[67] Terry J, Linford MR, Wigren C, Cao R, Pianetta P, Chidsey CED. Determination of the bonding of alkyl monolayers to the Si(111) surface using chemical-shift, scanned-energy photoelectron diffraction. Appl Phys Lett 1997;71(8):1056−8.

[68] Saito N, Youda S, Hayashi K, Sugimura H, Takai O. Alkyl self-assembled monolayer prepared on hydrogen-terminated Si(111) through reduced pressure chemical vapor deposition: chemical resistivities in HF and NH$_4$F solutions. Chem Lett 2002;31(12):1194−5.

[69] Sano H, Maeda H, Ichii T, Murase K, Noda K, Matsushige K, et al. Alkyl and alkoxyl monolayers directly attached to silicon: chemical durability in aqueous solutions. Langmuir 2009;25(10):5516−25.

[70] Silverstein RM, Bassler GC, Morrill TC. Spectrometric identification of organic compounds. New York: Wiley; 2005.

[71] Gorostiza P, de Villeneuve CH, Sun QY, Sanz F, Wallart X, Boukherroub R, et al. Water exclusion at the nanometer scale provides long-term passivation of silicon (111) grafted with alkyl monolayers. J Phys Chem B 2006;110(11):5576−85.

[72] Carpick RW, Salmeron M. Scratching the surface: fundamental investigations of tribology with atomic force microscopy. Chem Rev 1997;97(4):1163−94.

[73] Bhushan B, Kulkarni A, Koinkar VN. Microtribological characterization of self assembled and Langmuir−Blodgett monolayers by atomic and friction force microscopy. Langmuir 1995;11(8):3189−98.

[74] DePalma V, Tillman N. Friction and wear of self-assembled trichlorosilane monolayer films on silicon. Langmuir 1989;5(3):868−72.

[75] Hayashi K, Sugimura H, Takai O. Rictional properties of organosilane self-assembled monolayer in vacuum. Jpn J Appl Phys 2001;40(6B):4344−8.

[76] Sugimura H, Shiroyama H, Takai O, Nakagiri N. Differential friction force spectroscopy of micropatterned organosilane self-assembled monolayers. Appl Phys A 2001;72 (Suppl):S285−9.

[77] Cha KH, Kim DE. Investigation of the tribological behavior of octadecyltrichlorosilane deposited on silicon. Wear 2001;251(1−12):1169−76.

[78] Masuko M, Miyamoto H, Suzuki A. Tribological characteristics of self-assembled monolayer with siloxane to Si surface. Tribol Int 2007;40(10−12):1587−96.

[79] Srinivasan U, Houston MR, Howe RT, Maboudian R. Alkyltrichlorosilane-based self-assembled monolayer films for stiction reduction in silicon micromachines. J Microelectromech Syst 1998;7(2):252−60.

[80] de Boer MP, Mayer TM. Tribology of MEMS. MRS Bull 2001;26(April):302−4.

[81] Yang X, Perry SS. Friction and molecular order of alkanethiol self-assembled monolayers on Au(111) at elevated temperatures measured by atomic force microscopy. Langmuir 2003;19(15):6135−9.

[82] Xiao X, Hu J, Charych DH, Salmeron M. Chain length dependence of the frictional properties of alkylsilane molecules self-assembled on mica studied by atomic force microscopy. Langmuir 1996;12(2):235−7.

[83] Zhao J, Uosaki K. Electron transfer through organic monolayers directly bonded to silicon probed by current sensing atomic force microscopy: effect of chain length and applied force. J Phys Chem B 2004;108(44):17129−35.

[84] Seitz O, Böcking T, Salomon A, Gooding JJ, Cahen D. Importance of monolayer quality for interpreting current transport through organic molecules: alkyls on oxide-free Si. Langmuir 2006;22(16):6915−22.

[85] Zhao J, Huang Z, Yin X, Wang D, Uosaki K. Theoretical analysis of the potential distribution and transportation behavior of the ordered alkyl monolayer-silicon junction. Phys Chem Chem Phys 2006;8(48):5653−8.

[86] Li Q, Mathur G, Homsi M, Surthi S, Misra V. Capacitance and conductance characterization of ferrocene-containing self-assembled monolayers on silicon surfaces for
memory applications. Appl Phys Lett 2002;81(8):1494−6.

[87] Roth KM, Yasseri AA, Liu ZR, Dabke RB, Malinovskii V, Schweikart KH, et al.
Measurements of electron-transfer rates of charge-storage molecular monolayers on
Si(100): toward hybrid molecular/ semiconductor information storage devices. J Am
Chem Soc 2003;125(2):505−17.

[88] Bunimovich Y, Ge G, Beverly KC, Ries RS, Hood L, Heath JR. Electrochemically
programmed, spatially selective biofunctionalization of silicon wires. Langmuir
2004;20(24):10630−8.

[89] Masuda T, Uosaki K. Construction of organic monolayers with electron transfer function
on a hydrogen terminated Si(111) surface via silicon−carbon bond and their electrochemical characteristics in dark and under illumination. Chem Lett 2004;33(7):788−9.

[90] Fabre B, Bassani DM, Liang CK, Ray D, Hui F, Hapiot P. Anthracene and anthracene: C60 adduct-terminated monolayers covalently bound to hydrogen-terminated
silicon surfaces. J Phys Chem C 2011;115(30):14786−96.

[91] Yzambart G, Fabre B, Lorcy D. Multiredox tetrathiafulvalene-modified oxide-free
hydrogen- terminated Si(100) Surfaces. Langmuir 2012;28(7):3453−9.

[92] Fabre B. Ferrocene-terminated monolayers covalently bound to hydrogen-terminated
silicon surfaces. Toward the development of charge storage and communication
devices. Acc Chem Res 2010;43(12):1509−18.

[93] Han J, Lee KH, Fujii S, Sano H, Kim YJ, Murase M, et al. Scanning capacitance
microscopy for alkylsilane monolayer covered-Si substrate patterned by scanning
probe lithography. Jpn J Appl Phys 2007;46(8B):5621−5.

[94] Nakagiri N, Sugimura H, Ishida Y, Hayashi K, Takai O. Effects of an adsorbed water
layer and self-assembled organosilane monolayers on scanning probe microscopy of
silicon pn structures. Surf Sci 2003;532−535:999−1003.

[95] Tajimi N, Sano H, Murase K, Lee KH, Sugimura H. Thermal immobilization of ferrocene derivatives on (111) surface of n-type silicon-parallel between vinylferrocene
and ferrocenecarboxaldehyde. Langmuir 2007;23(6):3193−8.

[96] Sano H, Zhao M, Kasahara D, Murase K, Ichii T, Sugimura H. Formation of uniform
ferrocenyl-terminated monolayer covalently bonded to Si using reaction of hydrogen-
terminated Si(111) surface with vinylferrocene/n-decane solution by visible-light
excitation. J. Colloid Interf Sci 2011;361(1):259−69.

[97] Gowda G, Mathur G, Misra V. Valence band tunneling model for charge transfer of
redox-active molecules attached to n and p-silicon substrates. Appl Phys Lett 2007;90
(14):142113.

[98] Sze SM. Semiconductor device—physics and technology. 2nd ed. New York: Wiley;
1981.

[99] Pommerehne J, Vestweber H, Guss W, Mahrt RF, Bässler H, Porsch M, et al.
Efficient two layer LEDs on a polymer blend basis. Adv Mater 1995;7(6):551−4.

[100] Zhang G, Zhang H, Sun M, Liu Y, Pang X, Yu X, et al. Substitution effect on the geometry and electronic structure of the ferrocene. J. Comput Chem 2007;28(14):2260−74.

[101] Grigorov MG, Waber J, Vulliermet N, Chermette H, Tronchet JMJ. Numerical evaluation of the internal orbitally resolved chemical hardness tensor: second order chemical reactivity through thermal density functional theory. J. Chem Phys 1998;108
(21):8790−8.

[102] Seki K. Ionization energies of free molecules and molecular solids. Mol Cryst Liq
Cryst 1989;171(1):255−70.

6 Nanocomposite Dielectric Materials for Organic Flexible Electronics

Ye Zhou, Su-Ting Han and V.A.L. Roy

Department of Physics and Materials Science and Center of Super-Diamond and Advanced Films (COSDAF), City University of Hong Kong, Hong Kong SAR, PR China

6.1 Introduction

The possibility of using flexible materials for applications in the electronics and semi-conductor industry has become an important issue for low-cost, large-area printed electronics [1,2]. Organic materials are gaining much attention as a new approach to flexible electronics for various applications, such as sensors, memories, and radio frequency identification (RFID) tags [3−6]. Organic materials can be also applied in flexible electronic products include backplane circuitries, which can be used for the fabrication of bendable displays for e-paper and flexible computers. The development of small molecules and polymeric materials with high performance is fundamental to the progress of organic thin-film transistors (OTFTs) based electronic technologies [7,8]. While tremendous efforts have been focused on increasing the mobility of organic semiconductor materials for better device performance, an alternative approach is to enhance the dielectric properties of the gate insulators while maintaining easy processability and flexibility of the dielectric materials.

The choice of gate dielectric materials is essential which directly influence the electrical performance in OTFTs. A key challenge for realizing practical applications lies in developing dielectrics with low leakage current, high dielectric constant, and good mechanical stability [9,10]. The high-k dielectric materials help to enhance the field-induced carrier density in the semiconductor/insulator interface and reduce the operating voltage of the transistor. Most of the high-k inorganic dielectrics have the disadvantages, such as high vacuum or high temperature processing, limited tunability of the dielectric properties, poor processability, and incompatible with the flexible substrate. The polymer dielectrics can achieve low-cost fabrication and excellent mechanical flexibility, but the device performances are severely limited due to the high leakage current and low capacitance. Organic/inorganic multilayer dielectrics have got a lot of interest to gain simultaneous improvement of electrical and mechanical properties for OTFTs [11−14], but they still suffer from limited processability in high throughput printable technology.

Nanocrystalline Materials. DOI: http://dx.doi.org/10.1016/B978-0-12-407796-6.00006-3

To realize printability and flexibility in advanced electronic devices, it is necessary to develop solution-processable high-k dielectric materials that are suitable for roll-to-roll or other related printing technologies. The introduction of high-k inorganic nanoparticles into polymer matrix represents one of the most promising ways to fabricate high-k dielectrics [9]. The dielectric constant of the nanocomposite dielectrics can be tuned through changing the concentration of the nanoparticle fillers, which offers relatively easy processability for fabricating the dielectric layer in organic electronics. Most of the current studies on nanocomposite dielectrics are focused on the devices on rigid substrate, such as silicon wafers and glasses, the nanocomposite dielectric properties and associated transistor behaviors on flexible substrates need to be studied systematically.

This chapter focuses on the development and implementation of high-k nanocomposite dielectrics for organic electronics applications. We begin by the history of organic electronics and then describe the basics and the theory of dielectrics followed by the operating principles of OTFT devices to understand the motivations behind enhancing the dielectric constant. Next, an overview of the state-of-the-art OTFT performance achieved using various nanocomposite high-k dielectrics will be presented. The functionalization of the nanoparticles and the property characterization will be described. We will also discuss the recent experimental data analyzing the mechanical properties of the organic electronic devices on nanocomposite dielectrics.

6.2   The Development of Organic Electronics

Organic electronics is a branch of modern electronics, and it deals with organic materials, such as polymers or small molecules. The materials used in this kind of technology are carbon based, which is the same as the molecules of living things. When considering the properties of organic materials, we may not generally feel that they are electrical conductors. Organic materials are thought to be excellent insulators for a lot of technological applications. The discovery of the electrical conduction in organic materials can be traced to 1862 when Henry Letheby obtained a partly conductive material by anodic oxidation of aniline in sulfuric acid.

Heeger, MacDiarmid, and Shirakawa discovered in the 1970s that the polymer polyacetylene, after certain modifications, can be made conductive [15,16]. Doped polyacetylene can be used to form a new group of conducting polymers and the electrical conductivity of the material can be systematically tuned over a range of 11 orders of magnitude. When the dopant concentrations are near 1%, the metal-to-insulator transition happens which is verified by transport studies and far-infrared transmission measurements. In 2000, the Nobel prize in Chemistry was awarded to Heeger, MacDiarmid, and Shirakawa for their discoveries.

The discovery of electrically conductive polymers drastically changed our views on polymer materials and formed a basis of the future organic electronics. Thirty years later, a lot of success has been achieved in this field and the discovery of conductive polymers has led to interdisciplinary development among physicists, chemists, and engineers in terms of industrial applications.

For about 10 years after the discovery of conductive polyacetylene, research on conductive organic materials remained at the field of chemistry and solid-state physics. No application in electronics was involved at this stage. Then in the mid-1980s, the interest for organic conductive materials increased in the field of engineering and applied physics. Tang and VanSlyke [17] realized a major breakthrough by demonstrating a light-emitting diode (LED) based on two organic molecules, tris(8-hydroxyquinoline) aluminum (Alq3) and diamine. The fabrication of this device is considered as the real start of organic electronics and organic light-emitting diodes (OLEDs) were in fact the first application that attracted the attention of industrial research groups. Tsumura et al. [18] presented the first organic solid-state field-effect transistor (FET). Polythiophene was used as the semiconductor layer in the transistor. Later in 1988, Burroughes et al. [19] from Cavendish Laboratory, Cambridge, published the results on semiconductor devices made from polyacetylene, which is a conjugated polymer. The charge is stored in localized soliton-like excitations of the polymer chain, which are introduced by a surface electric field. The formation of charged solitons can introduce optical absorption below the band gap.

At the end of the 1980s, a new class of conductive organic materials was introduced. Several research groups presented organic devices using a small molecule, or can be called oligomer, rather than a polymer. In contrast to polymeric materials, small molecules are quite insoluble in solvents. Hence, deposition methods based on vacuum sublimation were introduced for depositing thin films of a semiconducting small molecule. Peng et al. fabricated α-sexithiophene (α-6T)-based OTFTs with different kinds of organic polymer insulators. With the help of polymer dielectrics, including polyvinyl alcohol (PVA) and cyanoethylpullulan (CYEPL), the electrical performance of the OTFTs has been enhanced compared with that on SiO_2 layer. They found a strong correlation between the dielectric constant of the dielectric layer and the mobility [20]. In 1992, Horowitz et al. [21] reported OTFTs made with various π-conjugated oligomers, including oligothiophene series and polyacenes. Furthermore, various inorganic and organic insulators were used in the study. The mobility increases as the conjugation length of the oligomer increased. They also found that the charge transport of the device is mainly governed by the interface between the semiconductor and the dielectric, which is a pioneer research in this area.

Garnier et al. [22] fabricated OTFTs from semiconducting polymers by printing approaches in 1994. The performances of the devices are insensitive to mechanical treatments, such as bending or twisting. Their reported technology opened the way for low-cost and large-area flexible electronics. Later, Bao et al. [23] reported the printed organic transistor in which all the key components can be easily screen-printed. This transistor had a polyimide dielectric layer, a regioregular poly(3-alkythiophene) (P3AT) semiconducting layer, and two silver electrodes, all of which are printed on an indium tin oxide (ITO)-coated plastic substrate.

All organic semiconductors used in OTFTs mentioned up to now are p-type semiconductors. In 1995, Haddon et al. [24] reported n-channel OTFTs with excellent device characteristics fabricated by utilizing C_{60} as the active layer.

At that time research in the field of organic electronics, and OTFTs in particular, remained focusing on the physical characterization of discrete components, as the technology was not sufficiently mature yet for fabricating integrated circuits. Brown et al. [25] from Philips Research Laboratories published the world's first results on organic integrated circuits. They fabricated OTFTs with solution-processed polymer semiconductors and demonstrated that these circuits can perform logic operations. This breakthrough marked the start of a new era in the field of organic electronics, as it became clear that OTFTs might provide a useful technology for many kinds of applications, accepting a relatively low performance, but requiring low cost. More groups started working on OTFTs to adopt them in electronic device, rather than focusing on the purely physical research.

During the 1990s, some novel device concepts were introduced by Bell Labs groups, especially the integration of n- and p-type semiconducting films. This technology led to some novel and interesting devices. In 1995, Dodabalapur et al. [26,27] from AT&T Bell Laboratories reported heterostructure OTFTs using two active materials: α-6T and C_{60}. It was shown that both electron and hole current could be found in this device structure. Organic light-emitting transistors are essentially relying on the basic principle of the organic heterostructure devices fabricated by Bell Laboratories.

Another concept which was investigated by the same group is the complementary organic technology. This technology integrated n-type and p-type OTFTs in one circuit. In 1996, Dodabalapur et al. [28] reported hybrid complementary inverters where the n-channel is Si and the p-channel is organic semiconductor. Later, they fabricated all organic complementary inverters [29]. In 1998, Sirringhaus et al. [30] from Cavendish Laboratory, University of Cambridge demonstrated an all-polymer semiconductor integrated device with regioregular poly(hexylthiophene) TFT driving a polymer LED of similar size. The device represented a step toward all-polymer optoelectronic integrated circuits such as active-matrix polymer LED displays. Later, Lin et al. [31] from Bell Labs reported organic ring oscillators based on complementary technologies. The n-type material is copper hexadecafluorophthalocyanine ($F_{16}CuPc$) and p-type material is an oligothiophene derivative. Crone et al. [32] also from Bell Laboratories showed much larger scales of integration with up to 864 transistors per circuit.

From 1999, high-k dielectric materials started to be used in OTFTs by researchers. Dimitrakopoulos et al. [33,34] from IBM research group first interpreted the gate voltage dependence of the mobility in pentacene-based transistors and this finding is based on the interaction of charge carriers with localized trap levels in the band gap. In the work, they demonstrated high-performance low-voltage organic transistors on transparent plastic substrates.

Tate et al. [35] demonstrates the combined use of two methods for fabricating organic transistors and that can operate at low voltages: anodization for thin (similar to 50 nm), high-capacitance gate dielectrics and microcontact printing on electroless silver for high-resolution (similar to 1 μm) source/drain electrodes. The near room temperature techniques are attractive and compatible with organic semiconductors and flexible substrates, meanwhile they are suitable for roll-to-roll printing technology. Sirringhaus et al. [36] demonstrated direct inkjet printing of complete transistor circuits in the same year. They used substrate surface energy patterning to direct the flow of conducting polymer inkjet droplets, which can enable high-resolution definition of the channel lengths of the transistors.

The complexity and performance of the OTFT-based circuits increased as time going. Someya et al. [37] from University of Tokyo developed large-area pressure sensors made with organic transistors which were proposed for electronic artificial skin applications. Subramanian et al. [38] from University of California, Berkeley reported that they realized transistors with mobilities of 10^{-1} cm^2/Vs using inkjet printer to print conductors, dielectric layers, and organic semiconductor materials. The performances of the transistors were adequate for 135 kHz RFID technology. Klauk et al. [39] from Max Planck Institute for Solid State Research fabricated integrated organic circuits that operated with low supply voltages (between 1.5 and 3 V) and the static power consumption was <1 nW per logic gate. Smits et al. [40] demonstrated self-assembled monolayer (SAM) FETs while there was long-range intermolecular $\pi-\pi$ coupling in the monolayer. They also fabricated the SAM logic circuits. Jung et al. [41] reported roll-to-roll printable antenna, rectifiers, and ring oscillators on plastic films and successfully demonstrated RFID tags operated at 13.56 MHz.

6.3   Basic Concept of Dielectric Materials in Organic Electronics

Dielectrics are widely used in various electrical or electronic applications. The fundamental usage of dielectrics is an insulating material against electrical conduction. An insulator material must have a large band gap compared with the semiconductor material. Therefore, there are no states available into which the electrons from the valence band can be excited. Nonetheless, there is always a breakdown voltage that will impart sufficient energy to the electrons to be excited into this band. Once this breakdown voltage is surpassed, the dielectric materials will lose the insulating properties. The other two major applications of dielectrics are capacitors and transistors, both of which are essential components of electronic logic circuits. It is well known that the role of the device insulator layer is as important as that of the semiconductor layer. In OTFTs, the device performance is very much dependent on the physical property and chemical nature of the dielectric surface.

A capacitor is a passive electrical component consisting of a dielectric sandwiched between two electrodes as shown in Figure 6.1. The application of a voltage at electric fields lower than the breakdown field of the dielectric across this material will induce a charge separation across the insulating layer forming the capacitor. An ideal capacitor is characterized by a single constant parameter, the capacitance C. Higher C values indicate that more charge may be stored for a given voltage. Real capacitors are not complete insulators and allow a small amount of current flowing through, terming as leakage current.

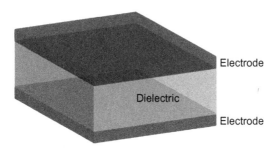

Figure 6.1 Schematic illustration of a capacitor.

Insulators or dielectric materials are characterized by the absence of charge transport. When an electric field (E) is applied, there is a shift in charge (Q) distribution in these materials. This field-induced polarization leads to dielectric behavior and hence to capacitance C. If we imagine two electrodes separated by a distance d in a vacuum, the application of a voltage (V) between them creates an electric field that is described as $E = V/d$. The charge created per unit area is proportional to this electric field, as given by Eq. (6.1).

$$Q = \varepsilon_0 E = \varepsilon_0 \frac{v}{d} \tag{6.1}$$

The proportionality constant between the applied voltage and the charge is called the capacitance C, and it is described as

$$C = \frac{Q}{v} = \frac{\varepsilon_0}{d} \tag{6.2}$$

where ε_0 is the vacuum permittivity. When a dielectric material is inserted between the electrodes, the capacitance is increased (by a factor of k, relative dielectric constant) due to the polarizability of the dielectric. In this case, the capacitance is described as

$$C = \frac{k\varepsilon_0}{d} \tag{6.3}$$

The transistor is the most important fundamental component in electronic devices. OTFTs can be fabricated on either rigid or flexible substrate. The common device configurations used in OTFTs are shown in Figure 6.2. OTFT is made of three parts: an insulator layer, a semiconducting layer, and three electrodes (gate, source, and drain). The gate configurations can be either bottom gate or top gate. For the drain and source configurations, two different structures can be used, bottom contact or top contact, depending on the position of the source and drain electrodes. The source and drain electrodes are in direct contact with the

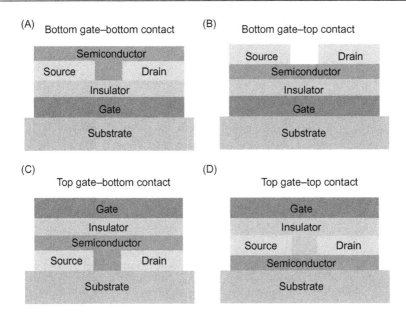

Figure 6.2 (A) Transistor with bottom gate−bottom contact configuration. (B) Transistor with bottom gate−top contact configuration. (C) Transistor with top gate−bottom contact configuration. (D) Transistor with top gate−top contact configuration.

semiconductor layer to form the channel and the gate electrode is isolated from the semiconductor layer by the dielectric layer. Each of the configurations has their advantages and disadvantages. The electrodes in bottom contact structure can be patterned by photolithography technologies which can achieve short channel length. The contact resistance in top contact device is usually lower than bottom contact device.

An organic semiconductor is said to be p-type when hole injection is easier than electron injection, which occurs when the highest occupied molecular orbital (HOMO) is closer to the Fermi level than the lowest unoccupied molecular orbital (LUMO). Symmetrically, an organic semiconductor is said to be n-type when LUMO is closer to the Fermi level than the HOMO. In this case, the electron injection is easier than the hole injection.

In principle, the characteristic that best defines an OTFT is the presence of an electric field that controls and modulates the conductivity of the channel between the source and drain. This electric field is created by the voltage applied between the source electrode and gate electrode, the gate−source voltage (V_{GS}). A positive/negative gate voltage will induce negative/positive charges at the insulator/semiconductor interface, and the number of these accumulated charges depends on V_{GS} and on the capacitance C of the insulator.

When no voltage is applied between the source and gate ($V_{GS} = 0$), the device is "off." On increasing both V_{GS} and V_{DS} (drain−source voltage), a linear current regime is initially observed at low drain voltages ($V_{DS} < V_{GS}$) (Eq. (6.4)) followed by a saturation regime at higher V_{DS} values (Eq. (6.5)).

$$I_{DS\text{-linear}} = \frac{W}{L} \times C \times \mu \times \left(V_{GS} - V_T - \frac{V_{DS}}{2} \right) V_{DS} \tag{6.4}$$

$$I_{DS\text{-saturate}} = \frac{W}{2L} \times C \times \mu \times (V_{GS} - V_T)^2 \tag{6.5}$$

In these equations, $I_{DS\text{-linear}}$ is the drain current in the linear regime, $I_{DS\text{-saturate}}$ is the drain current in the saturation regime, μ is the field-effect carrier mobility of the semiconductor, W is the channel width of the transistor, L is the channel length of the transistor, C is the capacitance per unit area of the insulator layer, V_T is the threshold voltage, V_{DS} is the drain−source voltage, and V_{GS} is the gate−source voltage.

The above equations indicate that the current between the source and drain can be increased by increasing either V_{GS} or V_{DS}. Nonetheless, these two parameters can be increased to only a certain extent. Another viable approach to minimizing V_{GS} and/or increasing the electrical current is adjusting the capacitance of the gate dielectric C. This relationship makes it clear that by either increasing the dielectric constant (k) or decreasing the thickness (d), the device current can be enhanced. There are specific requirements for gate dielectrics to be used in thin-film transistors. Apart from a high capacitance, high dielectric breakdown strength, high levels of purity, high on/off current ratios, materials processability, and device stability are also essential.

6.4   Nanocomposite Dielectrics with Unmodified Nanoparticles

One of the first reports based on nanocomposite insulators appeared in 1988 and TiO_2 nanoparticle−polystyrene (PS) composites were used as dielectric materials [42]. However, this first attempt resulted in an inhomogeneous system with problems of porosity in the composite material which needs to be addressed later. Chen et al. [43] prepared nanocomposite dielectric layers in 2004 using cross-linked poly-4-vinylphenol (PVP) and TiO_2 nanoparticles, which could be deposited by spin coating method form the nanocomposite solution. Figure 6.3 shows the application of the nanocomposite dielectrics in the transistor structure.

They tested different formulations of this dielectric by changing the nanoparticle concentrations. The vapor-deposited pentacene-based organic field-effect transistors (OFETs) using patterned ITO on a glass substrate as the gate electrode were fabricated and an additional PEDOT layer on the ITO could enhance the on/off current ratio from 10^3 to 10^4. The dielectric constant increased from 3.5 to 3.9 after inserting the additional PEDOT layer, and it reached 5.4 for the nanocomposite film with 7 wt

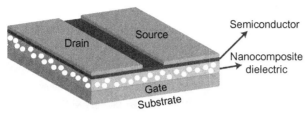

Figure 6.3 Schematic illustration of the thin-film transistor with nanocomposite dielectric layer.

% of TiO_2 nanoparticles. The authors also found that the transistors with 7 wt% TiO_2 nanoparticles in PVP as the gate insulator layer yield almost twice the drain–source current at the same gate voltage and the field-effect mobility could be increased from 0.20 to 0.24 cm^2/Vs. However, one of the drawbacks that the nanocomposite dielectrics present is the increased leakage current as compared to pure polymer, resulting in a decreased on/off current ratio. Furthermore, the device performance was still limited by the dispersibility of the TiO_2 nanoparticles in the polymer solution, leading to nanoparticle agglomeration at higher concentrations of fillers.

One strategy to reduce the leakage current of the nanocomposite dielectric was proposed by Kim et al. [44]. A solution-processed nanocomposite dielectric layer was fabricated by dispersing TiO_2 nanoparticles in a Nylon-6 polymer matrix. Nevertheless, it was found that even if the TiO_2 nanoparticles were uniformly dispersed in the polymer matrix, an additional polymer buffer layer was required to suppress the leakage current and to create a smoother surface between the semiconductor layer and the dielectric layer. In their experiment, an enhancement in mobility from 0.1 to 0.7 cm^2/Vs for pentacene-based transistors was achieved.

Bai et al. [45] fabricated a ceramic powder polymer composite using a P(VDF-TrFE) 50/50 mol% copolymer as the polymer matrix and $Pb(Mg_{1/3}Nb_{2/3})$ $O_3-PbTiO_3$ as the ceramic powder. The dielectric constant increased with the volume fraction of ceramic filler as expected. Furthermore, considering that previous research work on P(VDF-TrFE) copolymers indicates that high energy irradiation with proper dosage can increase the dielectric constant of the nanocomposite film [46], they irradiated the composites with the ceramic powder volume percentages ranging from 10% to 60% at 120°C by a 2.55 MeV electron source with different dosages. For a composite with a 50% ceramic volume content, the dielectric constant was quite high ($k \approx 250$) but with poor mechanical properties. Furthermore, the authors reported that adjusting the dosage of the irradiation allows tuning of the dielectric characteristics of the nanocomposite materials.

Yildirim et al. [47] fabricated ferroelectric P(VDF-TrFE) copolymer nanocomposites using barium titanate (BT) nanopowders. They obtained flexible insulating films with dielectric constants up to 51.5 when increasing the loading of the nanofillers (from 0 to 50 vol%). Low-voltage OFETs with these nanocomposites were also fabricated and they demonstrated ferroelectric hysteresis and good memory retention. Dang et al. [48] reported a nanocomposite composed of $BaTiO_3$ and Ni powders in a PVDF matrix. The authors compared $BaTiO_3$/PVDF and Ni-$BaTiO_3$/PVDF composites and found that only a small increase in the dielectric constant occurred upon introducing the Ni particles unless the concentration of the metallic particle was very close to the percolation threshold [49]. A dielectric constant above 800 could be achieved in their work. The advantages of these three-phase composites included easy processability, mechanical flexibility, and good dielectric behavior. In a following paper, another three-component composite with a dielectric constant of 120 was reported, which incorporated conductive carbon fibers [50]. The same group also investigated the dielectric behavior of Li- and Ti-doped NiO (LTNO)/PVDF composites [51]. They found a remarkable difference in the dielectric constant of the composites having different LTNO fillers and the dielectric constant could be achieved as high as ~ 600.

Schroeder et al. [52] reported high-capacitance gate insulators processed from aqueous solution. In this case, the nanocomposites were composed of $BaTiO_3$ nanoparticles dispersed in PVA or in a PVA random copolymer, poly(vinyl alcohol)-co-poly(vinyl acetate)-co-poly(itaconic acid) (PVAIA). Both types of nanocomposites exhibited leakage currents below 10^{-5} A/cm^2 with an electric field of ± 2 MV/cm. The dielectric constants for the PVAIA nanocomposite dielectrics ranged between 9 and 12, depending on the concentration of nanoparticles incorporated in the polymer matrix. The highest mobility for a pentacene-based transistor was 0.4 cm^2/Vs for the device using the lowest k dielectric versus 0.12 cm^2/Vs for the highest k one, which was ascribed to the surface roughness differences. For the $BaTiO_3$/PVA nanocomposite, a dielectric constant of 10.9 and a mobility of 0.35 cm^2/Vs were obtained.

Zhou et al. [53] investigated the effects of incorporating different concentrations of barium zirconate (BZ) nanoparticles into PVP dielectrics for flexible OTFTs. A dielectric constant of 6.7 could be achieved for the nanocomposite dielectrics with 12 wt% BZ nanoparticles. The effect of dielectric composition on the mobilities (μ), threshold voltages (V_T), on/off ratios, and subthreshold swings (S) of the transistors was discussed. The morphologies of dielectric and semiconductor films were also investigated and shown to correlate closely with the device performances.

6.5 Nanocomposite Dielectrics with Modified Nanoparticles

Simple solution processing of nanoparticles in a polymer host usually results in poor film quality and inferior insulating property which is mainly caused by the agglomeration of nanoparticles. To avoid the large leakage current and low breakdown strength of the nanocomposite, control over the particle−particle interaction and particle−polymer interface is essential to achieve better dispersion of the dielectric filler particles. When the size of the filler particles decreases to the nanometer scale, the properties of the polymer−filler interface would become dominant over the bulk properties of the constituents [54]. In a nanocomposite dielectric, the unique properties of the interface are amplified by the high surface area of the filler.

The large contrast in permittivity between two phases can give rise to highly inhomogeneous electric fields in the nanocomposite films. Incompatibility between the organophilic polymer matrix and the hydrophilic nanoparticle filler impedes the formation of a homogenous composite. Thus, a major research direction in this field is focused on modifying the nanoparticle filler surface to compatibilize the nanoparticle with the polymer matrix. Highly inhomogeneous fields and structural inhomogeneity generally lead to a significant reduction in the effective breakdown field strength of the composite.

In 2005, Maliakal et al. [55] fabricated TiO_2-PS polymer shell nanocomposites via a ligand exchange reaction between oleic acid-stabilized titanium oxide nanoparticles ($TiO_2-OLEIC$) and diethyl phosphonate terminated PS, leading to effective dispersion and enhanced dielectric properties. Capacitors were fabricated by spin coating TiO_2-PS (18.2% volume TiO_2) suspensions from chlorobenzene solution, yielding films of thicknesses ranging from 0.5 to 1.25 μm and exhibiting up to a 3.6 times enhancement in k versus PS. Pentacene-based TFT devices exhibited mobilities of 0.2 cm^2/Vs and low threshold voltage of -2 V, indicating a low trap density and dielectric compatibility with pentacene film growth. Later, the authors analyzed the blending of the aforementioned TiO_2-PS nanocomposites into PS thin films to investigate permittivity effects on OTFT performance while keeping the surface energy essentially constant [56]. Maliakal et al. found that the dielectric constant of these blends increases roughly linearly with the TiO_2-PS nanocomposite loading and the mobility of pentacene transistors also increases from 0.015 to 1.3 cm^2/Vs, with the latter being the highest value recorded for 100% TiO_2-PS, without PS blending. This value was much larger than that previously reported due to optimization of the solvent evaporation conditions. The significant enhancement in mobility with high-k TiO_2-PS was ascribed to morphological changes of pentacene, suggesting that the high mobility of pentacene on TiO_2-PS might be due to more efficient charge transfer between better connected small grains.

In 2007, Marks et al. [57] reported a method to disrupt nanoparticle agglomeration by in situ polymerization using metallocene catalysts supported on ferroelectric oxide nanoparticles. By coating the nanoparticles with methylalumoxane, followed by in situ propylene polymerization, they obtained homogeneously dispersed $BaTiO_3$ and TiO_2 nanoparticles within the matrix of a high-strength polypropylene. Transmission electron microscopy images evidenced the homogeneous dispersion of the nanoparticles in the polypropylene matrices, while scanning electron microscopy images were consistent with polymer chain growth from the nanoparticles. The insulators showed low leakage current ($\sim 10^{-6}-10^{-9}$ A/cm^2 at 200 V) and high breakdown strengths (~ 4 MV/cm), consistent with homogeneous dispersion of the metal oxide nanoparticles in polypropylene.

Lee et al. [58] achieved improved dispersion stability of TiO_2 nanoparticles by adding surfactants (polysorbate 80, Tween 80) to the solution mixtures and studied the dispersion and the sedimentation time of TiO_2 particles in a PVP solution. The surfactant aided TiO_2 dispersion in the polymer matrix but some aggregation was still found that increased leakage currents and eroded the device performance as compared to the pure PVP.

Chon et al. [59] reported organic−inorganic hybrid materials with high electronic polarizability through a chemical cross-linking reaction. To enhance the dispersion stability of BT nanoparticles in the matrix, the chemical nature of their surfaces was modified using diethyl 3-(trimethoxysilyl)propyl phosphonate, which was synthesized by the Michaelis—Arbuzov reaction of chloropropyltrimethoxysilane with 3-triethyl phosphate. The dielectric properties of the nanocomposite films could be significantly improved by a high degree of dispersion of the BT nanoparticles. The approach of adjusting the chemical composition allowed tailoring of the dielectric and film properties of the final composite materials.

Addition of surfactants could improve the dispersion of nanoparticles in polymers and consequently the film quality, but residual free surfactant may lead to high leakage current and dielectric loss. Thus, methods to modify nanoparticles through robust chemical bonds are highly desirable. Phosphonic acids could couple on the surface of metal oxides either by heterocondensation with surface hydroxyl groups or coordination to metal ions on the nanoparticle surface. The robust surface layer covering on the nanoparticles could decrease the aggregation of the particle and improve the dielectric properties of the nanocomposite films.

Recently, Roy et al. [60] used n-octadecylphosphonic acid (ODPA) to modify aluminum titanate (AT) nanoparticles for forming nanocomposite dielectrics. Figure 6.4A and B shows typical X-ray photoelectron spectroscopy (XPS) survey spectra of AT nanoparticles and ODPA-modified AT nanoparticles. Signals coming from aluminum, titanium, oxygen, carbon, and phosphorus were observed. The O1s, Ti2p, C1s, P2p, and Al2p peaks are illustrated in Figure 6.4C—G, separately. ODPA-modified AT nanoparticles show an O1s peak which is similar to that of unmodified AT nanoparticles. No evidence of P—O and P—O$^-$ bonding states was found, indicating a tridentate-bonded phosphorus in the ODPA head group. The intensities of Ti2p and Al2p slightly decrease due to the attenuation by the ODPA layer. The unmodified AT nanoparticles show two C1s peaks, which can be attributed to adventitious carbon acquired during the air transfer. A single and narrow C1s peak was observed in ODPA-modified AT nanoparticles, confirming the presence of ODPA layer on the surface of AT nanoparticles. The appearance of P2p peaks after ODPA modification also clearly indicates the formation of ODPA on the surface of AT nanoparticles.

The surface modification of BaTiO$_3$ (BT) nanoparticles with a phosphonic acid (PEGPA) to fabricate high volume fraction (up to 37 vol%) PEGPA-BT:cross-linked PVP nanocomposites was reported by Kim et al. [61]. The nanocomposite films were deposited by spin coating, followed by a soft baking at 100°C for 1 min and thermal cross-linking under vacuum at 160°C for 72 h. The nanocomposites showed significantly reduced leakage current density as compared to the ones obtained without phosphonic treatment of the nanoparticles due to improved dispersion, and differing dielectric constants were obtained for different particle concentrations. Nevertheless, the authors found that the increasing roughness of the nanocomposite films from nanoparticle loading greatly influences the morphology of the deposited pentacene layer and the OTFT performance. Further combining high volume fraction nanocomposites with a thin planarization layer of pure PVP could improve the electronic properties of the OTFTs, yielding a mobility of 0.17 cm^2/Vs, a threshold voltage of -1.1 V, a small subthreshold slope of 0.3 V/decade, and a large on/off current ratio of 10^5.

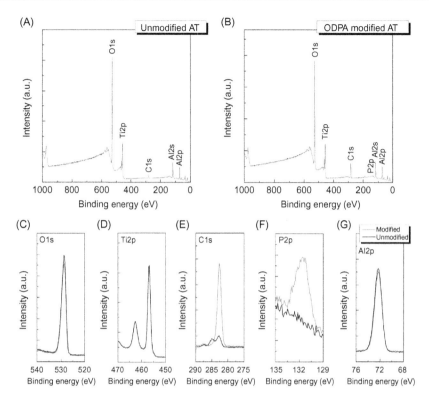

Figure 6.4 (A) XPS wide-scan spectrum of unmodified AT nanoparticles. (B) XPS wide-scan spectrum of ODPA-modified AT nanoparticles. (C) C1s, (D) Ti2p, (E) C1s, (F) P2p, and (G) Al2p spectra for unmodified AT nanoparticles and ODPA-modified nanoparticles. *Source*: Adapted from Ref. [60] with permission from The Royal Society of Chemistry.

Noh et al. [62] treated Al_2O_3 particles with a c-glycidoxypropyl−trimethoxysilane coupling agent and then dispersed them in a PVP matrix. In these functionalized nanocomposite dielectrics, the mechanical and electrical stabilities as well as the surface roughness were significantly improved compared with pure polymer. The dielectric constant increased monotonically from 4.9 for pure PVP to 7.2 for the nanocomposite dielectric with 24 vol% Al_2O_3. The mobilities measured for pentacene were decreased as compared to the pure PVP dielectric due to greater surface roughness; however, the on/off current ratio was increased by 3 orders of magnitude (from 10^2 to 10^5).

Huang et al. [63] studied the surface treatment of Al nanoparticles with octyl-trimethoxysilane coupling agent dispersed in a linear low density polyethylene (LLDPE) matrix. They found that the incorporation of nonpolar octyl groups onto the surface of Al nanoparticles not only increased the percolation threshold and the resistivity but also improved the dielectric properties as compared to the composites filled with unsurface-treated nanoparticles. It is possible to easily control the frequency and concentration dependences of dielectric constant and reduce the dielectric loss of the nanocomposites by the surface treatment. This approach is of great significance from the viewpoint of practical application of the polymer/metal nanocomposites in the electrical and electronic industries. The improved electrical properties could be directly ascribed to the good dispersion and special electrical feature of the surface-treated nanoparticles in the polymer matrix. A comparison between surface-treated and nonsurface-treated Al nanoparticles was also presented [64]. Because of the self-passivating nature of Al nanoparticles, both the treated and untreated particles contain an oxide shell around the metallic core that has a dramatic influence on the electrical properties of the composites. The surface treatment could lead better dispersion of Al nanoparticles in the LLDPE matrix and increase the nanofiller concentration, which translates into an improvement of the mechanical and thermal properties and stable dielectric characteristics.

6.6   Characterization of the Nanocomposite Dielectrics

Usually the surface morphology, leakage current density, and dielectric constant need to be considered for the dielectrics applied in thin-film electronic devices. The surface morphology can be studied by atomic force microscopy (AFM). AFM images of the PVP nanocomposite films with 2 vol% modified and unmodified AT nanoparticles are shown in Figure 6.5A and B. The nanocomposite films fabricated from unmodified AT nanoparticles exhibited large-scale aggregates, and the films with modified AT nanoparticles were found to be comparatively uniform by contrast. Figure 6.5C shows the root mean square (RMS) roughnesses of the dielectric films with respect to different nanoparticles concentration, both for modified and unmodified nanoparticles, as measured by AFM in tapping mode for a 2 μm $\times$ 2 μm area. These results reveal that the surface morphology is dependent on the nanoparticles concentration, with a general trend of larger roughness at higher concentration.

The leakage current density and capacitance can be studied by fabricating a capacitor with nanocomposite films sandwiched between two electrodes. The nanocomposite dielectric layer prepared using surface-modified AT nanoparticles shows reduced leakage current density than that of unmodified AT nanoparticles as shown in Figure 6.5D. The lower leakage current densities of the nanocomposite layer can be attributed to the decreased degree of AT nanoparticles aggregation in PVP matrix and reduction of direct particle−particle interaction by using ODPA surface modification. ODPA is also thought to passivate the ionizable hydroxyl group to eliminate the possible charge conduction pathways and minimize the concentration of charge carriers normally associated with the surface. Better interfacial adhesion of ODPA-modified AT nanoparticles can result in improved electrical properties of nanocomposite dielectrics. Figure 6.5E shows the dielectric constants of the dielectric films with different concentrations of modified AT nanoparticles measured in the range of 100 Hz to 200 kHz by a HP 4284 LCR meter. Larger effective k values are found for nanocomposite layer containing higher concentrations of

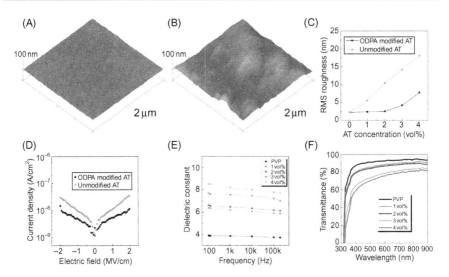

Figure 6.5 AFM images of the nanocomposite film with (A) 2 vol% modified and (B) 2 vol% unmodified AT nanoparticles. (C) RMS roughnesses of the dielectric films with respect to different nanoparticles concentration both for modified and unmodified nanoparticles.
(D) Leakage current densities of composite films with 1 vol% modified and unmodified AT nanoparticles as a function of applied electrical field. (E) Dielectric constant measured from the various composite films. (F) Optical transmission spectra of the nanocomposite films.
Source: Reproduced from Ref. [60] with permission from The Royal Society of Chemistry.

nanoparticles, and this can be explained by the higher dielectric permittivity of the filler relative to the polymer matrix. At a blending amount of 4 vol%, the dielectric constant of the nanocomposite film is 8.2 at 1 kHz, about twice the dielectric constant of pristine PVP. For the samples with low concentration of nanoparticles, the variation in the dielectric constant is smaller than those with higher nanoparticle concentration. For higher concentration of AT nanoparticle a trend of falling dielectric permittivity at high frequencies could be observed and explained by the greater space charge. Figure 6.5F shows the optical transmission spectra of the nanocomposite films. The average transmittance of the nanocomposite film with 2 vol% modified AT nanoparticles in the visible range (400−700 nm) is ~80%. The transmittance of the nanocomposite film decreases with increasing nanoparticles concentration, therefore suggesting an inhomogeneous structure for PVP host filled with large doping amount of nanoparticles.

6.7 Electrical Properties of the Devices on Nanocomposite Dielectrics

For preparing OTFTs and inverters with nanocomposite dielectric layer, step-by-step shadow-mask process which is an all-dry manufacturing process is usually used.

Figure 6.6A and B shows the energy band diagrams of p-channel (pentacene) and n-channel (F_{16}CuPc), respectively. For simplicity, the energy levels are drawn as straight lines and band bending occurs at the semiconductor/insulator interface upon an applied gate bias. For p-type operation, there is a barrier (~0.1 eV) for hole injection from the Au electrodes into the HOMO level of pentacene. For n-type operation, there is a barrier (~0.2 eV) for electron injection from gold electrodes into the LUMO level of F_{16}CuPc.

Figure 6.6C and D shows X-ray diffraction (XRD) patterns of pentacene and F_{16}CuPc films deposited on the PVP nanocomposite film with 2 vol% modified AT nanoparticles. The pentacene film produces first-order diffraction peaks at 5.72°, which demonstrate thin-film phases, with corresponding d_{001} spacing of 15.4 Å. The XRD pattern of F_{16}CuPc film shows first-order diffraction peaks at 5.98°, corresponding to d_{200} spacing of 14.7 Å. The insets depict the surface morphologies of 40 nm pentacene and 25 nm F_{16}CuPc films on the nanocomposite film with 2 vol% modified AT nanoparticles. Pentacene film exhibits relatively larger grains with a size of ~200 nm and a grain density of ~25 μm^{-2}. F_{16}CuPc film consists of numerous

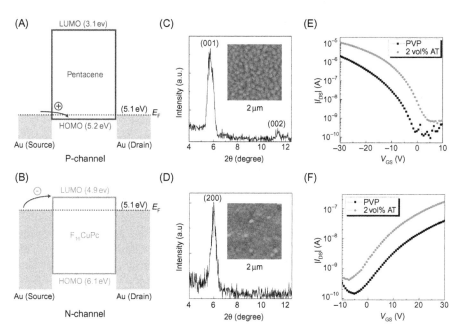

Figure 6.6 (A) Band diagram of pentacene in contact with gold electrodes. (B) Band diagram of F_{16}CuPc in contact with gold electrodes. XRD patterns and AFM images of (C) pentacene and (D) F_{16}CuPc films deposited on the nanocomposite film with 2 vol% modified AT nanoparticles. (E) Transfer characteristics of pentacene TFT with pristine PVP and nanocomposite dielectric layer containing 2 vol% ODPA-modified AT. (F) Transfer characteristics of F_{16}CuPc TFT with pristine PVP and nanocomposite dielectric layer containing 2 vol% ODPA-modified AT.
Source: Reproduced from Ref. [60] with permission from The Royal Society of Chemistry.

grains covering the dielectric layer, with a grain density of $\sim$300 μm^{-2} and a grain diameter of $\sim$60 nm. These results suggest that the structure of the nanocomposite dielectric films is compatible with pentacene and F_{16}CuPc film growth.

The electrical characteristics of the p-channel pentacene transistor and n-channel F_{16}CuPc transistor are shown in Figure 6.6E and F. The channel length is 50 μm and the channel width is 500 μm. The drain−source current (I_{DS}) of the transistor with nanocomposite dielectric layer is greater than that of the control device with PVP dielectric. This is attributed to the large charge density induced in the channel by high capacitance of the nanocomposite insulator. Saturation mobility could be determined from the slope of the transfer characteristics $|I_{DS}|^{1/2}$ versus gate−source voltage (V_{GS}) at drain−source voltage (V_{DS}) = −30 V by using the equation, $I_{DS} = (WC_{OX}/2L)\mu(V_{GS} - V_T)^2$. For the nanocomposite device, the pentacene-based transistor shows a carrier mobility (μ) of $\sim$0.4 cm^2/Vs, a threshold voltage (V_T) of $\sim$ − 5 V, a transconductance of $\sim$0.85 μS, and an on/off current ratio of $\sim$2 $\times$ 10^4. For the control device, the pentacene TFT has a μ of $\sim$0.3 cm^2/Vs, a V_T of $\sim$ − 14 V, a transconductance of $\sim$0.28 μS, and an on/off current ratio of $\sim$2 $\times$ 10^4. The F_{16}CuPc transistor demonstrates a μ of $\sim$5 $\times$ 10^{-3} cm^2/Vs, a V_T of $\sim$2 V, a transconductance of $\sim$0.013 μS, and an on/off current ratio of $\sim$10^3. For the control device with PVP dielectric, the F_{16}CuPc TFT has a μ of $\sim$2 $\times$ 10^{-3} cm^2/Vs, a V_T of $\sim$4 V, a transconductance of $\sim$0.003 μS, and an on/off current ratio of $\sim$5 $\times$ 10^2. All the devices show improved electrical properties compared with the transistors with pure PVP dielectric layers. Further increase in the mobilities of semiconductors could be achieved by increasing the substrate temperature when depositing the materials or flattening the interface between the semiconductor layer and dielectric layer.

An inverter or NOT gate is a logic gate which can make logic inversion. The inverter can give an output voltage representing the opposite logic level to the input voltage. There are usually four kinds of inverters in organic electronics which are called saturated load inverter, depleted load inverter, resistor load inverter, and complementary inverter as illustrated in Figure 6.7.

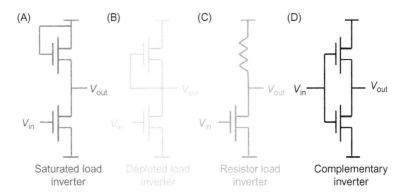

Figure 6.7 Schematic illustration of the four kinds of inverter configurations.

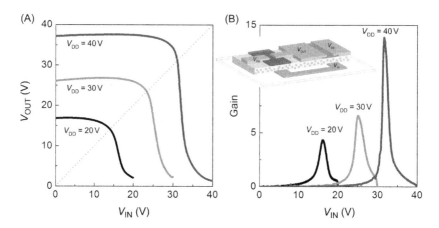

Figure 6.8 (A) Voltage transfer characteristics of the complementary inverters. (B) Signal gain of the inverters and schematic representation of the inverter architecture.
Source: Reproduced from Ref. [60] with permission from The Royal Society of Chemistry.

The quality of the inverter is often measured using the voltage transfer curve (VTC). The VTC is a plot of the input voltage with respect to the output voltage. Ideally, the VTC should be an inverted step function. However, a gradual transition region always exists in real devices. For low input voltage, the inverter outputs high voltage. For high input voltage, the inverter outputs low voltage. The gain of the inverter is expressed by the first-order derivative of an output voltage with respect to the input voltage: $-dV_{out}/dV_{in}$.

Using p-channel pentacene and n-channel $F_{16}CuPc$, the organic complementary circuits can be fabricated. During the patterning procedure, the gates of the p-channel and n-channel transistors are connected to form the input electrode, while the drains of the two transistors are connected to form the output electrode. The source of the p-channel acts as the power supply electrode and the source of the n-channel acts as the ground electrode. The VTC of the inverters with PVP nanocomposite filled with 2 vol% AT nanoparticles for the supply voltages (V_{DD}) of 20, 30, and 40 V is shown in Figure 6.8A. Inverter switching is clearly seen in the sweeping range of 0–40 V. The switching voltage (V_S) could be obtained from the intersection of the VTC with the line output voltage (V_{OUT}) = input voltage (V_{IN}). The measured V_S are 14, 23, and 31 V corresponding to V_{DD} = 20, 30, and 40 V. The V_S of a complementary inverter in the quadratic model is defined as

$$V_S = \frac{\beta_n V_{tn} + \beta_p (V_{DD} + V_{tp})}{\beta_n + \beta_p} \tag{6.6}$$

where $\beta_n = (\mu_n C_{OX} W_n/L_n)^{1/2}$ and $\beta_p = (\mu_p C_{OX} W_p/L_p)^{1/2}$.

The calculated V_S is determined to be 13.1, 21.6, and 30.1 V at V_{DD} = 20, 30, and 40 V, respectively. The calculated result is consistent with the measured value. Figure 6.8B shows the signal gain and it increases with increasing V_{DD}.

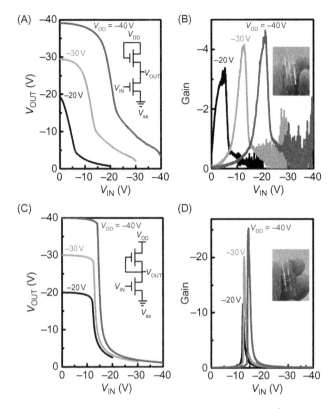

Figure 6.9 (A) Transfer characteristics of the saturated load inverter. (B) The signal gain of the saturated load inverter. (C) Transfer characteristics of the depleted load inverter. (D) The signal gain of the depleted load inverter. The insets show circuit schematics and optical images of the two different inverters.
Source: Adapted from Ref. [53] with permission from The Royal Society of Chemistry.

Figure 6.9 shows the electrical characteristics and circuit schematics of the saturated load inverter and depleted load inverter fabricated on the PVP nanocomposite dielectric films filled with 6 wt% BZ nanoparticles on PET substrate. Both inverters show good switching behavior and voltage amplification for the logic operation. The gain of the inverters increases as the V_{DD} augments. The depleted load inverter has the better static performance showing large signal gain and sharp switching with rail-to-rail output swings in the VTC.

6.8 Bending Experiments of the Flexible Devices

To evaluate the flexibility of the OTFTs, the PET substrates were bent along the channel transport axis of the transistor, as illustrated in Figure 6.10A. The strain

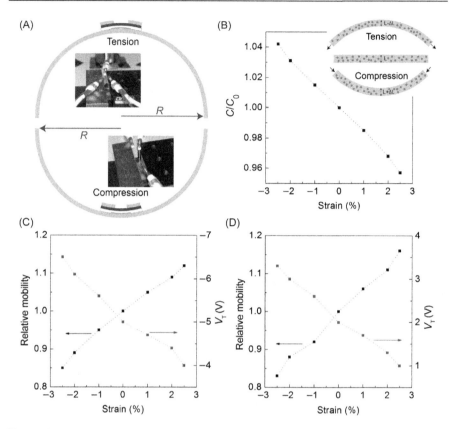

Figure 6.10 (A) Schematic illustration of compressive and tensile bending states of the transistors on PET substrate. (B) Relative changes in capacitances with respect to various strains and schematic illustration of change in film thickness induced by the Poisson effect. (C) Relative mobility and threshold voltage of pentacene as a function of bending strain. (D) Relative mobility and threshold voltage of $F_{16}CuPc$ as a function of bending strain. *Source*: Adapted from Ref. [60] with permission from The Royal Society of Chemistry.

S induced on the surface of a substrate by bending the substrate into a radius of R is given as

$$S = \frac{(t_L + t_S)(1 + 2\eta + \chi\eta^2)}{2R(1 + \eta)(1 + \chi\eta)} \tag{6.7}$$

where $\eta = t_L/t_S$, t_L is the thickness of the layer and t_S is the thickness of the substrate; $\chi = Y_L/Y_S$, Y_L is the Young's modulus of the layer and Y_S is the Young's modulus of the substrate [65].

The maximum compressive (positive) strain that applied was 2.5% (4 mm concave bending radii) and the maximum tensile (negative) strain was −2.5% (4 mm convex bending radii). To confirm that the changes in the electrical performance of OTFTs are not mostly caused by a strain-induced change in the dielectric capacitance, the capacitors were evaluated under the same compressive and tensile strain. Figure 6.10B shows the relative changes of capacitance (C/C_0) with respect to the applied strain of a dielectric film with 2 vol% ODPA-modified AT nanoparticles. The capacitance decreased a little under compressive strain and increased slightly under tensile strain, and this phenomenon could be explained by the change in film thickness induced by the Poisson effect. According to the Poisson model, the changes in capacitance are expressed by Eq. (6.8), where v is Poisson ratio of the dielectric film (0.4 was chosen for the composite film) [66]. The calculated result ($C/C_0 = 1.04$ at $S = -2.5\%$) agrees well with the experimental data in the experiment.

$$C/C_0 = 1 - S/(1 - v) \tag{6.8}$$

The electrical characteristics of the pentacene and $F_{16}CuPc$ transistors were recorded during the systematic application of different compressive or tensile strain at the flexible substrate. Figure 6.10C shows the relative mobility and threshold voltage of pentacene transistor as a function of bending strain. In the case of pentacene compared to the flat state, the μ is increased by 12% at $S = 2.5\%$ and decreased by 15% at $S = -2.5\%$, and the V_T is shifted positively by compressive strains and negatively by tensile strains. Figure 6.10D shows the relative mobility and threshold voltage of $F_{16}CuPc$ transistor with respect to the bending strain. For $F_{16}CuPc$ transistor, the μ is increased by 16% at $S = 2.5\%$ and decreased by 17% at $S = -2.5\%$, and the V_T is shifted negatively by compressive strains and positively by tensile strains. All the changes in the μ and V_T of the transistors are reversible during the bending test. This phenomenon could be explained by the hopping model for organic molecular systems. For pentacene and $F_{16}CuPc$, energy barriers for hole/electron hopping decrease owing to the smaller spacing between molecules under compressive strain and increase due to the larger spacing under tensile strain. The improved/deteriorated in-phase intermolecular coupling or phase transition between the bulk phase and thin-film phase by bending stress may also explain this result [67,68].

Figure 6.11 shows the investigation of the bending stability of the inverters. The signal gain at $V_{DD} = 30$ V increased by 22% at a strain of 2.5% and decreased by 18% at a strain of −2.5%, respectively. Figure 6.11A shows the signal gain before bending, bending at strain of 2.5%, and after bending. There is no noticeable change in the electrical performance after the bending test. Figure 6.11B summarizes the signal gain respect to different bending strain. The gain of the inverter increases with compressive strain and decreases with tensile strain, which is well accordance with the performance change in each individual transistor of the inverter.

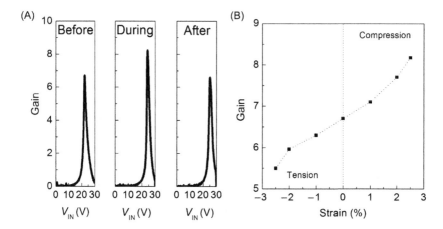

Figure 6.11 (A) Signal gain before bending, bending at strain of 2.5% and after bending. (B) Signal gain as a function of different bending strain.
Source: Reproduced from Ref. [60] with permission from The Royal Society of Chemistry.

6.9 Conclusion

With the development of organic flexible electronics, it becomes clear that new low-cost dielectric materials need to be investigated. For the nanocomposite dielectric materials, the dielectric constant can be tuned by changing the concentration of nanoparticles, which allows fine tuning of the gate insulators for flexible electronics applications. We discuss the dielectric characterization and the film morphology of the nanocomposites. The electrical performance of the capacitors, transistors, and inverters under different applied strain on flexible substrates is also described. Polymer nanocomposite dielectric layer should be quite promising for advanced flexible electronics.

References

[1] Facchetti A, Yoon MH, Marks TJ. Gate dielectrics for organic field-effect transistors: new opportunities for organic electronics. Adv Mater 2005;17(14):1705−25.

[2] Rogers JA, Someya T, Huang Y. Materials and mechanics for stretchable electronics. Science 2010;327(5973):1603−7.

[3] Sun J, Zhang B, Katz HE. Materials for printable, transparent, and low-voltage transistors. Adv Funct Mater 2011;21(1):29−45.

[4] Di C-A, Zhang F, Zhu D. Multi-functional integration of organic field-effect transistors (OFETs): advances and perspectives. Adv Mater 2013;25(3):313−30.

[5] Heremans P, Gelinck GH, Müller R, Baeg KJ, Kim DY, Noh YY. Polymer and organic nonvolatile memory devices. Chem Mater 2010;23(3):341−58.

[6] Han ST, Zhou Y, Wang C, He L, Zhang W, Roy VAL. Layer-by-layer-assembled reduced graphene oxide/gold nanoparticle hybrid double-floating-gate structure for low-voltage flexible flash memory. Adv Mater 2013;25(6):872−7.

[7] Guo Y, Yu G, Liu Y. Functional organic field-effect transistors. Adv Mater 2010;22 (40):4427−47.

[8] Klauk H. Organic thin-film transistors. Chem Soc Rev 2010;39(7):2643−66.

[9] Ortiz RP, Facchetti A, Marks TJ. High-k organic, inorganic, and hybrid dielectrics for low-voltage organic field-effect transistors. Chem Rev 2010;110(1):205−39.

[10] Veres J, Ogier S, Lloyd G, de Leeuw D. Gate insulators in organic field-effect transistors. Chem Mater 2004;16(23):4543−55.

[11] Hwang DK, Kim CS, Choi JM, Lee K, Park JH, Kim E, et al. Polymer/YOx hybrid−sandwich gate dielectrics for semitransparent pentacene thin-film transistors operating under 5 V. Adv Mater 2006;18(17):2299−303.

[12] Cao Q, Xia MG, Shim M, Rogers JA. Bilayer organic−inorganic gate dielectrics for high-performance, low-voltage, single-walled carbon nanotube thin-film transistors, complementary logic gates, and p−n diodes on plastic substrates. Adv Funct Mater 2006;16(18):2355−62.

[13] Lee NE, Seol YG, Noh HY, Lee SS, Ahn JH. Mechanically flexible low-leakage multilayer gate dielectrics for flexible organic thin film transistors. Appl Phys Lett 2008;93 (1):013305.

[14] Chung Y, Verploegen E, Vailionis A, Sun Y, Nishi Y, Murmann B, et al. Controlling electric dipoles in nanodielectrics and its applications for enabling air-stable n-channel organic transistors. Nano Lett 2011;11(3):1161−5.

[15] Chiang CK, Fincher Jr CR, Park YW, Heeger AJ, Shirakawa H, Louis EJ, et al. Electrical conductivity in doped polyacetylene. Phys Rev Lett 1977;39(17):1098−101.

[16] Shirakawa H, Louis EJ, MacDiarmid AG, Chiang CK, Heeger AJ. Synthesis of electrically conducting organic polymers: halogen derivatives of polyacetylene, (CH)$_x$. J Chem Soc Chem Commun 1977;16:578−80.

[17] Tang CW, VanSlyke SA. Organic electroluminescent diodes. Appl Phys Lett 1987;51 (12):913−5.

[18] Tsumura A, Koezuka H, Ando T. Macromolecular electronic device: field-effect transistor with a polythiophene thin film. Appl Phys Lett 1986;49(18):1210−2.

[19] Burroughes JH, Jones CA, Friend RH. New semiconductor device physics in polymer diodes and transistors. Nature 1988;335(6186):137−41.

[20] Peng X, Horowitz G, Fichou D, Garnier F. All-organic thin-film transistors made of alpha-sexithienyl semiconducting and various polymeric insulating layers. Appl Phys Lett 1990;57(19):2013−5.

[21] Horowitz G, Peng XZ, Fichou D, Garnier F. Role of the semiconductor/insulator interface in the characteristics of π-conjugated-oligomer-based thin-film transistors. Synth Met 1992;51(1−3):419−24.

[22] Garnier F, Hajlaoui R, Yassar A, Srivastava P. All-polymer field-effect transistor realized by printing techniques. Science 1994;265(5179):1684−6.

[23] Bao Z, Feng Y, Dodabalapur A, Raju VR, Lovinger AJ. High-performance plastic transistors fabricated by printing techniques. Chem Mater 1997;9(6):1299−301.

[24] Haddon RC, Perel AS, Morris RC, Palstra TTM, Hebard AF, Fleming RM. C[sub 60] thin film transistors. Appl Phys Lett 1995;67(1):121−3.

[25] Brown AR, Pomp A, Hart CM, de Leeuw DM. Logic gates made from polymer transistors and their use in ring oscillators. Science 1995;270(5238):972−4.

[26] Dodabalapur A, Katz HE, Torsi L, Haddon RC. Organic heterostructure field-èffect transistors. Science 1995;269(5230):1560−2.

[27] Dodabalapur A, Katz HE, Torsi L, Haddon RC. Organic field-effect bipolar transistors. Appl Phys Lett 1996;68(8):1108−10.

[28] Dodabalapur A, Baumbach J, Baldwin K, Katz HE. Hybrid organic/inorganic complementary circuits. Appl Phys Lett 1996;68(16):2246−8.

[29] Dodabalapur A, Laquindanum J, Katz HE, Bao Z. Complementary circuits with organic transistors. Appl Phys Lett 1996;69(27):4227−9.

[30] Sirringhaus H, Tessler N, Friend RH. Integrated optoelectronic devices based on conjugated polymers. Science 1998;280(5370):1741−4.

[31] Lin YY, Dodabalapur A, Sarpeshkar R, Bao Z, Li W, Baldwin K, et al. Organic complementary ring oscillators. Appl Phys Lett 1999;74(18):2714−6.

[32] Crone B, Dodabalapur A, Lin YY, Filas RW, Bao Z, LaDuca A, et al. Large-scale complementary integrated circuits based on organic transistors. Nature 2000;403 (6769):521−3.

[33] Dimitrakopoulos CD, Purushothaman S, Kymissis J, Callegari A, Shaw JM. Low-voltage organic transistors on plastic comprising high-dielectric constant gate insulators. Science 1999;283(5403):822−4.

[34] Dimitrakopoulos CD, Kymissis I, Purushothaman S, Neumayer DA, Duncombe PR, Laibowitz RB. Low-voltage, high-mobility pentacene transistors with solution-processed high dielectric constant insulators. Adv Mater 1999;11(16):1372−5.

[35] Tate J, Rogers JA, Jones CDW, Vyas B, Murphy DW, Li W, et al. Anodization and microcontact printing on electroless silver: solution-based fabrication procedures for low-voltage electronic systems with organic active components. Langmuir 2000;16 (14):6054−60.

[36] Sirringhaus H, Kawase T, Friend RH, Shimoda T, Inbasekaran M, Wu W, et al. High-resolution inkjet printing of all-polymer transistor circuits. Science 2000;290 (5499):2123−6.

[37] Someya T, Kato Y, Sekitani T, Iba S, Noguchi Y, Murase Y, et al. Flexible, large-area networks of pressure and thermal sensors with organic transistor active matrixes. Proc Natl Acad Sci USA 2005;102(35):12321−5.

[38] Subramanian V, Chang PC, Lee JB, Molesa SE, Volkman SK. Printed organic transistors for ultra-low-cost RFID applications. IEEE Trans Comp Pack Technol 2005;28 (4):742−7.

[39] Klauk H, Zschieschang U, Pflaum J, Halik M. Ultralow-power organic complementary circuits. Nature 2007;445(7129):745−8.

[40] Smits ECP, Mathijssen SGJ, van Hal PA, Setayesh S, Geuns TCT, Mutsaers K, et al. Bottom-up organic integrated circuits. Nature 2008;455(7215):956−9.

[41] Minhun J, Jaeyoung K, Jinsoo N, Namsoo L, Chaemin L, Gwangyong L, et al. All-printed and roll-to-roll-printable 13.56-MHz-operated 1-bit rf tag on plastic foils. IEEE Trans Electron Dev 2010;57(3):571−80.

[42] Khastgir D, Maiti HS, Bandyopadhyay PC. Polystyrene−titania composite as a dielectric material. Mater Sci Eng 1988;100(0):245−53.

[43] Chen FC, Chu CW, He J, Yang Y, Lin JL. Organic thin-film transistors with nanocomposite dielectric gate insulator. Appl Phys Lett 2004;85(15):3295−7.

[44] Kim CH, Bae JH, Lee SD, Choi JS. Fabrication of organic thin-film transistors based on high dielectric nanocomposite insulators. Mol Cryst Liq Cryst 2007;471(1):147−54.

[45] Bai Y, Cheng ZY, Bharti V, Xu HS, Zhang QM. High-dielectric-constant ceramic-powder polymer composites. Appl Phys Lett 2000;76(25):3804−6.

[46] Zhang QM, Bharti V, Zhao X. Giant electrostriction and relaxor ferroelectric behavior in electron-irradiated poly(vinylidene fluoride-trifluoroethylene) copolymer. Science 1998;280(5372):2101−4.

[47] Yildirim FA, Ucurum C, Schliewe RR, Bauhofer W, Meixner RM, Goebel H, et al. Spin-cast composite gate insulation for low driving voltages and memory effect in organic field-effect transistors. Appl Phys Lett 2007;90(8):083501.

[48] Dang Z-M, Shen Y, Nan CW. Dielectric behavior of three-phase percolative Ni-BaTiO[sub 3]/polyvinylidene fluoride composites. Appl Phys Lett 2002;81(25): 4814−6.

[49] Pecharromán C, Esteban-Betegón F, Bartolomé JF, López-Esteban S, Moya JS. New percolative BaTiO$_3$−Ni composites with a high and frequency-independent dielectric constant ($\epsilon_r \approx 80,000$). Adv Mater 2001;13(20):1541−4.

[50] Dang ZM, Fan LZ, Shen Y, Nan CW. Study on dielectric behavior of a three-phase CF/(PVDF + BaTiO$_3$) composite. Chem Phys Lett 2003;369(1−2):95−100.

[51] Dang ZM, Wu JB, Fan LZ, Nan CW. Dielectric behavior of Li and Ti co-doped NiO/ PVDF composites. Chem Phys Lett 2003;376(3−4):389−94.

[52] Schroeder R, Majewski LA, Grell M. High-performance organic transistors using solution-processed nanoparticle-filled high-k polymer gate insulators. Adv Mater 2005;17(12):1535−9.

[53] Zhou Y, Han ST, Xu ZX, Roy VAL. Polymer−nanoparticle hybrid dielectrics for flexible transistors and inverters. J Mater Chem 2012;22(9):4060.

[54] Lewis TJ. Interfaces: nanometric dielectrics. J Phys D Appl Phys 2005;38(2):202.

[55] Maliakal A, Katz H, Cotts PM, Subramoney S, Mirau P. Inorganic oxide core, polymer shell nanocomposite as a high k gate dielectric for flexible electronics applications. J Am Chem Soc 2005;127(42):14655−62.

[56] Jung C, Maliakal A, Sidorenko A, Siegrist T. Pentacene-based thin film transistors with titanium oxide−polystyrene/polystyrene insulator blends: high mobility on high K dielectric films. Appl Phys Lett 2007;90(6):062111.

[57] Guo N, DiBenedetto SA, Kwon D-K, Wang L, Russell MT, Lanagan MT, et al. Supported metallocene catalysis for *in situ* synthesis of high energy density metal oxide nanocomposites. J Am Chem Soc 2007;129(4):766−7.

[58] Lee K-H, Park BJ, Choi HJ, Park J, Choi JS. Effect of surfactant on preparation of poly(4-vinylphenol)/titanium dioxide composite for a gate insulator of organic thin film transistors. Mol Cryst Liq Cryst 2007;471(1):173−9.

[59] Chon J, Ye S, Cha KJ, Lee SC, Koo YS, Jung JH, et al. High-κ dielectric sol − gel hybrid materials containing barium titanate nanoparticles. Chem Mater 2010;22 (19):5445−52.

[60] Zhou Y, Han S-T, Xu ZX, Yang X-B, Ng HP, Roy VAL. Functional high-k nanocomposite dielectrics for flexible transistors and inverters with excellent mechanical properties. J Mater Chem 2012;22(28):14246−53.

[61] Kim P, Zhang X-H, Domercq B, Jones SC, Hotchkiss PJ, Marder SR, et al. Solution-processible high-permittivity nanocomposite gate insulators for organic field-effect transistors. Appl Phys Lett 2008;93(1):013302.

[62] Noh HY, Seol YG, Kim SI, Lee NE. Mechanically flexible low-leakage nanocomposite gate dielectrics for flexible organic thin-film transistors. Electrochem Solid-State Lett 2008;11(8):H218−21.

[63] Huang X, Kim C, Jiang P, Yin Y, Li Z. Influence of aluminum nanoparticle surface treatment on the electrical properties of polyethylene composites. J Appl Phys 2009;105(1):014105.

[64] Huang XY, Jiang PK, Kim CU. Electrical properties of polyethylene/aluminum nano-composites. J Appl Phys 2007;102(12):124103.

[65] Suo Z, Ma EY, Gleskova H, Wagner S. Mechanics of rollable and foldable film-on-foil electronics. Appl Phys Lett 1999;74(8):1177−9.

[66] Sekitani T, Kato Y, Iba S, Shinaoka H, Someya T, Sakurai T, et al. Bending experiment on pentacene field-effect transistors on plastic films. Appl Phys Lett 2005;86 (7):073511.

[67] Tan HS, Kulkarni SR, Cahyadi T, Lee PS, Mhaisalkar SG, Kasim J, et al. Solution-processed trilayer inorganic dielectric for high performance flexible organic field effect transistors. Appl Phys Lett 2008;93(18):183503.

[68] Yang C, Yoon J, Kim SH, Hong K, Chung DS, Heo K, et al. Bending-stress-driven phase transitions in pentacene thin films for flexible organic field-effect transistors. Appl Phys Lett 2008;92(24):243305.

7 Nanomaterials for Drug Delivery

Li Yan and Xianfeng Chen

Center of Super-Diamond and Advanced Films (COSDAF) and Department of
Physics and Materials Science, City University of Hong Kong, Kowloon Tong,
Hong Kong SAR, PR China

7.1 Introduction

Recently the demand for advanced drug delivery systems has been expanding due
to the fast growth in fields of medicine, biology, materials science, and other scientific sectors. With dramatic progress in biotechnology and biomedical engineering
fields, great amount of new and potentially promising drugs are discovered and
designed. However, they are often limited by poor stability, high toxicity, short cir-
culating half-life, tendency of aggregation, or their transport and delivery may be
hindered by biobarriers like blood—brain barriers or cell membranes. In fact, existing drugs also often have many problems, such as low efficacy, serious side effects,
high cost, and low controllability. To address these issues, researchers from multi-
disciplinary backgrounds, such as biology, materials science, pharmacy, and chemistry, have been gathering together to develop new drug delivery systems. This is
an important field of research, as an effective drug delivery system can develop
bridges to bring new medicines to clinical applications as well as renew classical
drugs and expand their market, all of which are intriguing to pharmaceutical
industry.

What is a drug delivery system? Generally, a drug delivery system is to control
the releasing profile of a drug at a designed rate in a targeted area to ultimate the
function of the drug. By having the concept of drug delivery systems, it is necessary to know what properties drug delivery systems should ideally have. The
desired properties include: (a) *enhanced permeability*: drug delivery systems can
facilitate drugs across cell membrane or other biobarriers; (b) *localized or targeted
delivery*: drug delivery systems enable drugs to localize in or target specific areas
like local cancers; (c) *increased stability*: drug delivery systems can protect drugs
from macrophage clearance or self-degradation; (d) *controllability*: drug release
may be controlled upon environmental change, such as temperature and pH or
external signal (e.g., near-infrared (NIR) light and ultrasound); and (e) *reduced toxicity and enhanced therapeutic efficacy*: drug delivery systems can minimize side
effect of drugs but in the meanwhile maintain or even enhance their therapeutic
effect. To achieve drug delivery systems with these properties, many approaches

Nanocrystalline Materials. DOI: http://dx.doi.org/10.1016/B978-0-12-407796-6.00007-5

have been developed. Among them, nanomaterials are of particular interest and have been widely researched for developing a new generation of drug delivery systems.

Nanomaterials are extremely small-sized materials with at least one dimension ranging from 1 to 100 nm. According to the number of the dimensions which are in the range of 1−100 nm, nanomaterials can be classified to 0D (e.g., nanoparticle), 1D (e.g., nanorod (NR) and nanowire), and 2D (e.g., quantum well, nanosheet, and nanofilm). The very small dimensions of nanomaterials lead to many superior physical, optical, electrical, and biological properties that are suitable for a wide range of applications including drug delivery. Herein are typical examples: a decrease of melting temperature and an increase of catalytic activity appear in nanomaterials due to their huge surface area, high ratio of surface atoms to total number of atoms, increased surface energy, etc. High surface to volume ratio enables more drugs to be attached onto the surface of nanomaterial, thus increasing the ability of drug loading. Ease of surface modification allows nanomaterial to attach different kinds of conjugate biomolecules to achieve target delivery. Drugs can be protected by nanomaterials to avoid degradation during transport. Nanomaterials can be fluorescent or responsive under UV/Vis/NIR excitation or electrical or magnetic field and therefore the drug delivery can be tracked and the release of drugs be precisely controlled. Nanomaterials can increase the cellular uptake of drugs because of the surface charge, size, etc., and control to the preference of specific types of cells. Owing to these properties, there has been a fast growing interest of using nanomaterials in a broad range of applications including drug delivery.

The aim of this chapter is to provide typical and detailed examples to describe how different types of nanomaterials are employed to achieve the desired properties of drug delivery systems.

7.2 Nanomaterials for Drug Delivery

7.2.1 Semiconductor Quantum Dots

Quantum dots (QDs) have attracted great attention in the past few decades, because of their unique optical and electronic properties. Compared with organic fluorescent dyes, QDs show size-dependent emission, which enables ease of tuning emission from visible to infrared wavelength by simply controlling their sizes [1−4]. Furthermore, QDs have other advantages, such as high photostability, narrow emission wavelength, strong luminescence intensity, broad excitation window, and ease of preparation and surface modification. Therefore, QDs have been extensively used for drug delivery and bioimaging [5].

High quality semiconductor nanoparticles can be prepared through organometallic approaches. Many excellent methods have been summarized in the literature [6−10]. However, the nanoparticles synthesized through these routes are not soluble in water and do not have the appropriate moieties to conjugate biomolecules. To make them ready for drug delivery, surface modifications need to be carried out and two general approaches were developed for this purpose [11,12]. The size of QDs can be tuned between 2 and 10 nm and it is convenient to link versatile functionalities to the QDs [13].

One of the greatest advantages of using nanomaterials for drug delivery is that these ultrasmall materials can greatly enhance the intracellular delivery of a broad range of molecules for improved efficacy. As one example, QDs can be used for siRNA delivery. Interference RNA has great potential for biomedical applications because of its powerful ability to target specific mRNAs, resulting in mRNA destruction and gene silencing [14,15]. However, the lack of stability and difficulty in cellular entry and endosome escape are the major problems that limit its application. Taking advantage of easy surface modification, QDs can be used for highly efficient and safe delivery of siRNA for mRNA interference [16]. High quality semiconductor QDs were synthesized through a TOPO (tri-n-octylphosphine oxide) route. In brief, a precursor cadmium oxide and stearic acid were mixed to form a clear solution; subsequently TOPO and hexadecylamine solution were added; finally, selenium solution was quickly injected into the above mixture to form QDs. The prepared hydrophobic TOPO-stabilized QDs were then transferred to aqueous solution by attaching an amphiphilic polymer containing carboxylic acid group through hydrophobic interaction. Next, N-(3-mimethylaminopropyl)-N'-ethylcarbodiimide hydrochloride (EDAC) and N,N-dimethylethylenediamine were added to functionalize the surface of the QDs with amine groups which provide positive charge for electrostatical attraction of siRNA molecules for gene delivery (Figure 7.1A) [16]. The fabricated 6 nm QDs (measured from transmission electron microscopy (TEM) images) have an average hydrodynamic diameter of about 13 nm. After siRNA adsorption to the surface of the QDs, the hydrodynamic size increases to 17 nm. The pristine QDs have positive charge with +19.4 mV, which decreases to +8.5 mV after siRNA loading. These QDs are capable of improving cellular uptake of siRNA. Once the QDs are in the endosomes in cells, the carboxyl group (or amine group in some applications) can cause proton-sponge effect and lead to endosome escape. Briefly, proton absorption occurs in acid organelles, which increases osmotic pressure across the organelle membrane [17]; subsequently, the increased osmotic pressure causes acidic endosome to swell and rupture, releasing the inside materials to cytoplasm (Figure 7.1C) [16]. Figure 7.1B illustrates the various steps of using siRNA-QDs for gene therapy, which includes cellular entry by endocytosis, endosome escape due to proton-sponge effect and gene silencing by siRNA [16].

Figure 7.2 shows the RNA silence efficacy of cyclophilin B protein by the above-described proton-sponge coated QDs (Figure 7.1) and three typical commercial transfection reagents (Lipofectamine, TransIT, and JetPEI) [16]. Lipofectamine is a cationic-lipid transfection reagent. TransIT is a lipopolypelex which is a lipid—polymer mixture. JetPEI is a linear polyethylenimine (PEI) derivative. These three transfection reagents can efficiently aid the intracellular delivery of siRNA or DNA to cytoplasm or nucleus for taking effect. In this study, QDs delivered siRNA showed much better RNA interference efficacy than commercial transfection reagents, which was demonstrated by cyclophilin B protein suppression. The results

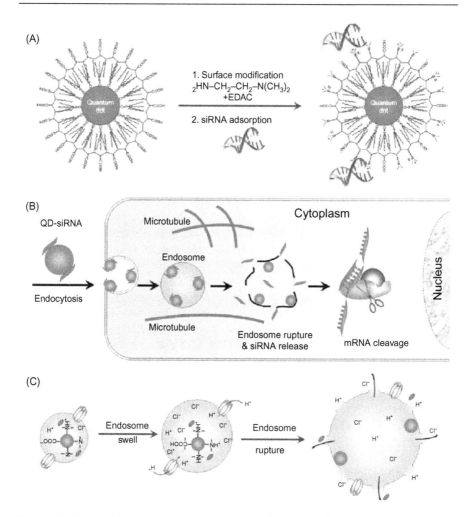

Figure 7.1 Rational design of proton-sponge coated QDs and their use as a multifunctional nanoscale carrier for siRNA delivery and intracellular imaging. (A) Chemical modification of polymer-encapsulated QDs to introduce tertiary amine groups and adsorption of siRNA on the particle surface by electrostatic interactions. (B) Schematic diagram showing the steps of siRNA-QD in membrane binding, cellular entry, endosomal escape, capturing by RNA binding proteins, loading to RNA-induced silencing complexes (RISCs), and target degradation. (C) Schematic illustration of the proton-sponge effect showing the involvement of the membrane protein ATPase (proton pump), osmotic pressure buildup, and organelle swelling and rupture. For optimized silencing efficiency and cellular toxicity, the QD surface layer is composed of 50% (molar) carboxylic acids and 50% tertiary amines. The optimal number of siRNA molecules per particle is approximately 2 (as shown in the diagram). *Source*: Reproduced with permission from Ref. [16]. Copyright 2008, American Chemical Society.

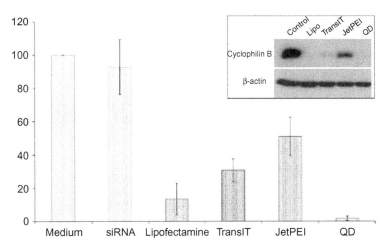

Figure 7.2 Comparison of gene silencing efficiencies between proton-sponge coated QDs and commercial transfection reagents by Western blotting. The level of cyclophilin B protein expression was reduced to 1.81 ± 1.47% by QDs, to 13.45 ± 9.48% by Lipofectamine, to 30.58 ± 6.90% by TransIT, and to 51.00 ± 11.46% by JetPEI (corresponding to efficiency improvements of 7.4-, 16.9-, and 28.2-fold, respectively). The quantitative values were obtained from the Western blot (inset). On the average, the proton-sponge coated QDs are 18 times as efficient as the three transfection agents commonly used.
Source: Reproduced with permission from Ref. [16]. Copyright 2008, American Chemical Society.

are shown in Figure 7.2. Naked siRNA had negligible gene silencing effect, which was indicated by protein expression close to 100%. When using JetPEI, TransIT, and Lipofectamine for siRNA delivery, the protein expression was reduced to 51.00 ± 11.46%, 30.58 ± 6.90%, and 13.45 ± 9.48%, respectively. Notably, when QDs were used, the protein expression was dramatically dropped to 1.81 ± 1.47%, which suggests efficiency improvements of 7.4-fold, 16.9-fold, and 28.2-fold when compared to Lipofectamine, TransIT, and JetPEI, respectively. These data clearly show that surface-modified QDs can be used for siRNA delivery to significantly improve its efficacy. The high gene silencing effect of the QDs delivered siRNA also indicates well-enhanced intracellular delivery and subsequent successful endosome escape of the QD-siRNA complex, because siRNA can only take effect after being released from endosome to cytoplasm [16].

Another advantage of using nanomaterials for drug delivery is that drugs can be specifically delivered to certain types of cells, such as cancer cells for improved efficacy and reduced side effect. Many biomarkers are expressed on the surface of cell membranes. To realize targeted drug delivery, one can conjugate certain types of biomolecules which have affinity with the biomarkers of specific kinds of cells. For example, folate receptors (FRs) are expressed at low levels in normal tissues but overexpressed in several types of tumors. Folic acid (FA) is found to have high affinity to FRs. Therefore, for tumor-targeted delivery, it is possible to attach FA

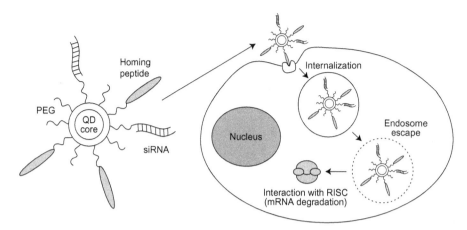

Figure 7.3 Design of an MFNP for siRNA delivery. Because of their photostable fluorescence and multivalency, QDs are suitable vehicles for carrying siRNA into live cells *in vitro* and *in vivo*. Conjugation of homing peptides (along with the siRNA cargo) to the QD surface allows targeted internalization in tumor cells. Once internalized, these particles must escape the endolysosomal pathway and reach the cytoplasm to interact with the RISC, which leads to degradation of mRNA homologous to the siRNA sequence.
Source: Reproduced with permission from Ref. [19]. Copyright 2007, American Chemical Society.

molecules to the surface of nanomaterials which can therefore preferably bind to tumor cells instead of normal ones. Cell surface necleolin (a tumor-homing peptide) and F3 peptides are other candidates for tumor-targeted delivery [18,19]. Many ligand—receptor interactions have been developed and reported. More examples can be found from one recently published excellent review [20].

Figure 7.3 shows the design of QDs surface modified with polyethylene glycol (PEG), which further covalently attaches to F3 peptide and siRNA [19]. PEG is one of most used polymers for steal effect: suppression of nonspecific interaction within body and resulting reduced blood clearance [21]. Steal effect can help nanoparticles avoid recognition by immune system, increase circulation time, and thus increase drug delivery efficacy. Moreover, PEG polymer can reduce nanoparticle's tendency of aggregation and increase their water solubility and stability [21]. F3 peptide is a targeting molecule which helps QDs-siRNA nanomedicine to be specifically delivered to tumor cells. The overwhelming advantage of targeted delivery can be demonstrated by the results shown in Figure 7.4 [19]. In Figure 7.4A, it can be seen that much greater (2 orders of magnitude higher) amount of QDs were internalized by HeLa cells in comparison with that of the QDs without attachment of targeting molecules (e.g., siRNA-QD and PEG-QD). As another control, QDs were attached with KAREC, which is a pentapeptide. The result showed that KAREC is not capable of increasing the cellular uptake of the QDs to tumor cells. To further confirm the targeting function of F3 peptide, free F3 and KAREC peptides (as a control) were

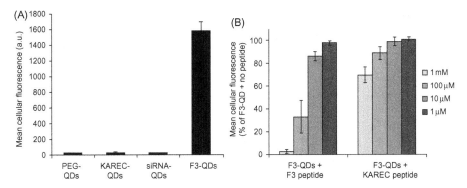

Figure 7.4 Attachment of F3 peptide leads to QD internalization in HeLa cells. (A) Thiolated peptides (F3 and KAREC control) and siRNA were conjugated to PEG-amino QD705 particles using sulfosuccinimidyl 4-(*N*-maleimidomethyl)cyclohexane-1-carboxylate (sulfo-SMCC). Particles were filtered to remove excess peptide or siRNA and incubated with HeLa cell monolayers for 4 h. (B) Flow cytometry indicated the F3 peptide is required for cell entry. The addition of free F3 peptide inhibits F3 QD uptake, while KAREC peptide does not, suggesting the F3 peptide and F3-labeled particles target the same receptor. *Source*: Reproduced with permission from Ref. [19]. Copyright 2007, American Chemical Society.

individually added to cell suspension containing F3-attached QD. The results showed that the addition of F3 peptide, particularly at high concentrations of 100 μM and 1 mM, significantly inhibited the QD internalization by cells, while the introduction of KAREC peptide did not have much effect on the cellular uptake of the QDs even at a high concentration of 1 mM (Figure 7.4B). The reason is that free F3 peptide is absorbed onto the cell membrane surface and significantly prevents the targeted delivery of the F3-peptide modified QDs. This indicates that the F3 peptide plays a very important role in QDs' targeted delivery.

There are many other researches of developing QDs as drug delivery systems. Yamomoto et al. [22] reported successful conjugation of captopril (antihypertensive drug) to QDs, which was found to effectively reduce rat blood pressure. Another work described using QDs to attach RNA aptamers and doxorubicin (DOX, chemotherapy drug for cancer treatment). It was demonstrated that targeting effect due to RNA aptamers could specifically bind to prostate-specific membrane antigen (PSMA) [23].

Besides being used for siRNA and drug delivery, QDs have also been widely used for delivering plasmid DNA [24].

7.2.2 Magnetic Nanoparticles

It has been of great interest in applying magnetic nanoparticles in magnetic resonance imaging (MRI), targeted drug/gene delivery, bioseparation [25−32], etc. Superparamagnetism is a unique property of magnetic nanoparticles.

Superparamagnetic property refers to the ability of magnetic nanoparticles to be magnetized under an applied magnetic field and such magnetization disappears once magnetic field is removed [33]. This property is able to prevent the agglomeration of magnetic nanoparticles after losing magnetic field, and thus prolong their circulation half-time [33]. With rapid development of nanotechnology, magnetic nanoparticle-based drug delivery systems have been emerging for many cancer therapies [28,29,34]. Magnetic nanoparticles conjugated with methotrexate (MTX, a chemotherapy anticancer drug) can increase the internalization by FRs overexpressed cancer cells, and therefore improve the drug efficacy in comparison with free MTX drug [35]. Engineered magnetic nanoparticles can also load cisplatin complexes (therapeutic drug that binds and interrupts DNA) and studies showed that the magnetic conjugates were more toxic to cancer cells than free cisplatin [36,37]. Porous hollow magnetic nanoparticles like Fe_3O_4 were also founded to be effective for targeted delivery. For instance, cisplatin was loaded into the hollow core of the magnetic nanoparticles which were bonded with targeting molecules (Herceptin) [38]. Moreover, magnetic nanoparticles coated with polyamidoamine (PAMAM) and PEI are promising for gene delivery as well [39,40]. Although effective, PEI molecules may have high toxicity due to their disruption to cell membranes [41,42]. To address this limitation, magnetic nanoparticles can be further attached with PEG to reduce the toxicity and increase the circulation time [43].

Next, one example is given in detail to explain how magnetic nanoparticles are applied in a drug delivery system to achieve desired properties. In the study, a self-protecting magnetic nanomedicine (SPMNM) was used for drug delivery. In the nanomedicine, highly magnetic nanocarrier (HMNC) with Fe_3O_4 core and poly [N-(1-one-butyric acid)]aniline (SPAnH) shell were used to load 1,3-bis(2-chloroethyl)-1-nitrosourea (BCNU) drug followed by an outmost layer of O-(2-aminoethyl)polyethylene glycol (EPEG) to protect BCNU drug (Figure 7.5) [44]. High magnetization Fe_3O_4 core was fabricated by conventional coprecipitation method. Briefly, NaOH was added into a solution containing $FeCl_3$ and $FeCl_2$ under nitrogen atmosphere. SPAnH shell was formed by adding an acid (HCl solution) to the Fe_3O_4 core and poly[aniline-co-sodium N-(1-one-butyric acid)aniline] (SPAnNa). In the acid environment, SPAnNa formed SPAnH to encapsulate the Fe_3O_4 core. The prepared Fe_3O_4 (core)−SPAnH (shell) was then attached with BCNU

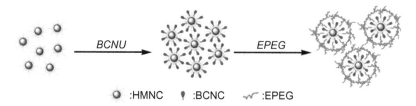

Figure 7.5 Schematic illustration of BCNU drug and EPEG protecting layer attaching onto HMNC surface.
Source: Reproduced with permission from Ref. [44]. Copyright 2011, Elsevier.

(a commercially available chemotherapeutic agent for malignant brain tumor treatment) [45] and an EPEG layer. The EPEG layer acts as a protect layer to increase the stability of BCNU drug [21]. In addition, with the EPEG coating, the magnetic nanoparticles exhibited greatly reduced removal rate by macrophages in the reticuloendothelial system [21]. This indicates that the macrophages were not activated by the magnetic nanoparticles in presence of EPEG molecules. Thus, circulation time and therapeutic effects were significantly increased.

External magnetic field can lead magnetic nanoparticle to accumulate to a certain area of body. This can be used to increase local drug density and reduce drug side effect. The effect of a magnetic field on intracellular delivery of the above prepared nanomedicine was investigated and the results are shown in Figure 7.6 [44].

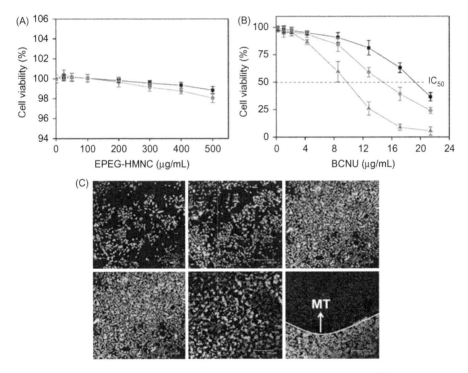

Figure 7.6 (A) Viability of U87 cells after incubation with EPEG—HMNC (■), EPEG—HMNC and exposure to a magnetic field of 800G (●). (B) Cytotoxicity of free BCNU (■), SPMNM (●), and SPMNM subjected to an 800-G magnetic field (▲) against U87 cells after 36 h incubation. Values are the mean ± SD ($n = 8$). (C) (Top) Fluorescence micrographs of U87 cells exposed to EPEG—HMNC for 2 days (left), 3 days (middle), and 4 days (right). (Bottom) Fluorescence micrographs of U87 cells treated with EPEG—HMNC (left), SPMNM (middle), and SPMNM subjected to an 800-G external magnetic field above the white line for 8 h (right). Live and dead cells display green and red fluorescence, respectively. (For interpretation of the references to color in this figure legend, the reader is referred to the web version of this book.)
Source: Reproduced with permission from Ref. [44]. Copyright 2011, Elsevier.

Figure 7.6A shows that magnetic nanoparticles have negligible cytotoxicity. Figure 7.6B shows that, with the aid of magnetic field, BCNU-coated magnetic nanoparticles perform much better efficacy in comparison with those without magnetic field. This is indicated by the reduced concentration of IC_{50} for the nanoparticles working under a magnetic field. The top three figures in Figure 7.6C show the fluorescence images of U87 cells at 2 (left), 3 (middle), and 4 (right) days after exposure to EPEG−HMNC. These images indicate that almost all cells were alive and kept growing. The bottom three images in Figure 7.6C are the fluorescence micrographs of U87 cells treated with EPEG−HMNC (left), SPMNM without magnetic field (middle), and SPMNM with an 800-G external magnetic field above the white line (right) for 8 h. These images indicate that: (a) the vast majority of cells were still alive if treated with EPEG−HMNC (magnetic nanocarrier without anticancer drug); (b) a small number of cells were killed (shown in red staining) if treated with SPMNM but without magnetic field; and (c) however, with the treatment by magnetic nanomaterials (SPMNM), if an 800-G external magnetic field was applied to the system, nearly all of the cells which were in the magnetic field were killed (shown in red staining in the area above the white line) while other area of cells free of magnetic field remain alive. This indicates that a target delivery was achieved by applying an external magnetic field.

To further demonstrate that such magnetic field induced targeted delivery may be used in clinical in the future, an *in vivo* mouse model study was carried out and the results are shown in Figure 7.7 [44]. Figure 7.7A shows that, with a magnetic field, the magnetic nanoparticles were guided to localize and concentrate to tumor tissue sites. In line with this targeted delivery, the therapeutic efficacy should be enhanced. To confirm this, experiments were performed and the results are

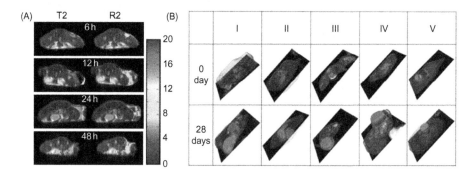

Figure 7.7 (A) *In vivo* imaging of MNM distribution in hypodermic tumors with magnetic targeting (MT) (left, T2-weighted images; right, combined R2 maps and T2-weighted images). (B) Magnetic resonance images and corresponding 3D reconstructions of rat brains with tumors induced by intracranial injection of U87 cells. Animals were treated intravenously with: I, control; II, free BCNU; III, EPEG−HMNC with MT; IV, SPMNM alone; and V, SPMNM with MT. The images were obtained on the day of treatment (Day 0) and 4 weeks later (Day 28).
Source: Reproduced with permission from Ref. [44]. Copyright 2011, Elsevier.

presented in Figure 7.7B. Figure 7.7BI indicates the negative control in the study. Without treatment, the tumor size dramatically increased after 28 days. Figure 7.7BII shows the tumor growth with treatment by free anticancer drug. The initial tumor size was very small, but it grew to a large size. Figure 7.7BIII is another negative control. Although magnetic nanoparticles were used, drug was not contained, so the tumor growth was not inhibited. The main comparison was between Figure 7.7BIV and BV. Same batch of magnetic nanoparticles carrying anticancer drugs was used in the experiments. The difference was that magnetic field was not used in the experiment for Figure 7.7BIV. Obviously, with the aid of magnetic field, magnetic nanomedicine was capable of achieving significantly better antitumor efficacy.

7.2.3 Layered Double Hydroxides

Layered double hydroxides (LDHs) are a class of layered materials with a general formula of $[(M^{II})_{1-x}(M^{III})_x(OH)_2]^{x+}(A^{m-}_{x/m}) \cdot nH_2O]$ and a structure as shown in Figure 7.8 [46]. The layers contain divalent (M^{II}) and trivalent (M^{III}) metal ions and the interlayer region is occupied with charge-balancing anions. Divalent metal ions are often Ca^{2+}, Mg^{2+}, Fe^{2+}, Co^{2+}, Mn^{2+}, Ni^{2+}, Cu^{2+}, or Zn^{2+}. Trivalent metal ions are commonly Al^{3+}, Fe^{3+}, Co^{3+}, and Ni^{3+}. The anions, such as CO_3^{2-}, NO_3^-, and Cl^-, in the interlayer galleries can be readily replaced [47]. Therefore, drugs may be conveniently loaded to LDHs simply by ion exchange [48,49]. By loading drugs in the interlayer regions, LDH is capable of protecting the drug molecules. LDHs have very large specific surface area and therefore drug loading quantity is high. LDHs at nanoscale are readily dispersible in aqueous media. Furthermore, LDHs are positively charged while cell membrane is negatively charged, which is beneficial for enhanced cellular uptake. These advantages have been demonstrated in extensive studies of using LDH nanoparticles for delivery of a range of drug molecules [50−53].

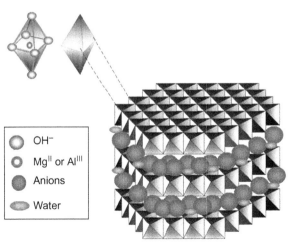

Figure 7.8 Structure of a hydrotalcite-like compound. *Source*: Reproduced with permission from Ref. [46]. Copyright 2011, Elsevier.

- ⬤ OH⁻
- ⬤ Mg^{II} or Al^{III}
- ⬤ Anions
- ⬤ Water

In one study, $Mg_{0.68}Al_{0.32}(OH)_2(NO_3^-)_{0.32} \cdot 1.2H_2O$ (Mg_2Al-NO_3^-LDH) was firstly prepared by a coprecipitation of mixed metal ions (Mg^{2+}:Al^{3+} = 2:1) in a base solution (0.1 N NaOH, pH $\approx$ 10) under nitrogen [51]. Different materials were then loaded to the interlayer regions of the LDH nanoparticles. Before loading of these materials, the basal spacing of the LDH nanoparticles was 8.7 Å. After being loaded with herring testis DNA, adenosine triphosphate (ATP), fluorescein 5-isothiocyanate (FITC), and c-*myc* antisense oligonucleotide (As-*myc*) through simple ion-exchange reaction, the spacing was enlarged to 23.9, 19.4, 18.8, and 17.1 Å, respectively, calculated from X-ray diffraction results. FITC-loaded LDH nanoparticles were used to perform cellular uptake study and the results revealed that the uptake of the LDH−FITC hybrids was well enhanced with increased concentration and incubation time. Confocal microscopy observation showed that the fluorophore was detected in cells within 1 h of incubation and the signal continuously increased for over 8 h. The fluorophores in the cells were primarily distributed in the peripheral and cytosol regions with some in nucleus. This was in line with the TEM observation in a separate report [54]. For the cells treated with LDH−FITC hybrids at 1 μM concentration, fluorescence could be observed in cells. In comparison, for the cells treated with 5 μM FITC alone, no fluorescence signal was observed inside cells. It is evident that LDH helped the intracellular uptake of FITC molecules. In another study, [γ-^{32}P]-labeled ATP was employed to conveniently compare the cellular uptake efficiency of ATP molecules alone and LDH−ATP hybrids. The experimental data clearly verified that greatly enhanced transfer efficiency could be obtained for LDH−ATP in comparison with ATP alone [55]. Following the confirmation that LDH's function in aiding intracellular delivery, LDH nanoparticles and LDH−As-*myc* hybrids were incubated with human promyelocytic leukemia (HL-60) cells. At a concentration below 1 mg/mL and incubation time for up to 4 days, LDH nanoparticles did not show noticeable cytotoxicity. In contrast, at a concentration of the hybrids of 20 μM and incubation for 4 days, 65% growth inhibition of the HL-60 cells was observed when compared to untreated cells [51,56].

The use of LDH nanoparticles for improving DNA vaccine delivery in the treatment of melanoma in mice was also recently studied. The results demonstrated that with LDH nanoparticles, the DNA vaccine could induce well-enhanced serum antibody response compared with naked DNA [57].

With the success of using LDH nanoparticles for DNA delivery, the application was extended to anticancer drug delivery. MTX is widely used for human cancer treatment. However, MTX has a very short plasma half-life and therefore it is often used at a high dose, which may lead to drug resistance and nonspecific toxicities in normal cells. Researchers have carried out studies attempting to use LDH nanoparticles to solve this problem. MTX was hybridized with LDH nanoparticles by ion-exchange reaction. Before MTX loading, the basal spacing was 8.4 Å. After loading, the spacing dramatically expanded to 19.9 Å. This indicates that MTX entered the interlayers of the LDH nanoparticles, which was further confirmed by comparing the FTIR spectra of the LDH−MTX hybrids, LDH nanoparticles, and MTX. The samples were then tested in one normal cell line (human tendon fibroblast) and another cancer cell line (human osteosarcoma SaOS-2). The results showed that LDH−MTX hybrids, MTX, and LDH nanoparticles did not have significant effect on the proliferation of normal cells. However, LDH−MTX hybrids and MTX suppressed the proliferation of tumor cells while LDH nanoparticles still did not have effect. When comparing the drug efficiency of different samples at various periods of times, it can be concluded that, during the first 3 days, LDH−MTX worked much more efficiently than MTX. The mechanism was investigated by extracting genomic DNA from the cancer cells after being treated by MTX with a concentration of 500 μg/mL and carrying out electrophoresis analysis of the DNA. In the cancer cell group treated by LDH−MTX hybrids, internucleosomal cleavage by DNA apoptosis was observed while no noticeable DNA apoptosis was detected in the groups treated by individual LDH and MTX. Overall, the study suggested that LDH−MTX hybrids damaged the DNA of cancer cells more efficiently than MTX and correspondingly had well-enhanced drug efficacy [58]. The reason may be as follows. First, LDH protected MTX from rapid decomposition in blood flow. Second, LDH increased the cellular uptake of MTX by cancer cells [58]. LDH nanoparticles were also used for siRNA delivery. It was demonstrated that LDH-mediated siRNA delivery effectively silenced neuronal gene expression [59].

Except aiding in intracellular delivery of a range of molecules, LDHs were also used to reduce harm of drugs to patients. Indomethacin is an anti-inflammatory drug with very bad gastrointestinal side effects. It can cause peptic ulcers, which may result in serious bleeding and perforation. To eliminate or at least reduce the side effects, the drug was incorporated with LDHs followed by *in vivo* study in mouse model. The result showed that the LDH−indomethacin hybrids caused much less gastric damage compared to oral intake of the drug alone [60].

Furthermore, controlled and sustained release could be realized by using LDHs. In one study, various drugs including antibiotics tetracycline, DOX, 5-fluorouracil, vancomycin (VAN), sodium fusidate (SF) and antisense oligonucleotides were bound to LDH clay particles and then dispersed in poly(lactic-co-glycolic acid) films. The control groups were drugs bound to only LDH clay particles. Without poly(lactic-co-glycolic acid), the drugs showed a large burst phase of release (except DOX) with little extended drug release for 4 days. With the polymer films, the drug systems had a reduced burst phase of release followed by a slow continuous release for over 12 days [61]. Very recently, another study in rabbit model was reported in which LDH nanoparticles were impregnated with antibiotic ciprofloxacin and coated on Bioverit® II implants to act as a sustained drug delivery (release) system against an infection with *Pseudomonas aeruginosa* in the middle ear [62]. The results indicated that LDH nanoparticles may be very useful for controlled release applications because they are capable of protecting drugs from being degraded and releasing effective drug doses over a longtime periods.

7.2.4 Mesoporous Silica Nanomaterials

In recent years, silica nanomaterials with tunable mesoporous structures have attracted great attention as potential drug delivery vectors, because mesoporous silica has many advantages, such as facile fabrication and surface modification, tunable pore volume, high capability of drug loading and low cost [63−67]. Mesoporous silica pore diameter can be readily controlled, leading to different release profiles, loading capacities, and drug categories: Vallet-Regi et al. [68] observed that bovine serum albumin loading was improved from 15 to 27% if pore size was increased from 8.2 to 11.4 nm; Horcajada et al. [69] and Izquierdo-Barba et al. [70] reported that a decrease of pore size of mesoporous silica led to a drop in releasing rate of ibuprofen and erythromycin. The structure and shape of mesoporous silica nanomaterials can be easily tuned to attach different types of guest molecules, such as proteins, pharmaceutical drugs, gene (DNA and siRNA) for drug delivery [71−74]. Generally, drug molecules are attached to mesoporous silica by weak noncovalent interaction, such as hydrogen bonding, physical adsorption, and electrostatic interaction [75]. Surface-modified mesoporous silica, which provides versatile functional groups, is very attractive for enhancing drug loading or adding more biomolecules [76−78]. For example, without surface modification, ibuprofen can only be loaded by hydrogen bond between silanol group (in mesoporous silica) and carboxylic acid group (ibuprofen); while mesoporous silica surface modified with amine group provides stronger electrostatic interaction, which can increase the loading capacity and slower release profile [76].

Mesoporous silica nanoparticles (MSNPs) can be fabricated having hollow or tattle structures, which provide larger empty space to upload guest molecules and increase drug loading capacity. For example, mesoporous silica can be fabricated to form much larger pore size (20 nm) for large molecule (DNA) delivery [79,80]. Figure 7.9 shows typical TEM images of silica nanoparticles with hollow or tattle structures [81]. Shi and coworkers explored that hollow mesoporous structure had twice loading capacity of ibuprofen (744.5 mg/g) compared with conventional mesoporous silica (358.6 mg/g) [82]. Another research reported that silica nanoparticles were loaded with as high as 18.2% of water-soluble DOX [83]. This delivery system showed a sustained pH-sensitive release profile: for the first hour, 13 and 8% of drug released at pH 4 and pH 7.4, respectively; by 88 h, 50 and 11% of drug released at pH 4 and pH 7.4. In this system, drug release was lasted for up to 3 days and faster drug release was observed at low pH (pH = 4, close to lysosomes pH) [83]. This property could favor nanoparticle drug release to cytoplasm in cancer cell by endocytosis. In the following, some typical studies are selected to describe how mesoporous silica can be designed to be a smart material for drug delivery.

Conventional chemotherapeutic drugs are widely distributed in human blood and tissues, which affect both healthy and cancer cells, leading to serious side effect [84]. To increase drug delivery efficacy and reduce side effect, it is attractive and desirable for drug delivery systems to release drug predominantly at disease foci. For this purpose, mesoporous silica nanomaterials can be carefully designed and modified to be a stimuli-responsive system for controlled release of drugs [63,85−87]. In one typical research work, MSNPs functionalized by 2-(propyldisulfanyl)ethylamine were loaded with various drugs in its interior pore structures and capped by covalently bonded mercaptoacetic acid-derivated cadmium sulfide (CdS) QDs by amidation reaction (Figure 7.10A) [88,89]. In brief, mesoporous

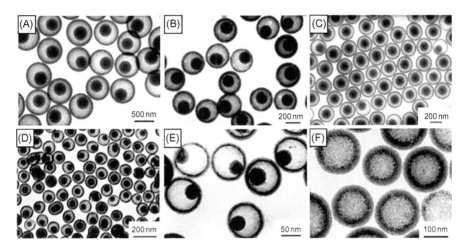

Figure 7.9 Silica nanorattle with mesoporous and hollow structure synthesized by selective etching strategy. It was used as nanoreactor for synthesis of gold nanoparticle with controllable size in the hollow interior. TEM images of silica nanorattles with different sizes and core/shell ratios: (A) 645 nm (core 330 nm, shell 50 nm), (B) 330 nm (core 160 nm, shell 40 nm), (C) 260 nm (core 140 nm, shell 20 nm), (D) 110 nm (core 53 nm, shell 13 nm), (E) 95 nm (core 46 nm, shell 15 nm), and (F) hollow silica spheres of 170 nm diameter (shell 25 nm).
Source: Reproduced with permission from Ref. [81]. Copyright 2009, Wiley-VCH Verlag GmbH & Co. KGaA, Weinheim.

silica was modified with 2-(propyldi-sulfanyl)ethylamine on the surface, which provided amine ending. Subsequently, drugs were loaded to the mesopores of the surface functionalized mesoporous silica. Then, carboxyl-stabilized QDs were added to the solution. In the presence of catalyst 1-(3-(dimethylamino)propyl)-3-ethylcarbodiimide hydrochloride (EDC), carboxyl group of the QDs bonded with the amine group on mesoporous silica to form covalent bonds, therefore capping the QDs to the mesoporous silica nanomaterials. From Figure 7.10A, it can be seen that disulfide bonds are in the linkage between the QDs and the MSNPs. These bonds are chemically labile and can easily detach in the presence of a reduction agent like dithiothreitol (DTT) for drug release. Figure 7.10B shows that the drug release was <1.0% in 10 mM Phosphate buffered saline (PBS) solution (pH 7.4) for over a period of 12 h, but when DTT (disulfide-reducing molecules) was added, the drug release was triggered and about 85% of the loaded drug released within 24 h. The results indicate that the drugs were initially successfully encapsulated within the MSNPs by covalent bonded CdS nanoparticles and the release could be triggered by the addition of disulfide-reducing molecules. The release was found to be dependent on the concentration of disulfide-reducing molecules, which demonstrates that the drug release rate was controlled by the detachment rate of the QDs.

To enhance drug efficacy, mesoporous nanoparticles can also be designed for targeted drug delivery [90−94]. In a targeting system, surface-attached targeting

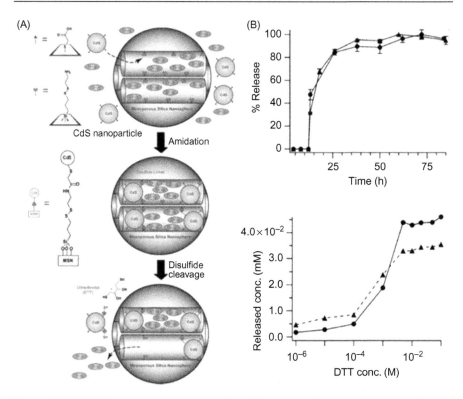

Figure 7.10 (A) Schematic representation of the CdS nanoparticle-capped MSN-based drug/ neurotransmitter delivery system. The controlled release mechanism of the system is based on chemical reduction of the disulfide linkage between the CdS caps and the MSN hosts. (B) The DTT-induced release profiles of VAN (●) and ATP (▲) from the CdS-capped MSN system.
Source: Reproduced with permission from Ref. [88]. Copyright 2003, American Chemical Society.

agents can help mesoporous nanoparticles selectively approach to overexpressed antigens or receptors on the surface of cancer cells [95,96]. Owing to ease of surface modification of mesoporous silica, a dual stimuli-responsive and targeting function can be simultaneously incorporated into MSNPs drug delivery systems [97]. To achieve this, disulfide bonds as described above were firstly induced onto the surface of MSNPs (Figure 7.11A-2): first, mesoporous silica was functionalized with sulfhydryl group by refluxing with 3-mercaptopropyltrimethoxysilane followed by reacting with 2-carboxyethyl-2-pyridyl disulfide. After loading a typical anticancer drug DOX into the mesopores inside silica nanoparticles, per-diamino-β-cyclodextrin (β-CD(NH$_2$)$_7$) was attached to block the mesopores and inhibit the release of the loaded drug by forming a complex structure which is denoted as MSNPs-S-S-CD in Figure 7.11A-3. Subsequently, two types of PEG including mPEG-Ad (methoxyl PEG polymer chain functionalized with adamanyane (Ad) at

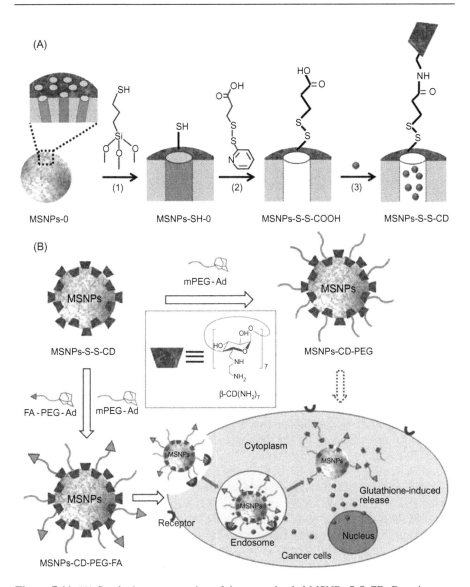

Figure 7.11 (A) Synthetic representation of the cargo-loaded MSNPs-S-S-CD. Reaction conditions: (1) 3-mercaptopropyltrimethoxysilane (MPTMS) in toluene; (2) 2-carboxyethyl-2-pyridyl disulfide in ethanol, followed by the removal of the surfactant cetyltrimethylammonium bromide (CTAB); (3) loading of cargo molecules, followed by additions of β-CD(NH$_2$)$_7$ and 1-(3-(dimethylamino)propyl)-3-ethylcarbodiimide hydrochloride (EDC). MSNPs-0 and MSNPs-SH-0 mean that the mesopores are occupied with the CTAB template. (B) Schematic illustration of multifunctional MSNPs-CD−PEG−FA for targeted and controlled drug delivery.
Source: Reproduced with permission from Ref. [97]. Copyright 2012, Wiley-VCH Verlag GmbH & Co. KGaA, Weinheim.

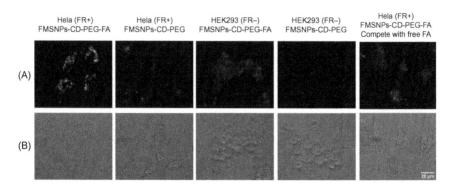

Figure 7.12 (A) Fluorescence and (B) bright-field images of HeLa cells and 293 cells after 3 h incubation with either FMSNPs-CD−PEG−FA or FMSNPs-CD−PEG as indicated. After trypan blue was used to quench the extracellular fluorescence, the cells were monitored live by fluorescence microscopy.
Source: Reproduced with permission from Ref. [97]. Copyright 2012, Wiley-VCH Verlag GmbH & Co. KGaA, Weinheim.

one end) and FA-PEG-Ad (methoxyl PEG polymer chain functionalized with FA at one end and with Ad at the other end) were attached onto MSNPs (Figure 7.11B). The Ad-ending in the PEG chain performs strong Ad/CD complexation in aqueous solution [98,99], resulting in PEG binding to MSNPs-CD. MSNPs surface modified with β-CD(NH$_2$)$_7$ and FA-mPEG-Ad (MSNPs-CD−PEG−FA) can enter cancer cells by receptor-mediated endocytosis (Figure 7.11B, bottom) because FA ending can be recognized by overexpressed FR on the cancer cell's surface. After entering cells, the nanoparticles are trapped in endosome in which the pH is about 5−6. Then the nanoparticles can escape from endosome and the drug is released in cytoplasm. To confirm the intracellular delivery ability of MSNPs-CD−PEG−FA, fluorescent microscopy images of HeLa cells cocultured with fluorescent MSNPs-CD−PEG−FA (FMSNPs-CD−PEG−FA) and fluorescent MSNPs-CD−PEG (FMSNPs-CD−PEG) are shown in Figure 7.12. With attachment of fluorescein, both FMSNPs-CD−PEG−FA and FMSNPs-CD−PEG can emit green fluorescence. Thus, the nanoparticles can be conveniently detected by fluorescent signals. From Figure 7.12A, FMSNPs-CD−PEG−FA treated HeLa cells (FR + : FR overexpressed cell) show very strong florescent signal in comparison with those treated by FMSNPs-CD−PEG, indicating FA can help nanoparticles effectively deliver to HeLa cells. However, when free FA and MSNPs-CD−PEG−FA were coincubated with HeLa cells, only very weak signal was detected [97]. This lowered intracellular delivery was because free FA also bound to the FR on cell membrane surface, which reduced the chance of MSNPs-CD−PEG−FA binding. This further indicates that the FA can aid the MSNPs to target HeLa cells (FR positive cells) and increase their intracellular delivery.

Besides the targeted delivery characteristic, these MSNPs-CD−PEG−FA have two stimuli-responsive attributes for controlled release [97]. One is responsive to reduction reagents which can cause cleavage of disulfide bond (Figure 7.13A). The

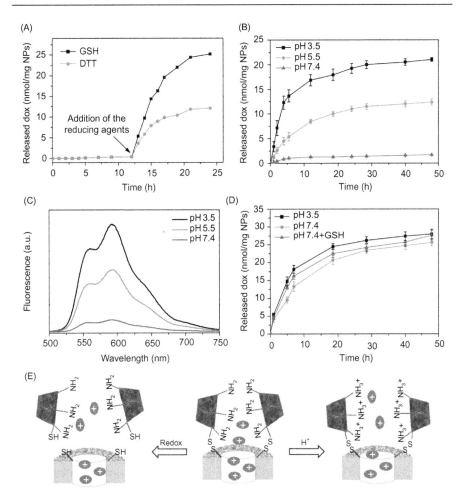

Figure 7.13 (A) Release profiles of DOX-loaded MSNPs-S-S-CD in PBS (pH 7.4) triggered by the addition of DTT (10 mM) or GSH (10 mM). (B) Release profiles of DOX-loaded MSNPs-S-S-CD at 37°C under different pH conditions. (C) Fluorescence spectra of DOX (λ_{ex} = 480 nm) after 40 h incubation of DOX-loaded MSNPs-S-S-CD at different pH values. (D) Release profiles of DOX-loaded MSNPs-S-S-COOH at 37°C under different pH conditions. The DOX release was not affected by the addition of GSH (10 mM). (E) Dual release mechanism of DOX-loaded MSNPs-S-S-CD. Drug release of DOX-loaded MSNPs-S-S-CD can be triggered by acidic pH or the addition of a reduction agent, such as DTT or GSH.

Source: Reproduced with permission from Ref. [97]. Copyright 2012, Wiley-VCH Verlag GmbH & Co. KGaA, Weinheim.

other factor is pH value change (Figure 7.13B). Two reduction reagents (DTT and glutathione (GSH)) were used as triggers for the cleavage of disulfide bonds, and correspondingly the block CD was detached to stimulate DOX anticancer drug release (Figure 7.13E, left). From Figure 7.13A, it can be seen that rapid release of DOX was observed after adding reduction agents to cut the disulfide bonds. For pH change triggered release, the mechanism can be described as below. Remaining amine groups at β-CD(NH$_2$)$_7$ can accept proton in acid environment to become positive charged and therefore generate electrostatic repulsion between neighbor β-CD (NH$_2$)$_7$, leading to opening of drug release channels, as shown in Figure 7.13E (right). The release profiles in Figure 7.13B show that significantly enhanced release of DOX was realized under acid environments, compared with that at neutral pH 7.4. In great contrast, for the MSNPs without β-CD(NH$_2$)$_7$ on the surface, the release kinetics at different pHs were very similar (Figure 7.13D), revealing the role of β-CD(NH$_2$)$_7$ in the pH triggered stimuli system. The two stimuli-responsive systems were carefully designed for controlled release and could be very useful in biomedical applications. For example, with the injection of reduction regents (DTT and GSH) in certain disease areas like tumor tissues, the drug can be controlled to release in the tumor regions by cleavage of disulfide bonds, while the drug remains within the MSNPs-CD$-$PEG$-$FA nanoparticles in other parts of the body. In addition, tumor tissues and endosomes have lower pH value than normal tissue, which can trigger pH stimuli-responsive system to release the drug.

7.2.5 Organic Nanoparticles

Organic nanoparticles have recently attracted growing interest in drug delivery [100−105]. They include natural organic nanoparticles (e.g., protein aggregates, lipid droplets, milk emulsions, and viruses) and synthetic nanoparticles, such as micelles [106−108], liposomes (LPs) [109,110], polymersomes [111−114], and polyplexes [115−117]. Organic nanoparticles have one important advantage over inorganic ones. It is that many organic nanoparticles are biodegradable, which can overcome the risk of chronic toxicity of the nonbiodegradable ones to cells or tissues [118,119]. In addition, natural materials like phospholipids can be employed to synthesize nanoparticles which have improved biocompatibility and ability to penetrate cell membrane [120,121].

Since many examples about drug delivery systems have been described earlier, in the following, study cases about using nanomaterials for immunization are discussed. Organic nanoparticles are commonly used for enhancing the immunogenicity of vaccines. In one report, vaccines were covalently conjugated carboxylated polystyrene microspheres ranging in size from 0.02 to 2 μm (0.02, 0.04, 0.1, 0.2, 0.5, 1, and 2 μm) and then a series of immunological study was carried out in mouse model [122]. First, ovalbumin (OVA) protein was tested: mice were immunized intradermally once or twice with 100 μg of OVA protein. Spleen T-cell responses to OVA protein or SIINFEKL were measured at 10 days postimmunization. The results clearly indicated that the 0.04 μm bead and OVA protein complex induced the highest immune response to OVA and to its MHC class I epitope

SIINFEKL no matter whether one or two immunizations were administered to mice. Except inducing the highest T-cell responses, 0.04 μm bead and OVA protein complex also led to the strongest antibody response. The author also compared the T-cell responses promoted by OVA protein, 0.04 μm bead/OVA protein covalent conjugation system, and 0.04 μm bead/OVA protein physical mixture. The results showed: (a) OVA protein generated extremely weak immune response; (b) by simply mixing with the 0.04 μm polymer beads, the T-cell response was greatly enhanced; and (c) by covalent bonding the vaccine to the 0.04 μm polymer beads, the immune response was further well improved. Collectively, these experimental outcomes demonstrated that conjugating vaccine to nanoparticles was capable to dramatically enhance immune responses and the size of the nanoparticles and conjugation nature had an important effect on the intensity of immune responses.

The fate of the vaccine-conjugated beads was also studied [122]. Fluorescent beads were used for the convenience of tracking. It was found that, at both 48 h and 14 days after intradermal injection, the 0.04 μm beads were found in more cells than the beads of other sizes. This finding suggested that 0.04 μm beads could effectively carry vaccine to the dendritic cells in the draining lymph nodes, which is very important for improving the immunogenicity of vaccines. The adjuvant effect of the beads was compared to that of a number of commonly used adjuvants. The results showed that the 0.04 μm beads could induce significantly stronger immune response than those promoted by Quil-A, MPL, DC-OVA, alum, and DC-SIINFEKL. Finally, the study further demonstrated that the 0.04 μm nanobeads were not only able to effectively protect mice against tumor growth for a certain period of time but also capable of treating established tumors. As one example, the results of the treatment of tumors are described here: two groups of mice were injected with EG7 tumor cells on day 0. On day 8, one group of mice was left untreated and the other group was injected with 100 μg of OVA/0.04 μm beads. The tumor growth was monitored. In the untreated group, all animals had substantial tumors on day 19. In great contrast, all animals in the treated group were healthy and had no tumors on days 19 and 21. This showed that a single dose of nanobeads/OVA-conjugated complex had the potential to eliminate tumors [122].

For immunization, vaccines are generally delivered to peripheral tissues, such as skin and muscle. However, these tissues contain low density of dendritic cells and these dendritic cells need to travel to lymph nodes after antigen uptake for inducing immune responses. In contrast, lymph nodes themselves contain much higher density of dendritic cells. Therefore, if nanovaccines can be directly and conveniently transported to lymph nodes, greatly improved immune responses would be produced. To achieve this, in a study, pluronic-stabilized polypropylene sulfide (PPS) nanoparticles with two different sizes, 25 and 100 nm were injected into mice intradermally. It was found that the transportation of the 25 nm nanoparticles to lymphatic capillaries and the draining lymph nodes was much more efficient than that of the 100 nm ones (with efficiency dropping to 10%) (Figure 7.14) [123].

Other than protein vaccines, the immunogenicity of DNA vaccines can also be dramatically improved. DNA vaccines are considered to be the third generation of vaccines, because they are easy to develop and produce as well as safe to use.

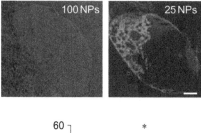

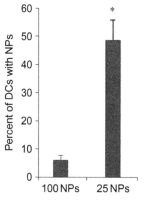

Figure 7.14 Ultrasmall nanoparticles accumulate in lymph nodes after intradermal injection, whereas slightly larger ones do not. The 25-nm, but not the 100-nm, nanoparticles are visible in mouse lymph node sections 24 h after injection. Cell nuclei shown in blue 4′,6-Diamidino-2-phenylindole dihydrochloride (DAPI); scale bar, 200 mm. Calculations of the fraction of dendritic cells that have internalized NPs after a coinjection supports the hypothesis that lymph node dendritic cells are effectively targeted after intradermal injection of smaller but not larger nanoparticles. (For interpretation of the references to color in this figure legend, the reader is referred to the web version of this book.)
Source: Reproduced with permission from Ref. [123]. Copyright 2007, Nature Publishing Group.

Despite of the promising outlook, their use has been seriously limited by the low immunogenicity. Therefore, it is essential to develop suitable vectors to deliver DNA vaccines for greatly enhanced immunogenicity. Extensive researches have been done to use cationic polymer PEI to form complex with DNA to significantly increase DNA's cellular uptake and transfection efficiency. The cytotoxicity and transfection efficiency of PEI/DNA complexes were found to be heavily dependent on the structure, size, and the molar ratio of the amine in cationic polymers to phosphate in DNA (N/P ratio). The cytotoxicity can be reduced and the transfection efficiency can be improved by cross-linking PEI with PEG [124]. Using this technique for DNA vaccines, one recent study showed that a family of PEI polymers, when forming a complex with plasmid DNA and being delivered to mice intravenously, were able to increase *in vivo* DNA expression 20−400 fold and enhance DNA induced epitope-specific CD8$^+$ T-cell response 10−25 fold in comparison with those of naked DNA [125]. Many other types of cationic polymers have also been used to form complexes with DNA for enhanced transfection efficiency. Except being applied to improve immunogenicity of vaccines, organic nanoparticles have also been widely used to promote delivery of siRNA and a broad range of other molecules [116,118].

7.2.6 Metal Nanomaterials

Many metal nanomaterials like gold nanostructures are able to convert optical energy to heat via nonradiative electron relaxation dynamics [126]. With this unique property, gold nanomaterials are broadly employed to be a noninvasive

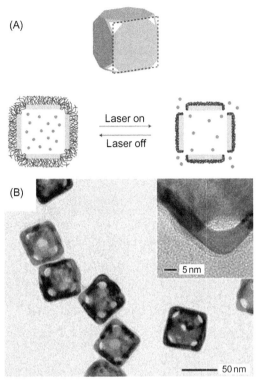

(A)

Laser on

Laser off

(B)

5 nm

50 nm

Figure 7.15 (A) Schematic illustrating how the system works. A side view of the Au nanocage is used for the illustration. On exposure to an NIR laser, the light is absorbed by the nanocage and converted into heat, triggering the smart polymer to collapse and thus release the preloaded effector. When the laser is turned off, the polymer chains will relax back to the extended conformation and terminate the release. (B) TEM images of Au nanocages for which the surface was covered by a pNIPAAm-*co*-pAAm copolymer with an LSCT at 39°C. The inset shows a magnified TEM image of the corner of such a nanocage.
Source: Reproduced with permission from Ref. [128]. Copyright 2009, Nature Publishing Group.

surgery to kill cancer cells and remove diseased tissues [127]. In addition to this, this photothermal effect can also be used to improve drug permeability and control drug release. In this section, we will focus on examples of using gold nanomaterials to enhance drug release controllability, targeted delivery, and drug permeability followed by brief summary of applying gold nanomaterials directly for drug delivery.

Gold nanocages have been demonstrated for controlled drug release. These nanocages are hollow gold cages with holes at the truncated corners (Figure 7.15) [128]. The surface of the nanocages is covered with a layer of a pNIPAAm-*co*-pAAm copolymer with a low critical solution temperature (LCST) of 39°C. Below 39°C, the polymer is soluble in water and the extended conformation blocks the preloaded drug in the nanocages to release. When the temperature is raised above 39°C, the polymer becomes hydrophobic due to a phase transition and collapses to release the preloaded drug (Figure 7.15A). In the research work, the release profile of a dye (alizarin-PEG) was firstly demonstrated. The dye was loaded to the nanocages by incubating the dye and the nanocages at 42°C for 12 h. Then the samples were cooled with an ice bath. To control the temperature change, the samples were exposed to an NIR laser to make use of the photothermal property of the gold nanocages to heat the polymer. When the laser power was fixed at 10 mW/cm², the majority of the dye molecule could be released within 16 min of laser illumination (Figure 7.16). Once the release of the dye molecule upon laser exposure was

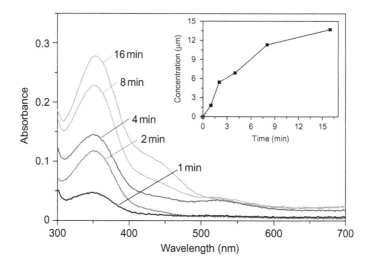

Figure 7.16 Absorption spectra of alizarin-PEG released from the copolymer-covered Au nanocages by heating at 42°C for 1, 3, 5, and 10 min by exposure to the NIR laser for 2 min at 10, 25, and 40 mW/cm^2.
Source: Reproduced with permission from Ref. [128]. Copyright 2009, Nature Publishing Group.

confirmed, the test was extended to the controlled release of DOX to kill cancer cells. The result showed that the DOX release profile was similar to that of the dye molecule and the release of DOX was capable of killing breast cancer cells.

Photo-induced heating of metal nanomaterials can also be used for controlled release of drugs from other types of containers like hollow polymer capsules. The walls of the capsules can be functionalized with metal nanoparticles. Once the capsules are exposed to light, metal nanoparticles can produce heat, which causes local ruptures of the capsule walls and release the inside drug molecule. This has been demonstrated on a single capsule level for controlled release of molecules from capsules inside living cells [129]. In the study, Alexa Fluor 488 (AF-488) dextran was encapsulated in a polyelectrolyte sodium poly(styrene sulfonate) (PSS)/poly (diallyldimethylammonium chloride) (PDADMAC) microcapsules containing gold and gold sulfide nanoparticles in the walls (Figure 7.17A). The fluorescence microscopy image of the microcapsule is shown in Figure 7.17A and B. The super-imposed fluorescence and TEM images of the cell and microcapsule are presented in Figure 7.17B. Figure 7.17B and C shows the fluorescence intensity profile across the line drawn in Figure 7.17A and B. It can be clearly seen that the capsule is filled with fluorescent dye before laser illumination. Figure 7.17B−F shows the same set of data of the same cell and microcapsule after laser exposure (830 nm with a power of 50 mW). The result indicates that the majority of the fluorescent molecules have been released from the capsules after laser exposure. This means that the release of the cargo molecules in the microcapsules can be remotely

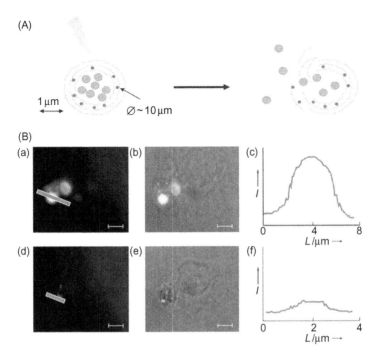

Figure 7.17 (A) Illustration of light-controlled opening of individual polymer capsules (drawn in grey) by local heating, mediated by Au nanoparticles. Gold nanoparticles are embedded in the walls of polyelectrolyte capsules. The capsule cavity is loaded with cargo molecules (drawn in orange). Upon illumination with light the heat created by the nanoparticles causes local ruptures in the capsule walls and thus release of the cargo. (B) Sequence of images showing the release of fluorescent AF-488 dextran inside a living MDA-MB-435S cell. (a) Fluorescence image of a filled capsule; (b) superimposed fluorescence and transmission images of the capsule inside the same cell; (c) fluorescence intensity I profile plotted along the length L of the red line in (a). (d−f) Similar data after exposure of the cell to a laser beam. The scale bars in all images correspond to 5 mm. (For interpretation of the references to color in this figure legend, the reader is referred to the web version of this book.)
(A) Reproduced with permission from Ref. [130]. Copyright 2008, Royal Society of Chemistry. (B) Reproduced with permission from Ref. [129]. Copyright 2006, Wiley-VCH Verlag GmbH & Co. KGaA, Weinheim.

controlled by laser illumination. In addition to gold nanostructures, silver nanoparticles can also be used to achieve a similar effect.

Other than using the photothermal effect of gold nanostructures for controlled release, gold nanomaterials have also been demonstrated to be capable of upregulating tumor cell receptor expression and increasing cell membrane permeability. For targeted delivery, the surfaces of nanomaterials are usually coated with ligands which can bind to the receptors of targeted cells. It can be therefore expected that upregulated receptor expression will lead to increased amount of nanomaterial delivery. In the meantime, increased cell membrane permeability can cause tumor cells to uptake more drugs. In this case, gold nanomaterials can cooperatively work together with many types of nanomedicine to enhance drug efficacy, which is illustrated in Figure 7.18A [131]. Figure 7.18B indicates that the tumor temperature increases upon extended exposure of the accumulated gold NRs to NIR laser (810 nm, ~0.75 W/cm^2). The PEG-coated gold NRs were passively accumulated to a MDA-MB-435 xenograft tumor as a result of leakiness of the tumor vasculature. Figure 7.18C and D shows the temperature-induced amplification of *in vivo* targeting of magnetic nanoworms (NWs) and LPs. In the figures, LyP1 indicates that, in the group, the nanomaterials were conjugated with LyP1 (a tumor targeting peptide which binds to the p32 receptors in the MDA-MB-435 tumor). Experiment showed that the p32 receptor expression can be upregulated by heating. From the results, it can be seen that, with the aid of gold NRs and laser exposure, the targeted delivery of LyP1-conjugated nanomaterials to tumor, indicated by the intensity of fluorescence for magnetic NWs or the amount of DOX accumulation in tumor for LPs, was dramatically enhanced. Furthermore, *in vivo* tumor treatment results demonstrated that therapeutic nanoparticles were much more efficient in treating the tumors whose local microenvironment had been engineered by gold NR heating under NIR illumination. Figure 7.18E shows that when combining with gold NR-medicated local heating, a single injection of LPs containing only 3 mg of DOX/kg is enough to achieve significant tumor regression or even elimination. This is much better than other reported targeted therapies with multiple high doses. Also important, all of the mice experienced anticancer treatment did not have significant body mass loss. This research indicated that cooperative use of dual or multiple nanomaterials has the potential to dramatically increase drug efficacy, mitigate side effects, and effectively eradicate drug-resistant diseases [131].

Besides playing a subsidiary role in drug delivery, as biocompatible and bioinert materials, it has also been very common to directly use gold nanomaterials for drug delivery. The administration of gold particles can be divided into two major approaches: gene gun, ballistic acceleration of gene-loaded gold nanoparticles to introduce gene into cells [132] and natural ingestion of drug/gene-loaded gold nanoparticles by cells [130]. There have been many reports studying gold nanoparticles for peptides, proteins, and nucleic acids (DNA or RNA) delivery. Gold nanoparticle surface modified with cationic quaternary ammonium can attach plasmid DNA by coulomb attraction [133], and this complex can prevent DNA from enzymatic digestion [134]. DNA-gold nanoparticles with noncovalent bonding were shown effective for gene delivery to mammalian 293T cell. This complex was even maximally eightfold more effective than using PEI for DNA delivery [135]. Klibanov and Thomas [136] coated 2 kDa PEI on the surface of gold nanoparticles leading to 12-fold increase in the gene transfection efficiency compared with that of having polymer only. In addition, nucleic acid can be covalently linked to nanoparticles as well. Mirkin et al. reported gold nanoparticles modified with oligonucleotide and these nanoparticle complexes were able to avoid enzymatic digestion. By chemically tailoring of density of DNA, oligonucleotide-modified gold

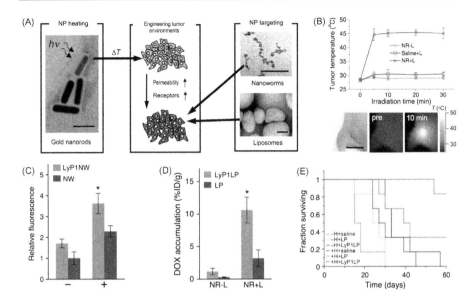

Figure 7.18 (A) Schematic showing the components of the two cooperative nanomaterial systems used in this study. The first component consists of gold NRs, which act as a photothermal sensitizer. The second component consists of either magnetic nanoworms (NW) or DOX-loaded liposomes (LP). Irradiation of the NR with an NIR laser induces localized heating that stimulates changes in the tumor environments. The NW or LP components decorated with LyP1 tumor targeting peptides bind to the heat-modified tumor environments more efficiently than to the normal tumor environments. TEM images of all three components are shown. Scale bars indicate 50 nm. (B) Temperature changes induced by localized laser irradiation (+L) of mice injected with NR alone (no NW or LP). Tumor-bearing mice were injected intravenously with either PEGylated NRs (NR) or saline (saline). Trace labeled "NR-L" is a control where NRs were injected but the tumor was not irradiated. Data and images obtained 72 h postinjection; infrared thermographic maps of average tumor surface temperature were obtained after laser exposure for the indicated times. Scale bar indicates 1 cm. (C) Fluorescence intensity from Cy7-labeled LyP1-conjugated magnetic nanoworms (LyP1NW) and Cy7-labeled control nanoworms (NW) in MDA-MB-435 tumor as a function of externally applied heat (30 min). Heated (45°C) and unheated (37°C) samples indicated with (+) and (−), respectively. The tissues were collected from the mice 24 h postinjection; NIR fluorescence images use Cy7 channel. * indicates $P < 0.05$ ($n = 3-4$). (D) Quantification of *in vivo* accumulation of DOX in tumors as a function of NR-mediated laser heating of LyP1-conjugated liposomes (LyP1LP) or control liposomes that contain no targeting peptide (LP). NR + L and NR − L indicate mice containing gold NRs that were or were not subjected to laser treatment, respectively. Amount of DOX present quantified by fluorescence microscopy to yield a percentage of injected dose per tissue mass. * indicates $P < 0.05$ ($n = 3-4$). (E) Survival rate in different treatment groups after a single dose (3 mg DOX/kg) into mice ($n = 6$) containing single MDA-MB-435 xenograft tumors. Error bars indicate standard deviations from ≥ 3 measurements.

Source: Reproduced with permission from Ref. [131]. Copyright 2010, National Academy of Sciences, USA.

nanoparticles could effectively control protein expression [137]. Gold nanoparticles can also serve as peptide and protein delivery vectors. For example, Pokharkar et al. [138] reported that gold nanoparticles whose surfaces were functionalized with chitosan (nontoxic biopolymer) were used for effective insulin transmucosal delivery.

7.2.7 Micro/Nanobubbles

Microbubbles are usually gas-filled bubbles with a size of a few micrometers. They are often used as a contrast agent for ultrasound imaging [139,140]. Recently, microbubbles started to be broadly applied to drug delivery [141,142]. One unique characteristic of these microbubbles is that high pressure ultrasound is able to cause the destruction of the microbubbles, which can lead to rupture of microvessels and cell membranes and subsequently increase the delivery of drug molecules or other nanomaterials to tumor sites or into cells. In one study, lipid-coated CdSe quantum dots (LQDs) of different sizes were intravenously injected into mice through the tail vein for tumor treatment [143]. In some of the groups, phospholipid-coated microbubbles (mean diameter of 2.5 μm, SonoVue®) were also intravenously injected before LQD injection. Figure 7.19A illustrates the nanoparticle delivery

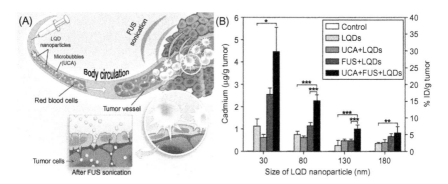

Figure 7.19 (A) A schematic representation of nanoparticle delivery through tumor microvasculature enhanced by FUS sonication in the presence of microbubbles (UCA). The yellow hexagons represent LQD nanoparticles, white circles represent microbubbles, and red circles represent red blood cells. Ultrasound-mediated microbubbles' oscillation and cavitation could disrupt tumor vascular walls, and this disruption would enhance the delivery of LQDs to interstitial spaces. (B) The amounts of Cd extracted from four different sizes of LQD nanoparticles deposited in the tumor tissues with five different treatment conditions: control, LQDs, UCA + LQDs, FUS + LQDs, and UCA + FUS + LQDs. The figure also shows the percentage values of the injected dose of LQD nanoparticles per gram of tumor tissues (for each condition, $n \geq 3$; and mean + SD). Statistical significance compared with LQD injections only; ***$P < 0.0001$, **$P \leq 0.0005$, and *$P < 0.005$. (For interpretation of the references to color in this figure legend, the reader is referred to the web version of this book.)
Source: Reproduced with permission from Ref. [143]. Copyright 2010, Elsevier.

through tumor microvasculature enhanced by focused ultrasound (FUS) sonication in the presence of microbubbles [143]. Ultrasound-mediated microbubbles can enhance the delivery. Figure 7.19B shows the amount of cadmium of the LQDs delivered in the tumor tissues under different conditions. The results indicate that the amount of any size LQDs delivered to the tumor tissues has been significantly enhanced when microbubbles (ultrasound contrast agent, UCA) and FUS are cooperatively used [143].

Although microbubbles have been employed for drug delivery, there are a number of limitations. For example, the diameter is generally in micrometer range which is too large for intravascular applications [144] and it is often difficult to functionalize the surface with targeting molecules. To overcome these limitations, much smaller bubble LPs have been developed. These LPs contain perfluoropropane (an ultrasound imaging gas) and the size can be controlled to be <200 nm. The effect of these bubble LPs to cells and plasmid DNA had been investigated [145]. The data show that the ultrasound used in the work had negligible damage to cells in the absence of bubble LPs and plasmid DNA was not degraded after 10−30 s of ultrasound with and without bubble LPs. Subsequently, the function of the bubble LPs to aid gene delivery was studied. Figure 7.20A clearly reveals that the LPs and ultrasound can successfully achieve gene delivery in cell lines. Many other cell lines were also tested and the results demonstrated that the strategy could

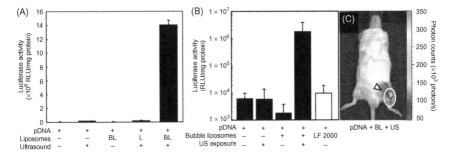

Figure 7.20 (A) Luciferase expression in COS-7 cells transfected by ultrasound with bubble LPs. COS-7 cells (1×10^5 cells/500 μL) mixed with pCMV-Luc (5 μg) and bubble LPs (60 μg) were exposed to ultrasound (frequency, 2 MHz; Duty, 50%; burst rate, 2 Hz; intensity, 2.5 W/cm^2; time 10 s). Data are shown as mean ± SD ($n = 3$). BL, Bubble liposomes; L, PEG liposomes. (B and C) Gene delivery to femoral artery with bubble liposomes. Each sample containing plasmid DNA 10 μg was injected into femoral artery. In the same time, ultrasound (frequency, 1 MHz; duty, 50%; burst rate, 2 Hz; intensity, 1 W/cm^2; time 2 min) was exposed to the downstream area of injection site. (B) Luciferase expression in femoral artery of the ultrasound exposure area at 2 days after transfection. Data are shown as means ± SD ($n = 5$) (LF2000: Lipofectamine 2000). (C) *In vivo* luciferase imaging at 2 days after transfection in the mouse treated with plasmid DNA, bubble liposomes, and ultrasound exposure. The photon counts are indicated by the pseudo-color scales. Arrow head shows injection site and circle shows ultrasound exposure area. BL, Bubble liposomes; US, ultrasound.
Source: Reproduced with permission from Ref. [145]. Copyright 2007, Elsevier.

be used for gene delivery to a range of cells. Finally, the ability of the bubble LPs to *in vivo* gene delivery was also evaluated. The bubble LPs with plasmid DNA were injected into femoral artery. At the same time, the downstream area of the injection site was exposed to ultrasound. It can be observed from Figure 7.20B that the combination of the bubble LPs and ultrasound led to efficient gene expression, which was even much higher than that achieved with the conventional lipofection method. Figure 7.20C indicates that gene expression was realized only at the area of ultrasound exposure. It means that it is possible to establish a noninvasive and tissue-specific gene delivery approach with systemic administration of bubble LPs.

7.2.8 Multifunctional Nanomaterials

With the rapid development of nanomaterials and nanotechnology, there is a growing interest in incorporating various nanomaterials as a whole entity for multifunctional applications [146−148]. Although some examples have been mentioned above, more representative studies will be introduced for multifunctional nanoparticle (MFNP) applications in biomedical sector to simultaneously achieve many functions, such as disease diagnosis, bioimaging, and targeted drug delivery.

The beauty of MFNP systems is that two or more biomedical applications can be achieved with a single treatment. Such a system not only can circumvent the limitation of using single type of nanoparticles but also is cost-effective and convenient. For example, recently, an MFNP system was designed for multimodal imaging and dual-targeted photothermal therapy [149]. The MFNPs were fabricated based on upconversion nanoparticles (UCNPs) as shown in Figure 7.21A. Scanning electron microscopy (SEM, Figure 7.21B−D) and TEM (Figure 7.21E−G) were employed for observation. First, poly(acrylic acid) (PAA)-stabilized UCNPs were prepared. These UCNPs had a diameter of about 160 nm (Figure 7.21B and E). Second, dopamine-modified ultrasmall superparamagnetic ferrite (Fe_3O_4) nanoparticles (IONPs) with a diameter of approximately 5 nm were attached to the surface of UCNPs by electrostatic attraction (Figure 7.21C and F). Third, negatively charged gold nanoparticles were attached to the surface of UCNP−IONP nanoparticles. Fourth, a thin layer of gold shell was grown on the surface by seed-mediated reduction of $HAuCl_4$ (Figure 7.21D and G). A TEM image of an individual MFNP was shown in Figure 7.21H. In the synthesis, a layer of IONPs was incorporated between the UCNPs and the gold shell because this layer of IONPs enables the MFNPs to have magnetic properties and dramatically reduces the luminescence quenching effects of the gold layer to the UCNPs. Finally, the MFNPs were coated with PEG followed by attaching FA on the surface. Well distribution of Y, Fe, and Au elements in the nanoparticle can be confirmed by line profiles of elemental composition (Figure 7.21I and J) and elemental mapping (Figure 7.21K−O) by energy-disperse X-ray spectroscopy (EDS) and high-angle annular dark-field scanning TEM (HAADF-STEM), respectively. Figure 7.21P shows the digital image of PEG-coated MFNPs dispersed in water under ambient light (left), 980 nm laser (middle), and magnetic field (right). The MFNPs disperse very well in water and exhibit upconversion luminescence (UCL), which is demonstrated by the green

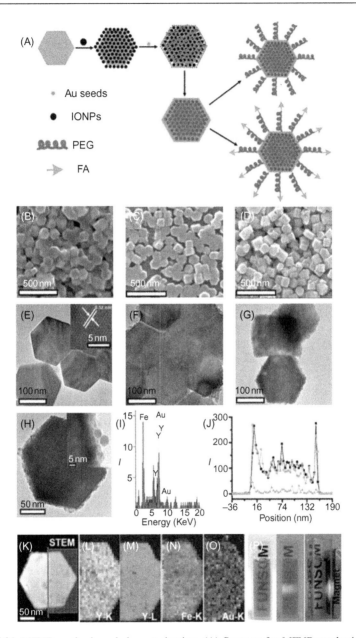

Figure 7.21 MFNP synthesis and characterization. (A) Strategy for MFNP synthesis and functionalization. (B—D) SEM images of the UCNPs (B), UCNP—IONP nanocomposites (C), and UCNP—IONP—Au MFNPs (D). (E—G) TEM images of UCNPs (E), the inset shows the high-resolution (HR)TEM image of an MFNP and the indicated *d* spacing is 0.52 nm; UCNP—IONP nanocomposites (F); and MFNPs (G). (H) A TEM image of a single MFNP with high magnification. Inset is a HRTEM image showing the edge of this MFNP.

fluorescence under 980 nm excitation. Under magnetic field, these nanoparticles can be attracted to a certain place.

This specially designed nanoparticle system, incorporating several nanomaterials into one system, enables multifunctional medical applications [149]. First, these MFNPs can achieve targeted delivery by FA-modified nanoparticles to FR positive human epidermoid carcinoma KB cells. As shown in Figure 7.22A and B, when the MFNPs were modified with FA, much stronger signal was detected in intracellular delivery to FA positive KB cells in comparison with that in the group of cells coculturing with the MFNPs without FA targeting molecules on the surface. In addition, if the cells were not FA positive, the intracellular delivery of FA−PEG−MFNPs was negligible. Second, the system can be used for MRI (Figure 7.22C) because of the incorporated IONPs. After incubation with MFNPs, cells were transferred into agarose gel pellets for MRI. It can be seen that dramatically decreased signal in T_2-weighted images was observed for the KB cells incubated with FA−PEG−MFNPs when compared with that in other groups. Third, due to the existence of a thin layer of gold shell in the MFNPs, the strong surface plasmon resonance scattering is useful for dark-field light scattering imaging. Figure 7.22D shows that, in the group of FA positive KB cells incubating with FA−PEG−MFNPs, much brighter scattered light signals can be observed.

In addition to bioimaging, this nanoparticle system can also be used for photothermal therapy [149]. Photothermal therapy is a method in which light energy can be converted to heat for increasing local temperature to destruct or ablate cancer tumors. Figure 7.23A shows that, after exposure to 808 nm NIR laser for 5 min, PEG−MFNP nanoparticle solution temperature significantly increases from 20 to 50°C, making the nanoparticles promising for killing cancer cells. In sharp contrast, the temperature of water and UCNP−IONP only slightly increases after laser irradiation. These data indicate the importance of a thin layer of gold shell, demonstrating its important role of increasing temperature upon laser exposure. Figure 7.23B shows that, with the help of targeting molecules, FA, the effect of photothermal treatment of nanoparticles was significantly increased. This is mainly because FA enables receptor-mediated endocytosis and improved intracellular delivery of MFNP, thereby increasing local temperature and enhancing efficacy dramatically. Besides FA targeting, this nanoparticle system has another targeting

---

◄ (I) EDS of the MFNPs under the STEM pattern. (J and K) A STEM image of a single MFNP (K) and the cross-sectional compositional line profile (J). ■ Y L edge, ● Fe L edge, ▲ Au K edge. (L−O) HAADF-STEM−EDS mapping images of an MFNP showing the yttrium K edge (L, yellow), yttrium L edge (M, green), iron K edge (N, orange), and gold K edge (O, blue). (P) Photos of a PEG-coated MFNP sample in an aqueous solution under ambient light (left), exposed to a 980 nm laser (middle), and with a neighboring magnet (right). (For interpretation of the references to color in this figure legend, the reader is referred to the web version of this book.)
Source: Reproduced with permission from Ref. [149]. Copyright 2011, Wiley-VCH Verlag GmbH & Co. KGaA, Weinheim.

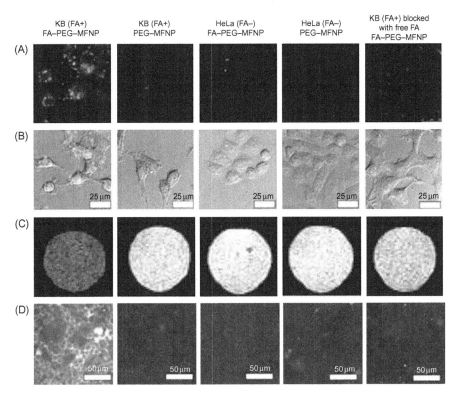

Figure 7.22 Multimodal-targeted *in vitro* imaging of cancer cells. (A and B) Confocal UCL emission (A) and merged UCL/bright field (B) images of KB or HeLa cells after incubation with FA−PEG−MFNP or PEG−MFNP as indicated. (C) T2-weighted MR images of cancer cells treated with MFNPs. The cells were suspended in 1% agarose gel for MR imaging. (D) Dark-field optical microscopy images of the corresponding cancer cell samples. Strong UCL signals, obvious T2-weighted MR darkening effect, and bright light scattering were observed from KB cells treated with FA−PEG−UCNP but not from other control samples. MFNPs at 0.05 mg were used in the above experiments.
Source: Reproduced with permission from Ref. [149]. Copyright 2011, Wiley-VCH Verlag GmbH & Co. KGaA, Weinheim.

approach through magnetic IONP, which is magnetic targeting. Figure 7.23C−J shows the magnetic targeting photothermal therapy: majority of the cells close to magnetic field are dead (Figure 7.23F and J); however, the cells far away from magnetic field remain unaffected (Figure 7.23D and H). Therefore, the cell killing efficacy is dependent on magnetic field. The reason is that magnetic field can effectively guide the accumulation of MFNP to a certain area. This can be used for targeted delivery such as tumors, reducing the possibility of nanoparticles concentrated in healthy area and circumventing side effects. Overall, the MFNP drug delivery system can provide multimodal bioimaging, dual targeting pathways, and photothermal therapy, all of which are based on *in vivo* tests.

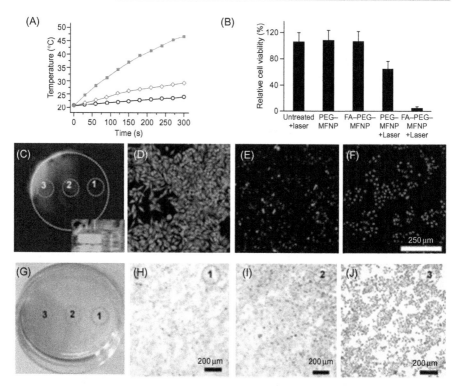

Figure 7.23 Dual-targeted photothermal ablation of cancer cells. (A) The heating curves of water (○), UCNP−IONP (1 mg/mL) nanocomposite (◇), and MFNP (■, 1 mg/mL in total or 0.034 mg/mL by Au content) under 808 nm laser irradiation at a power density of 1 W/cm^2. (B) Relative viabilities of PEG−MFNP or FA−PEG−MFNP treated KB cells with or without laser irradiation (808 nm, 1 W/cm^2, 5 min). (C) An UCL image of HeLa cells in a culture dish after incubation with PEG−MFNP in the presence of magnetic field taken by the Maestro *in vivo* imaging system (980 nm excitation). Inset: A photo showing the experimental setup. A magnet was placed close to the cell culture dish. (D−F) Confocal images of calcein AM (green, live cells) and propidium iodide (red, dead cells) costained cells after magnetic-targeted photothermal therapy (PTT). Images were taken at different locations in the culture dish: (1) far from the magnet (D), (2) in the middle (E), and (3) close to the magnet (F). (G) A digital photo of the cell culture dish after magnetic-targeted PTT and trypan blue staining. (H−J) Optical microscopy images of trypan blue stained cells after magnetic-targeted PTT. Images were taken at different locations in the culture dish: (1) far from magnet (E), (2) in the middle (F), and (3) close to the magnet (G). MFNPs at 0.05 mg/mL were used in the above *in vitro* experiments. (For interpretation of the references to color in this figure legend, the reader is referred to the web version of this book.)

Source: Reproduced with permission from Ref. [149]. Copyright 2011, Wiley-VCH Verlag GmbH & Co. KGaA, Weinheim.

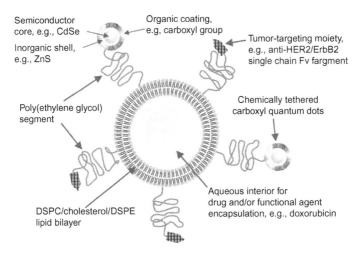

Figure 7.24 Schematic showing the structure of a QD-IL nanoparticle.
Source: Reproduced with permission from Ref. [150]. Copyright 2008, American Chemical Society.

Immunoliposome (IL, lipid bilayer structure) combined with QDs can also form an MFNP system for cellular and *in vivo* bioimaging, tumor targeting, and drug delivery [150]. As shown in Figure 7.24, LP surface is modified with QDs and tumor targeting moieties through PEG. In this multifunctional system, QDs can provide photoluminescent signal for bioimaging and drug tracing, while tumor targeting moiety can target specific cell and achieve targeted intracellular delivery by receptor-mediated endocytosis. Figure 7.25 shows the quantitative analysis of cellular uptake of cells by flow cytometry. With anti-her2 targeting molecules, QD-ILs' cellular uptake by HER2-overexpressing cells (SK-BR-3 and MCF-7/HER2) was significantly increased, compared with anti-her2 free QDs and quantum dot−liposome complex (QD-Ls). However, there is no obvious cellular uptake difference between QD-ILs, QDs, and QD-Ls for MCF-7 cell (low HER2-expressing human breast cancer cell line). Therefore, Figure 7.25 indicates that QD-ILs can successfully target the tumor cells and increase the intracellular delivery efficacy *in vitro*. Using QD-ILs nanoparticles, not only tumor cells can be tracked (Figure 7.26) but also targeted delivery of drug to tumor cells can be realized (Figure 7.27). The *in vitro* fluorescent microscopy images are in line with the flow cytometry results. Figure 7.26 shows the high targeting ability of QD-ILs nanoparticles (Figure 7.26D, red fluorescence—QD and blue fluorescence—nuclei), in which the payload (green fluorescence) is delivered into cytoplasm (Figure 7.26E and F). However, without anti-her2 targeting molecules, QD-Ls cannot achieve intracellular delivery (Figure 7.26C). Moreover, Figure 7.26I shows QD-ILs were not internalized into MCF-7 cell (low HER2-expressing human breast cancer cell line), confirming high selectivity of QD-ILs. In *in vivo* test (Figure 7.27), QD-ILs were accumulated in tumor tissue as well as mononuclear phagocytic system (MPS) after

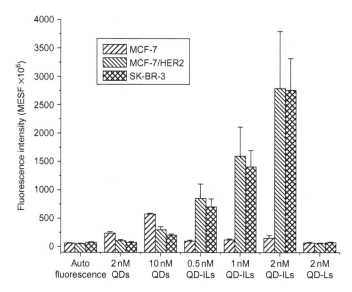

Figure 7.25 Quantitative flow cytometry results showing fluorescence intensities (mean ± SE; $n = 6$) of SK-BR-3, MCF-7/HER2 (both high HER2-overexpressing human breast cancer cell lines) and MCF-7 (low HER2-expressing human breast cancer cell line). Ten thousand cells were collected in each measurement. The columns, from left to right, represent autofluorescence, cells treated by carboxyl QDs at 2 and 10 nM, HER2-targeted QD-ILs at 0.5, 1, and 2 nM, and 2 nM nontargeted QD-Ls at 37°C for 1 h. The fluorescence intensities were calibrated by standard fluorescent microbeads and presented in units of molecules of equivalent soluble fluorochrome (MESF) of phycoerythrin.
Source: Reproduced with permission from Ref. [150]. Copyright 2008, American Chemical Society.

injection into nude mouse. In summary, the above data indicate that this system is well designed for tumor imaging and targeted intracellular delivery.

7.2.9 Toxicity and Hazards of Nanomaterials

As summarized in this chapter, nanomaterials can be used as advanced drug delivery systems with significant advantages, such as controlled release, targeted delivery, improved efficacy, and reduced side effects. Therefore, this is a very promising technique for a broad range of clinical applications. In fact, tens of nanomedicines have been approved for use in market [151]. With the development of nanoscience and nanotechnology, it is certain that more and more nanomaterials will be created for drug delivery systems. However, nanomaterials are very complex. Scientists still need time to have a thorough understanding of the system, particularly the long-term effect of nanomaterials on the health of human beings. Thus, when nanomaterials are used as nanomedicines for medication, full attention of toxicity issues toward human body is essential. For example, nanomaterials'

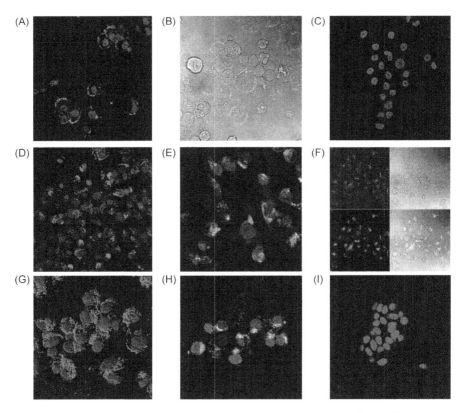

Figure 7.26 SK-BR-3 cells (panels a−g), MCF-7/HER2 (panel h), and MCF-7 cells (panel i) were treated with free QDs or QD-conjugated nanoparticles (red fluorescence) for 60 min at 37°C. Cells were stained by DAPI (blue fluorescence). An argon laser (488 nm) or titanium: sapphire (800 nm) was used for QD excitation. A HeNe laser (633 nm) was used for lipidic dye DiD (emission peak at 665 nm) excitation. (A and B) SK-BR-3 cells treated with free carboxyl QDs showing nonspecific association on the cell surface. In panel a, QD fluorescence (red) and nucleus DAPI staining (blue) are shown. In panel b, differential interference contrast (DIC) and QD fluorescence images are combined to reveal more vividly the binding of QDs along the outline of SK-BR-3 cells. (C) SK-BR-3 cells treated with nontargeted QD-liposomes (QD-Ls) showing virtually no detectable QDs. (D) QD-conjugated anti-HER2 ILs (QD-ILs) internalized in SK-BR-3 cells. (E) Dye-incorporated (green) QD-ILs internalized in SK-BR-3 cells. A 0.25−0.5 mol% amount of 1,1′-dioctadecyl-3,3,3′,3′ tetramethylindodicarbocyanine perchlorate (DiD oil) was included in the lipid mixture for dual labeling. (F) Subpanels of QDs (red, top left), DiD (green, bottom left), DIC (gray scale, top right), and composite (bottom right) images show the intracellular uptake of QD-ILs, in contrast to surface binding of free QDs in panels a and b. (G) Reconstructed 3D image of QD-ILs (red) internalized within SK-BR-3 cells (nucleus in blue). The z-stack constitutes 11 panels that span a total 30 μm elevation. (H) QD-ILs internalized within MCF-7/HER2 cells. (I) MCF-7 cells treated with QD-conjugated anti-HER2 ILs (QD-ILs), showing minimal uptake of QD-ILs. (For interpretation of the references to color in this figure legend, the reader is referred to the web version of this book.)
Source: Reproduced with permission from Ref. [150]. Copyright 2008, American Chemical Society.

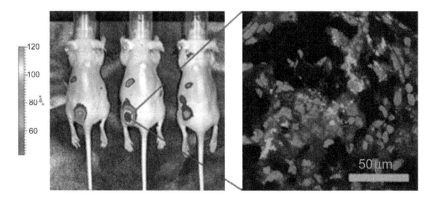

Figure 7.27 (Left panel) *In vivo* fluorescence imaging of three nude mice bearing MCF-7/ HER2 xenografts implanted in the lower back 30 h after i.v. injection with anti-HER2 QD-ILs. The image was acquired by using an excitation filter of 580−610 nm and an emission band-pass filter of 695−770 nm. Imaging showed that QD-ILs had localized prominently in tumors as well as in MPS organs. Units: Efficiency (the fractional ratio of fluorescence emitted per incident photon). (Right panel) A 5 μm section cut from frozen tumor tissues harvested at 48 h postinjection and examined by confocal microscopy by a 63 × oil immersion objective. The tumor section was examined in two-color scanning mode for nuclei stained by DAPI (blue) and QD-ILs (red). For the DAPI channel, an 800 nm (titanium:sapphire) two-photon excitation and 390−465 nm emission filter was used. For QD-ILs channel, 488 (argon) or 633 nm (HeNe) for excitation and 650 nm long pass for emission were used. (For interpretation of the references to color in this figure legend, the reader is referred to the web version of this book.)
Source: Reproduced with permission from Ref. [150]. Copyright 2008, American Chemical Society.

surface is the outmost layer which directly contacts with human environment, such as tissue, cell, and blood. Therefore, the surface property is crucial for biological responses of human physiological system to nanomaterials. However, nanomaterials have unique surface which is associated with their special chemical and physical properties. There is a need for investigating toxicological issues of such special surface properties of nanomaterials. Because of the unique but not fully investigated properties of nanomaterials, prediction of responses of human physiological system towards exposure to high amount of nanomaterials is very difficult. Thus nanomaterials hazards remain a big concern. For instance, nanomaterials may result in a change of the distribution and passage of blood−brain barrier or triggering blood coagulation pathways [152]. Moreover, because of the significant alternation of physical and chemical properties of nanomaterials, there is still a safety concern of nanomaterials even when the corresponding bulk materials are widely accepted as nontoxic. Here, taking gold nanomaterials as an example, bulk gold is commonly regarded having excellent biocompatibility and is a safe material, but gold nanoparticles are found to be "toxic" in many reports. For example, 2 nm cationic gold nanoparticles are toxic when they reach certain concentration. The toxicity may be resulted from their high permeability and damage to cell membranes [153]. In addition, 1.4 nm gold nanoparticles were found to cause necrosis, mitochondrial damage, and induce an oxidative stress on HeLa cell lines [154]. Furthermore, gold nanoparticle injected into blood can cause serious side effect, such as blood clotting and hemolysis [155]. As another example, silver has been widely used to make eating utensils and silver compounds are broadly used as antibacterials, but silver nanoparticles had been observed to interfere with spermatogonial stem cell proliferation, particularly when the cells were exposed to ultrafine silver nanoparticles [156]. Therefore, one should not rely on bulk materials' biological performance to predict the biocompatibility and safety of nanostructured materials. Instead, careful study of nanomaterials' biological behavior, including distribution within human body and kinetics of nanomedicines, is essential.

During the whole life cycle (manufacture, usage, and disposal) of engineered nanoparticles, there are also great concerns of nanoparticle exposure and release to the environment. These extremely small nanoparticles can easily and rapidly spread to the whole environment and bioaccumulate in many organisms with still unknown consequences [157]. One of the major environmental hazards of nanoparticles is their inhalation toxicity, since nanoparticles can easily travel long distance in air [158]. These airborne nanoparticles have serious adverse health effects. For example, they can exacerbate respiratory illness as well as cardiovascular diseases [159,160]. Moreover, inhaled nanoparticles can cross strong biological barriers like blood−brain barrier, which causes great safety concern. Some studies showed that nanoparticles, such as MnO_2 and ^{13}C, after inhalation can efficiently translocate to brain−central nervous system by olfactory neuronal pathway, which may result in inflammatory changes [161,162]. Last but not least, nanoparticles may directly interact with environment, resulting in hazardous effects or disruption to ecosystems. Reports had found that fullerenes have deleterious effect towards bacteria and aquatic species [163,164] and silver nanoparticles had negative effects on beneficial soil bacteria [165]. With the explored toxicity and hazards of nanomaterials, one can conclude that a detailed research of the positive and negative effects of nanomaterials on human beings and environment is essential in order to make them beneficial to us.

7.3 Conclusion

The applications of nanomaterials for drug delivery are very promising. Nanomaterials offer drug delivery systems with desirable properties for well-enhanced efficacy and alleviated side effects. A wide variety of nanomaterials can be engineered to have different shapes, structures, and sizes to meet clinical requirements of drug delivery systems. In this chapter, we provide experimental descriptions and examples in detail to explain the use of various nanomaterials for drug delivery systems with certain attractive characteristics, and also briefly summarize the general applications of specific types of nanomaterials. In spite of the promising future of nanomaterials for drug delivery, it is worth noting that the toxicity and safety of nanomaterials need to be thoroughly investigated. This knowledge will also advance the applications of nanomaterials for drug delivery.

References

[1] Gaponik N, Hickey SG, Dorfs D, Rogach AL, Eychmueller A. Progress in the light emission of colloidal semiconductor nanocrystals. Small 2010;6(13):1364−78.

[2] Bruchez M, Moronne M, Gin P, Weiss S, Alivisatos A. Semiconductor nanocrystals as fluorescent biological labels. Science 1998;281(5385):2013−6.

[3] Chan W, Nie S. Quantum dot bioconjugates for ultrasensitive nonisotopic detection. Science 1998;281(5385):2016−8.

[4] Medintz I, Uyeda H, Goldman E, Mattoussi H. Quantum dot bioconjugates for imaging, labelling and sensing. Nat Mater 2005;4(6):435−46.

[5] Gao X, Cui Y, Levenson R, Chung L, Nie S. In vivo cancer targeting and imaging with semiconductor quantum dots. Nat Biotechnol 2004;22(8):969−76.

[6] Murray C, Norris D, Bawendi M. Synthesis and characterization of nearly monodisperse Cde (E = S, Se, Te) semiconductor nanocrystallites. J Am Chem Soc 1993;115 (19):8706−15.

[7] Yin Y, Alivisatos A. Colloidal nanocrystal synthesis and the organic-inorganic interface. Nature 2005;437(7059):664−70.

[8] Qu L, Peng X. Control of photoluminescence properties of CdSe nanocrystals in growth. J Am Chem Soc 2002;124(9):2049−55.

[9] Talapin D, Rogach A, Kornowski A, Haase M, Weller H. Highly luminescent monodisperse CdSe and CdSe/ZnS nanocrystals synthesized in a hexadecylamine−trioctylphosphine oxide−trioctylphospine mixture. Nano Lett 2001;1(4):207−11.

[10] Chen X, Dobson PJ. Synthesis of semiconductor nanoparticles. In: Mikhail Soloviev, editor. Nanoparticles in biology and medicine: methods and applications. New York, NY: Humana Press; 2012. p. 103−23.

[11] Bailey R, Smith A, Nie S. Quantum dots in biology and medicine. Physica E 2004;25 (1):1−12.

[12] Dubertret B, Skourides P, Norris D, Noireaux V, Brivanlou A, Libchaber A. In vivo imaging of quantum dots encapsulated in phospholipid micelles. Science 2002;298 (5599):1759−62.

[13] Qi L, Gao X. Emerging application of quantum dots for drug delivery and therapy. Expert Opin Drug Deliv 2008;5(3):263−7.

[14] Dykxhoorn DM, Lieberman J. Knocking down disease with siRNAs. Cell 2006;126 (2):231−5.

[15] Dykxhoorn DM, Chowdhury D, Lieberman J. RNA interference and cancer: endogenous pathways and therapeutic approaches. Programmed cell death in cancer progression and therapy. Dordrecht: Springer; 2008. p. 299−329.

[16] Yezhelyev MV, Qi L, O'Regan RM, Nie S, Gao X. Proton-sponge coated quantum dots for siRNA delivery and intracellular imaging. J Am Chem Soc 2008;130 (28):9006−12.

[17] Bousif O, Lezoulch F, Zanta M, Mergny M, Scherman D, Demeneix B, et al. A versatile vector for gene and oligonucleotide transfer into cells in culture and in vivo polyethylenimine. Proc Natl Acad Sci USA 1995;92(16):7297−301.

[18] Christian S, Pilch J, Akerman M, Porkka K, Laakkonen P, Ruoslahti E. Nucleolin expressed at the cell surface is a marker of endothelial cells in angiogenic blood vessels. J Cell Biol 2003;163(4):871−8.

[19] Derfus AM, Chen AA, Min D, Ruoslahti E, Bhatia SN. Targeted quantum dot conjugates for siRNA delivery. Bioconjug Chem 2007;18(5):1391−6.

[20] Chen H, Zhen Z, Todd T, Chu PK, Xie J. Nanoparticles for improving cancer diagnosis. Mater Sci Eng R 2013;73(3):35−69.

[21] Knop K, Hoogenboom R, Fischer D, Schubert US. Poly(ethylene glycol) in drug delivery: pros and cons as well as potential alternatives. Angew Chem Int Ed 2010;49 (36):6288−308.

[22] Manabe N, Hoshino A, Liang Y, Goto T, Kato N, Yamamoto K. Quantum dot as a drug tracer *in vivo*. IEEE Trans Nanobiosci 2006;5(4):263−7.

[23] Bagalkot V, Zhang L, Levy-Nissenbaum E, Jon S, Kantoff PW, Langer R, et al. Quantum dot−aptamer conjugates for synchronous cancer imaging, therapy, and sensing of drug delivery based on Bi-fluorescence resonance energy transfer. Nano Lett 2007;7(10):3065−70.

[24] Srinivasan C, Lee J, Papadimitrakopoulos F, Silbart LK, Zhao M, Burgess DJ. Labeling and intracellular tracking of functionally active plasmid DNA with semiconductor quantum dots. Mol Ther 2006;14(2):192−201.

[25] Zhang L, Qiao S, Jin Y, Chen Z, Gu H, Lu GQ. Magnetic hollow spheres of periodic mesoporous organosilica and Fe_3O_4 nanocrystals: fabrication and structure control. Adv Mater 2008;20(4):805−9.

[26] Herr J, Smith J, Medley C, Shangguan D, Tan W. Aptamer-conjugated nanoparticles for selective collection and detection of cancer cells. Anal Chem 2006;78(9):2918−24.

[27] Mornet S, Vasseur S, Grasset F, Duguet E. Magnetic nanoparticle design for medical diagnosis and therapy. J Mater Chem 2004;14(14):2161−75.

[28] Dobson J. Magnetic nanoparticles for drug delivery. Drug Dev Res 2006;67(1):55−60.

[29] Veiseh O, Gunn JW, Zhang M. Design and fabrication of magnetic nanoparticles for targeted drug delivery and imaging. Adv Drug Deliv Rev 2010;62(3):284−304.

[30] Jain TK, Morales MA, Sahoo SK, Leslie-Pelecky DL, Labhasetwar V. Iron oxide nanoparticles for sustained delivery of anticancer agents. Mol Pharm 2005;2(3):194−205.

[31] Neuberger T, Schopf B, Hofmann H, Hofmann M, von Rechenberg B. Superparamagnetic nanoparticles for biomedical applications: possibilities and limitations of a new drug delivery system. J Magn Magn Mater 2005;293(1):483−96.

[32] Yi D, Lee S, Papaefthymiou G, Ying J. Nanoparticle architectures templated by SiO_2/Fe_2O_3 nanocomposites. Chem Mater 2006;18(3):614−9.

[33] Pankhurst Q, Connolly J, Jones S, Dobson J. Applications of magnetic nanoparticles in biomedicine. J Phys D Appl Phys 2003;36(13):R167−81.

[34] Nie S, Xing Y, Kim GJ, Simons JW. Nanotechnology applications in cancer. Annu Rev Biomed Eng 2007;9:257−88.

[35] Kohler N, Sun C, Fichtenholtz A, Gunn J, Fang C, Zhang M. Methotrexate-immobilized poly(ethylene glycol) magnetic nanoparticles for MR imaging and drug delivery. Small 2006;2(6):785−92.

[36] Xu C, Wang B, Sun S. Dumbbell-like Au-Fe_3O_4 nanoparticles for target-specific platin delivery. J Am Chem Soc 2009;131(12):4216−7.

[37] Xu C, Xie J, Ho D, Wang C, Kohler N, Walsh EG, et al. Au-Fe_3O_4 dumbbell nanoparticles as dual-functional probes. Angew Chem Int Ed 2008;47(1):173−6.

[38] Cheng K, Peng S, Xu C, Sun S. Porous hollow Fe_3O_4 nanoparticles for targeted delivery and controlled release of cisplatin. J Am Chem Soc 2009;131(30):10637−44.

[39] Pan B, Cui D, Sheng Y, Ozkan C, Gao F, He R, et al. Dendrimer-modified magnetic nanoparticles enhance efficiency of gene delivery system. Cancer Res 2007;67 (17):8156−63.

[40] Scherer F, Anton M, Schillinger U, Henkel J, Bergemann C, Kruger A, et al. Magnetofection: enhancing and targeting gene delivery by magnetic force *in vitro* and *in vivo*. Gene Ther 2002;9(2):102−9.

[41] Moghimi S, Symonds P, Murray J, Hunter A, Debska G, Szewczyk A. A two-stage poly(ethylenimine)-mediated cytotoxicity: implications for gene transfer/therapy. Mol Ther 2005;11(6):990−5.

[42] Hong S, Leroueil P, Janus E, Peters J, Kober M, Islam M, et al. Interaction of polycationic polymers with supported lipid bilayers and cells: nanoscale hole formation and enhanced membrane permeability. Bioconjug Chem 2006;17(3):728−34.

[43] Kievit FM, Veiseh O, Bhattarai N, Fang C, Gunn JW, Lee D, et al. PEI−PEG−chitosan-copolymer-coated iron oxide nanoparticles for safe gene delivery: synthesis, complexation, and transfection. Adv Funct Mater 2009;19(14):2244−51.

[44] Yang H, Hua M, Liu H, Huang C, Tsai R, Lu Y, et al. Self-protecting core−shell magnetic nanoparticles for targeted, traceable, long half-life delivery of BCNU to gliomas. Biomaterials 2011;32(27):6523−32.

[45] Chae G, Lee J, Kim S, Seo K, Kim M, Lee H, et al. Enhancement of the stability of BCNU using self-emulsifying drug delivery systems (SEDDS) and *in vitro* antitumor activity of self-emulsified BCNU-loaded PLGA water. Int J Pharm 2005;301 (1−2):6−14.

[46] Paredes SP, Valenzuela MA, Fetter G, Flores SO. TiO$_2$/MgAl layered double hydroxides mechanical mixtures as efficient photocatalysts in phenol degradation. J Phys Chem Solids 2011;72(8):914−9.

[47] Wang Q, O'Hare D. Recent advances in the synthesis and application of layered double hydroxide (LDH) nanosheets. Chem Rev 2012;112(7):4124−55.

[48] Ladewig K, Niebert M, Xu ZP, Gray PP, Lu GQM. Efficient siRNA delivery to mammalian cells using layered double hydroxide nanoparticles. Biomaterials 2010;31 (7):1821−9.

[49] Oh J, Choi S, Lee G, Han S, Choy J. Inorganic drug-delivery nanovehicle conjugated with cancer-cell-specific ligand. Adv Funct Mater 2009;19(10):1617−24.

[50] Choy J, Kwak S, Park J, Jeong Y, Portier J. Intercalative nanohybrids of nucleoside monophosphates and DNA in layered metal hydroxide. J Am Chem Soc 1999;121 (6):1399−400.

[51] Choy J, Kwak S, Jeong Y, Park J. Inorganic layered double hydroxides as nonviral vectors. Angew Chem Int Ed 2000;39(22):4042−5.

[52] Gu Z, Rolfe BE, Xu ZP, Thomas AC, Campbell JH, Lu GQM. Enhanced effects of low molecular weight heparin intercalated with layered double hydroxide nanoparticles on rat vascular smooth muscle cells. Biomaterials 2010;31(20):5455−62.

[53] Wong Y, Cooper HM, Zhang K, Chen M, Bartlett P, Xu ZP. Efficiency of layered double hydroxide nanoparticle-mediated delivery of siRNA is determined by nucleotide sequence. J Colloid Interface Sci 2012;369(1):453−9.

[54] Oh J, Park C, Choy J. Intracellular drug delivery of layered double hydroxide nanoparticles. J Nanosci Nanotechnol 2011;11(2):1632−5.

[55] Choy J, Kwak S, Park J, Jeong Y. Cellular uptake behavior of [gamma-P-32] labeled ATP−LDH nanohybrids. J Mater Chem 2001;11(6):1671−4.

[56] Kwak S, Jeong Y, Park J, Choy J. Bio-LDH nanohybrid for gene therapy. Solid State Ionics 2002;151(1−4):229−34.

[57] Li A, Qin L, Wang W, Zhu R, Yu Y, Liu H, et al. The use of layered double hydroxides as DNA vaccine delivery vector for enhancement of anti-melanoma immune response. Biomaterials 2011;32(2):469−77.

[58] Choy J, Jung J, Oh J, Park M, Jeong J, Kang Y, et al. Layered double hydroxide as an efficient drug reservoir for folate derivatives. Biomaterials 2004;25(15):3059−64.

[59] Wong Y, Markham K, Xu ZP, Chen M, Lu GQ, Bartlett PF, et al. Efficient delivery of siRNA to cortical neurons using layered double hydroxide nanoparticles. Biomaterials 2010;31(33):8770−9.

[60] Del Arco M, Cebadera E, Gutierrez S, Martin C, Montero M, Rives V, et al. Mg,Al layered double hydroxides with intercalated indomethacin: synthesis, characterization, and pharmacological study. J Pharm Sci 2004;93(6):1649−58.

[61] Chakraborti M, Jackson JK, Plackett D, Gilchrist SE, Burt HM. The application of layered double hydroxide clay (LDH)−poly(lactide-co-glycolic acid) (PLGA) film composites for the controlled release of antibiotics. J Mater Sci-Mater M 2012;23 (7):1705−13.

[62] Hesse D, Badar M, Bleich A, Smoczek A, Glage S, Kieke M, et al. Layered double hydroxides as efficient drug delivery system of ciprofloxacin in the middle ear: an animal study in rabbits. J Mater Sci-Mater M 2013;24(1):129−36.

[63] Giri S, Trewyn B, Stellmaker M, Lin V. Stimuli-responsive controlled-release delivery system based on mesoporous silica nanorods capped with magnetic nanoparticles. Angew Chem Int Ed 2005;44(32):5038−44.

[64] Slowing II, Trewyn BG, Giri S, Lin V. Mesoporous silica nanoparticles for drug delivery and biosensing applications. Adv Funct Mater 2007;17(8):1225−36.

[65] Slowing II, Vivero-Escoto JL, Wu C, Lin V. Mesoporous silica nanoparticles as controlled release drug delivery and gene transfection carriers. Adv Drug Deliv Rev 2008;60(11):1278−88.

[66] Vallet-Regi M, Balas F, Arcos D. Mesoporous materials for drug delivery. Angew Chem Int Ed 2007;46(40):7548−58.

[67] Zhu Y, Shi J, Shen W, Dong X, Feng J, Ruan M, et al. Stimuli-responsive controlled drug release from a hollow mesoporous silica sphere/polyelectrolyte multilayer core−shell structure. Angew Chem Int Ed 2005;44(32):5083−7.

[68] Vallet-Regi M, Balas F, Colilla M, Manzano M. Bone-regenerative bioceramic implants with drug and protein controlled delivery capability. Prog Solid State Chem 2008;36(3):163−91.

[69] Horcajada P, Ramila A, Perez-Pariente J, Vallet-Regi M. Influence of pore size of MCM-41 matrices on drug delivery rate. Micropor Mesopor Mater 2004;68(1-3):105−9.

[70] Izquierdo-Barba I, Martinez A, Doadrio A, Perez-Pariente J, Vallet-Regi M. Release evaluation of drugs from ordered three-dimensional silica structures. Eur J Pharm Sci 2005;26(5):365−73.

[71] Xia T, Kovochich M, Liong M, Meng H, Kabehie S, George S, et al. Polyethyleneimine coating enhances the cellular uptake of mesoporous silica nanoparticles and allows safe delivery of siRNA and DNA constructs. ACS Nano 2009;3 (10):3273−86.

[72] Zhao Y, Trewyn BG, Slowing II, Lin VSY. Mesoporous silica nanoparticle-based double drug delivery system for glucose-responsive controlled release of insulin and cyclic AMP. J Am Chem Soc 2009;131(24):8398−400.

[73] Fang Y, Gu D, Zou Y, Wu Z, Li F, Che R, et al. A low-concentration hydrothermal synthesis of biocompatible ordered mesoporous carbon nanospheres with tunable and uniform size. Angew Chem Int Ed 2010;49(43):7987−91.

[74] Meng H, Liong M, Xia T, Li Z, Ji Z, Zink JI, et al. Engineered design of mesoporous silica nanoparticles to deliver doxorubicin and P-glycoprotein siRNA to overcome drug resistance in a cancer cell line. ACS Nano 2010;4(8):4539−50.

[75] Tang S, Huang X, Chen X, Zheng N. Hollow mesoporous zirconia nanocapsules for drug delivery. Adv Funct Mater 2010;20(15):2442−7.

[76] Vallet-Regi M. Ordered mesoporous materials in the context of drug delivery systems and bone tissue engineering. Chem Eur J 2006;12(23):5934−43.

[77] Balas F, Manzano M, Horcajada P, Vallet-Regi M. Confinement and controlled release of bisphosphonates on ordered mesoporous silica-based materials. J Am Chem Soc 2006;128(25):8116−7.

[78] Song S, Hidajat K, Kawi S. Functionalized SBA-15 materials as carriers for controlled drug delivery: influence of surface properties on matrix−drug interactions. Langmuir 2005;21(21):9568−75.

[79] Liu J, Wang B, Hartono SB, Liu T, Kantharidis P, Middelberg APJ, et al. Magnetic silica spheres with large nanopores for nucleic acid adsorption and cellular uptake. Biomaterials 2012;33(3):970−8.

[80] Gao F, Botella P, Corma A, Blesa J, Dong L. Monodispersed mesoporous silica nanoparticles with very large pores for enhanced adsorption and release of DNA. J Phys Chem B 2009;113(6):1796−804.

[81] Chen D, Li L, Tang F, Qi S. Facile and scalable synthesis of tailored silica "Nanorattle" structures. Adv Mater 2009;21(37):3804−7.

[82] Zhu Y, Shi J, Chen H, Shen W, Dong X. A facile method to synthesize novel hollow mesoporous silica spheres and advanced storage property. Micropor Mesopor Mater 2005;84(1−3):218−22.

[83] Li L, Guan Y, Liu H, Hao N, Liu T, Meng X, et al. Silica nanorattle-doxorubicin-anchored mesenchymal stem cells for tumor-tropic therapy. ACS Nano 2011;5 (9):7462−70.

[84] Danhier F, Feron O, Preat V. To exploit the tumor microenvironment: passive and active tumor targeting of nanocarriers for anti-cancer drug delivery. J Control Release 2010;148(2):135−46.

[85] Chen C, Geng J, Pu F, Yang X, Ren J, Qu X. Polyvalent nucleic acid/mesoporous silica nanoparticle conjugates: dual stimuli-responsive vehicles for intracellular drug delivery. Angew Chem Int Ed 2011;50(4):882−6.

[86] He D, He X, Wang K, Chen M, Cao J, Zhao Y. Reversible stimuli-responsive controlled release using mesoporous silica nanoparticles functionalized with a smart DNA molecule-gated switch. J Mater Chem 2012;22(29):14715−21.

[87] Zhu S, Zhou Z, Zhang D. Control of drug release through the *in situ* assembly of stimuli-responsive ordered mesoporous silica with magnetic particles. Chemphys Chem 2007;8(17):2478−83.

[88] Lai C, Trewyn B, Jeftinija D, Jeftinija K, Xu S, Jeftinija S, et al. A mesoporous silica nanosphere-based carrier system with chemically removable CdS nanoparticle caps for stimuli-responsive controlled release of neurotransmitters and drug molecules. J Am Chem Soc 2003;125(15):4451−9.

[89] Colvin V, Goldstein A, Alivisatos A. Semiconductor nanocrystals covalently bound to metal surfaces with self-assembled monolayers. J Am Chem Soc 1992;114 (13):5221−30.

[90] Rosenholm JM, Peuhu E, Eriksson JE, Sahlgren C, Linden M. Targeted intracellular delivery of hydrophobic agents using mesoporous hybrid silica nanoparticles as carrier systems. Nano Lett 2009;9(9):3308−11.

[91] Rosenholm JM, Sahlgren C, Linden M. Towards multifunctional, targeted drug delivery systems using mesoporous silica nanoparticles-opportunities & challenges. Nanoscale 2010;2(10):1870−83.

[92] Rosenholm J, Sahlgren C, Linden M. Cancer-cell targeting and cell-specific delivery by mesoporous silica nanoparticles. J Mater Chem 2010;20(14):2707−13.

[93] Wang L, Wu L, Lu S, Chang L, Teng I, Yang C, et al. Biofunctionalized phospholipid-capped mesoporous silica nanoshuttles for targeted drug delivery: Improved water suspensibility and decreased nonspecific protein binding. ACS Nano 2010;4(8):4371−9.

[94] Wu H, Liu G, Zhang S, Shi J, Zhang L, Chen Y, et al. Biocompatibility, MR imaging and targeted drug delivery of a rattle-type magnetic mesoporous silica nanosphere system conjugated with PEG and cancer cell-specific ligands. J Mater Chem 2011;21 (9):3037−45.

[95] Peer D, Karp JM, Hong S, FaroKHzad OC, Margalit R, Langer R. Nanocarriers as an emerging platform for cancer therapy. Nat Nanotechnol 2007;2(12):751−60.

[96] Wagner E. Programmed drug delivery: nanosystems for tumor targeting. Expert Opin Biol Ther 2007;7(5):587−93.

[97] Zhang Q, Liu F, Nguyen KT, Ma X, Wang X, Xing B, et al. Multifunctional mesoporous silica nanoparticles for cancer-targeted and controlled drug delivery. Adv Funct Mater 2012;22(24):5144−56.

[98] Zhang B, Breslow R. Enthalpic domination of the chelate effect in cyclodextrin dimers. J Am Chem Soc 1993;115(20):9353−4.

[99] Davis ME. The first targeted delivery of siRNA in humans via a self-assembling, cyclodextrin polymer-based nanoparticle: from concept to clinic. Mol Pharm 2009;6 (3):659−68.

[100] Bajpai AK, Shukla SK, Bhanu S, Kankane S. Responsive polymers in controlled drug delivery. Prog Polym Sci 2008;33(11):1088−118.

[101] Gillies E, Frechet J. Dendrimers and dendritic polymers in drug delivery. Drug Discov Today 2005;10(1):35−43.

[102] Gombotz W, Pettit D. Biodegradable polymers for protein and peptide drug delivery. Bioconjug Chem 1995;6(4):332−51.

[103] Kabanov A, Batrakova E, Alakhov V. Pluronic (R) block copolymers as novel polymer therapeutics for drug and gene delivery. J Control Release 2002;82 (2−3):189−212.

[104] Patri A, Majoros I, Baker J. Dendritic polymer macromolecular carriers for drug delivery. Curr Opin Chem Biol 2002;6(4):466−71.

[105] Schmaljohann D. Thermo-and pH-responsive polymers in drug delivery. Adv Drug Deliv Rev 2006;58(15):1655−70.

[106] Kataoka K, Harada A, Nagasaki Y. Block copolymer micelles for drug delivery: design, characterization and biological significance. Adv Drug Deliv Rev 2001;47 (1):113−31.

[107] Lavasanifar A, Samuel J, Kwon G. Poly(ethylene oxide)-block-poly(L-amino acid) micelles for drug delivery. Adv Drug Deliv Rev 2002;54(2):169−90.

[108] Nishiyama N, Kataoka K. Current state, achievements, and future prospects of polymeric micelles as nanocarriers for drug and gene delivery. Pharmacol Ther 2006;112 (3):630−48.

[109] Ikehara Y, Kojima N. Development of a novel oligomannose-coated liposome-based anticancer drug-delivery system for intraperitoneal cancer. Curr Opin Mol Ther 2007;9(1):53−61.

[110] Khuller G, Kapur M, Sharma S. Liposome technology for drug delivery against mycobacterial infections. Curr Pharm Des 2004;10(26):3263−74.

[111] Jain JP, Ayen WY, Kumar N. Self assembling polymers as polymersomes for drug delivery. Curr Pharm Des 2011;17(1):65−79.

[112] Lee JS, Feijen J. Polymersomes for drug delivery: design, formation and characterization. J Control Release 2012;161(2):473−83.

[113] Meng F, Zhong Z, Feijen J. Stimuli-responsive polymersomes for programmed drug delivery. Biomacromolecules 2009;10(2):197−209.

[114] Onaca O, Enea R, Hughes DW, Meier W. Stimuli-responsive polymersomes as nanocarriers for drug and gene delivery. Macromol Biosci 2009;9(2):129−39.

[115] Bauhuber S, Hozsa C, Breunig M, Goepferich A. Delivery of nucleic acids via disulfide-based carrier systems. Adv Mater 2009;21(32−33):3286−306.

[116] Howard KA. Delivery of RNA interference therapeutics using polycation-based nanoparticles. Adv Drug Deliv Rev 2009;61(9):710−20.

[117] Jeong JH, Kim SW, Park TG. Molecular design of functional polymers for gene therapy. Prog Polym Sci 2007;32(11):1239−74.

[118] Panyam J, Labhasetwar V. Biodegradable nanoparticles for drug and gene delivery to cells and tissue. Adv Drug Deliv Rev 2003;55(3):329−47.

[119] Uhrich K, Cannizzaro S, Langer R, Shakesheff K. Polymeric systems for controlled drug release. Chem Rev 1999;99(11):3181−98.

[120] Fricker G, Kromp T, Wendel A, Blume A, Zirkel J, Rebmann H, et al. Phospholipids and lipid-based formulations in oral drug delivery. Pharm Res 2010;27(8):1469−86.

[121] van Hoogevest P, Liu X, Fahr A, Leigh MLS. Role of phospholipids in the oral and parenteral delivery of poorly water soluble drugs. J Drug Deliv Sci Tech 2011;21 (1):5−16.

[122] Fifis T, Gamvrellis A, Crimeen-Irwin B, Pietersz G, Li J, Mottram P, et al. Size-dependent immunogenicity: therapeutic and protective properties of nano-vaccines against tumors. J Immunol 2004;173(5):3148−54.

[123] Reddy ST, van der Vlies AJ, Simeoni E, Angeli V, Randolph GJ, O'Neill CP, et al. Exploiting lymphatic transport and complement activation in nanoparticle vaccines. Nat Biotechnol 2007;25(10):1159−64.

[124] Nimesh S, Goyal A, Pawar V, Jayaraman S, Kumar P, Chandra R, et al. Polyethylenimine nanoparticles as efficient transfecting agents for mammalian cells. J Control Release 2006;110(2):457−68.

[125] Grant EV, Thomas M, Fortune J, Klibanov AM, Letvin NL. Enhancement of plasmid DNA immunogenicity with linear polyethylenimine. Eur J Immunol 2012;42 (11):2937−48.

[126] Tong L, Wei Q, Wei A, Cheng J. Gold nanorods as contrast agents for biological imaging: optical properties, surface conjugation and photothermal effects. Photochem Photobiol 2009;85(1):21−32.

[127] Kennedy LC, Bickford LR, Lewinski NA, Coughlin AJ, Hu Y, Day ES, et al. A new era for cancer treatment: gold-nanoparticle-mediated thermal therapies. Small 2011;7 (2):169−83.

[128] Yavuz MS, Cheng Y, Chen J, Cobley CM, Zhang Q, Rycenga M, et al. Gold nanocages covered by smart polymers for controlled release with near-infrared light. Nat Mater 2009;8(12):935−9.

[129] Skirtach AG, Javier AM, Kreft O, Koehler K, Alberola AP, Moehwald H, et al. Laser-induced release of encapsulated materials inside living cells. Angew Chem Int Ed 2006;45(28):4612−7.

[130] Sperling RA, Rivera gil P, Zhang F, Zanella M, Parak WJ. Biological applications of gold nanoparticles. Chem Soc Rev 2008;37(9):1896−908.

[131] Park J, von Maltzahn G, Xu MJ, Fogal V, Kotamraju VR, Ruoslahti E, et al. Cooperative nanomaterial system to sensitize, target, and treat tumors. Proc Natl Acad Sci USA 2010;107(3):981−6.

[132] Chen D, Wendt CH, Pui DYH. A novel approach for introducing biomaterials into cells. J Nanopart Res 2000;2(2):133−9.

[133] McIntosh C, Esposito E, Boal A, Simard J, Martin C, Rotello V. Inhibition of DNA transcription using cationic mixed monolayer protected gold clusters. J Am Chem Soc 2001;123(31):7626−9.

[134] Han G, Martin C, Rotello V. Stability of gold nanoparticle-bound DNA toward biological, physical, and chemical agents. Chem Biol Drug Des 2006;67(1):78−82.

[135] Sandhu K, McIntosh C, Simard J, Smith S, Rotello V. Gold nanoparticle-mediated transfection of mammalian cells. Bioconjug Chem 2002;13(1):3−6.

[136] Thomas M, Klibanov A. Conjugation to gold nanoparticles enhances polyethylenimine's transfer of plasmid DNA into mammalian cells. Proc Natl Acad Sci USA 2003;100(16):9138−43.

[137] Rosi N, Giljohann D, Thaxton C, Lytton-Jean A, Han M, Mirkin C. Oligonucleotide-modified gold nanoparticles for intracellular gene regulation. Science 2006;312 (5776):1027−30.

[138] Bhumkar DR, Joshi HM, Sastry M, Pokharkar VB. Chitosan reduced gold nanoparticles as novel carriers for transmucosal delivery of insulin. Pharm Res 2007;24(8):1415−26.

[139] Ferrara K, Pollard R, Borden M. Ultrasound microbubble contrast agents: fundamentals and application to gene and drug delivery. Annu Rev Biomed Eng 2007;9:415−47.

[140] Klibanov A. Microbubble contrast agents-targeted ultrasound imaging and ultrasound-assisted drug-delivery applications. Invest Radiol 2006;41(3):354−62.

[141] Lentacker I, De Smedt SC, Sanders NN. Drug loaded microbubble design for ultrasound triggered delivery. Soft Matter 2009;5(11):2161−70.

[142] Mayer CR, Geis NA, Katus HA, Bekeredjian R. Ultrasound targeted microbubble destruction for drug and gene delivery. Expert Opin Drug Del 2008;5(10):1121−38.

[143] Lin C, Liu T, Chen C, Huang Y, Huang W, Sun C, et al. Quantitative and qualitative investigation into the impact of focused ultrasound with microbubbles on the triggered release of nanoparticles from vasculature in mouse tumors. J Control Release 2010;146(3):291−8.

[144] Lindner J. Microbubbles in medical imaging: current applications and future directions. Nat Rev Drug Discov 2004;3(6):527−32.

[145] Suzuki R, Takizawa T, Negishi Y, Hagisawa K, Tanaka K, Sawamura K, et al. Gene delivery by combination of novel liposomal bubbles with perfluoropropane and ultrasound. J Control Release 2007;117(1):130−6.

[146] Kim J, Kim HS, Lee N, Kim T, Kim H, Yu T, et al. Multifunctional uniform nanoparticles composed of a magnetite nanocrystal core and a mesoporous silica shell for magnetic resonance and fluorescence imaging and for drug delivery. Angew Chem Int Ed 2008;47(44):8438−41.

[147] Liong M, Lu J, Kovochich M, Xia T, Ruehm SG, Nel AE, et al. Multifunctional inorganic nanoparticles for imaging, targeting, and drug delivery. ACS Nano 2008;2 (5):889−96.

[148] Nasongkla N, Bey E, Ren J, Ai H, Khemtong C, Guthi JS, et al. Multifunctional polymeric micelles as cancer-targeted, MRI-ultrasensitive drug delivery systems. Nano Lett 2006;6(11):2427−30.

[149] Cheng L, Yang K, Li Y, Chen J, Wang C, Shao M, et al. Facile preparation of multifunctional upconversion nanoprobes for multimodal imaging and dual-targeted photothermal therapy. Angew Chem Int Ed 2011;50(32):7385–90.

[150] Weng KC, Noble CO, Papahadjopoulos-Sternberg B, Chen FF, Drummond DC, Kirpotin DB, et al. Targeted tumor cell internalization and imaging of multifunctional quantum dot-conjugated immunoliposomes *in vitro* and *in vivo*. Nano Lett 2008;8 (9):2851–7.

[151] Fuente JM, Grazu V, editors. Nanobiotechnology: inorganic nanoparticles vs organic nanoparticles. London: Elsevier; 2012.

[152] De Jong WH, PJA Borm. Drug delivery and nanoparticles: applications and hazards. Int J Nanomed 2008;3(2):133–49.

[153] Goodman C, McCusker C, Yilmaz T, Rotello V. Toxicity of gold nanoparticles functionalized with cationic and anionic side chains. Bioconjug Chem 2004;15 (4):897–900.

[154] Pan Y, Leifert A, Ruau D, Neuss S, Bornemann J, Schmid G, et al. Gold nanoparticles of diameter 1.4 nm trigger necrosis by oxidative stress and mitochondrial damage. Small 2009;5(18):2067–76.

[155] Dobrovolskaia MA, Aggarwal P, Hall JB, McNeil SE. Preclinical studies to understand nanoparticle interaction with the immune system and its potential effects on nanoparticle biodistribution. Mol Pharm 2008;5(4):487–95.

[156] Braydich-Stolle LK, Lucas B, Schrand A, Murdock RC, Lee T, Schlager JJ, et al. Silver nanoparticles disrupt GDNF/Fyn kinase signaling in spermatogonial stem cells. Toxicol Sci 2010;116(2):577–89.

[157] Stern ST, McNeil SE. Nanotechnology safety concerns revisited. Toxicol Sci 2008;101(1):4–21.

[158] Bailey M. The new Icrp model for the respiratory tract. Radiat Prot Dosimet 1994;53 (1–4):107–14.

[159] Dockery D, Pope C, Xu X, Spengler J, Ware J, Fay M, et al. An association between air pollution and mortality in 6 United States cities. N Engl J Med 1993;329 (24):1753–9.

[160] Pope C, Burnett R, Thurston G, Thun M, Calle E, Krewski D, et al. Cardiovascular mortality and long-term exposure to particulate air pollution—epidemiological evidence of general pathophysiological pathways of disease. Circulation 2004;109 (1):71–7.

[161] Elder A, Gelein R, Silva V, Feikert T, Opanashuk L, Carter J, et al. Translocation of inhaled ultrafine manganese oxide particles to the central nervous system. Environ Health Perspect 2006;114(8):1172–8.

[162] Oberdorster G, Sharp Z, Atudorei V, Elder A, Gelein R, Kreyling W, et al. Translocation of inhaled ultrafine particles to the brain. Inhal Toxicol 2004;16 (6–7):437–45.

[163] Lyon D, Fortner J, Sayes C, Colvin V, Hughes J. Bacterial cell association and antimicrobial activity of a C-60 water suspension. Environ Toxicol Chem 2005;24 (11):2757–62.

[164] Oberdorster E. Manufactured nanomaterials (fullerenes, C-60) induce oxidative stress in the brain of juvenile largemouth bass. Environ Health Perspect 2004;112 (10):1058–62.

[165] Colman BP, Arnaout CL, Anciaux S, Gunsch CK, Hochella Jr MF, Kim B, et al. Low concentrations of silver nanoparticles in biosolids cause adverse ecosystem responses under realistic field scenario. PLoS ONE 2013;8(2):e57189.

8 Processing and Deformation Characteristics of Metals Reinforced with Ceramic Nanoparticles

Sie Chin Tjong[1,2]

[1]Department of Physics and Materials Science, City University of Hong Kong, Tat Chee Avenue, Kowloon, Hong Kong, PR China, [2]Department of Physics, Faculty of Science, King Abdulaziz University, Jeddah, Saudi Arabia

8.1 Introduction

Metal matrix composites (MMCs) reinforced with continuous ceramic fibers exhibit high specific strength and specific elastic modulus over unreinforced metals/alloys [1]. The high cost of reinforcing fibers and high processing cost of fiber-reinforced MMCs render them uneconomical for technological applications. In contrast, MMCs reinforced with ceramic microparticles are isotropic, easier to manufacture, and lower cost in comparison to the continuous fiber-reinforced composites. MMCs inherent excellent properties from their matrix alloy and the reinforcing phase components, i.e., ductility and toughness of the metal matrix, high modulus and strength of the reinforcement. Therefore, particulate-reinforced MMCs exhibit high strength, superior creep, and wear resistances [2−6]. They find a broad range of applications from structural components in the aerospace, automotive, and transportation industries to the thermal management for electronic devices [7,8]. Generally, the metal matrices used in the particulate-reinforced MMCs include commercially available alloys based on aluminum, magnesium, and titanium. Among them, aluminum-based alloys with face-centered cubic structure are used extensively due to their low density, good workability, and tailored mechanical property. The mechanical strength of heat-treatable aluminum alloys can be increased dramatically through aging heat treatment due to the formation of fine precipitates, i.e., precipitation hardening. However, precipitation-hardened aluminum alloys suffer a large reduction in mechanical strength upon exposure to elevated temperature for long periods because of the precipitate coarsening. Ceramic particles are stable at temperatures up to the melting temperature of the matrix metal and do not coarsen at elevated temperatures.

Particulate-reinforced MMCs generally contain large volume fraction of reinforcement to achieve desired mechanical and physical properties. High reinforcement

Nanocrystalline Materials. DOI: http://dx.doi.org/10.1016/B978-0-12-407796-6.00008-7

loadings impair mechanical properties and increase the weight of resulting composites. Furthermore, ceramic particles with sizes of several micrometers often act as the preferential sites for crack initiation and propagation, resulting in low mechanical ductility and toughness. In recent years, the escalation cost of fossil fuel and the urgent need of reducing carbon dioxide emission have driven materials scientists to search for light-weight structural materials for use in aerospace and transportation industries.

With the advent of nanotechnology, novel nanocrystalline materials with unique chemical, physical, and mechanical properties have been developed and synthesized recently. Nanocrystalline materials exhibit much higher mechanical strength and modulus compared to their microcrystalline counterparts. Inorganic nanoparticles can be synthesized from a wide variety of techniques, including sol−gel, spray forming, chemical vapor deposition, and laser-induced gas phase reaction [9]. Thus, ceramic nanomaterials overcome the limitations of microcrystalline counterparts, showing great potential for use as reinforcing components for metals. For example, Tjong and coworkers introduced silicon nitride nanoparticles (10 nm) to the aluminum matrix, and reported that the tensile strength 1 vol% Si_3N_4/Al nanocomposite is comparable to that of 15 vol% SiC (3.5 μm)/Al microcomposite. The yield stress of 1 vol% Si_3N_4/Al nanocomposite is significantly higher than that of the 15 vol% SiC-reinforced microcomposite [10]. Furthermore, the additions of 1−2 vol% Si_3N_4 nanoparticles to aluminum also enhance its high-temperature creep resistance markedly [11].

In spite of several advantages of reinforcing ceramic nanoparticles, the fabrication of metal matrix nanocomposites (MMNCs) remains a big challenge for materials scientists. Ceramic nanoparticles of large surface areas often agglomerate into clusters since they are poorly wetted with metals during the composite fabrication. In this regard, several processing strategies have been adopted to fabricate MMNCs with homogeneous dispersion of nanoparticles in the metal matrix. These include compocasting, ultrasonic cavitation, ball milling, and friction stir processing (FSP). The microstructures and mechanical properties of MMNCs depend greatly on the processing technique employed. This chapter overviews the state-of-the-art development in the processing strategies and mechanical deformation characterization of MMNCs reinforced with ceramic nanoparticles. Particular attention is paid to their structure−property relationships.

8.2   Fabrication of MMNCs

Two processing routes are generally deployed for fabricating MMNCs, i.e., liquid- and solid-state processing. Liquid-state processing route includes melt stirring and compocasting. Liquid-state processing is widely known to be very cost-effective since it can produce bulk composites in large quantities using existing melting and casting facilities. However, particle agglomeration, poor wetting of ceramic nanoparticles with molten metal, and preferential formation of interfacial products hinder its extensive use for manufacturing MMNCs. The potential of bulk MMNCs cannot be fully realized for industrial applications unless nanocomposite structural components can be manufactured cost effectively using liquid-state processing route. Because of poor wettability between the metal matrix and ceramic particles, the reinforcing particulates tend to agglomerate into clusters in the matrix. Therefore, an external force field is required to break up the clusters and disperse them into the melt. This can be done by using high intensity ultrasonic probe for creating violent agitation in the melt, terming as the ultrasonic cavitation [12−15].

Solid-state processing route is typically a powder metallurgy (PM)-based process, in which the matrix powder and reinforcing materials are mixed together in a simple mechanical mixer followed by cold compaction and sintering, hot pressing, or spark plasma sintering (SPS) to form a bulk composite. Conventional furnace sintering requires high temperature and long heating time to obtain dense products. In contrast, SPS offers advantages of low processing temperature and very short heating time for consolidating composite powders. In certain cases, secondary mechanical processing such as hot extrusion, hot forging, hot rolling, or FSP is necessary to further improve mechanical properties of MMNCs by consolidating the compacts into full-dense products. PM processing has the advantages of better dispersion of reinforcing particles and near net-shape fabrication. The disadvantage is the high cost of powder materials.

8.2.1 Liquid-State Processing

Depending on the temperature at which the reinforcing particles are introduced into the melt, there exist two types of melting practices for fabricating composites. In the stir mixing/casting process, the particles are added to molten alloy above its liquidus temperature. On the contrary, ceramic nanoparticles are introduced into metal in the semisolid state in the compocasting process. Stir casting involves an initial melting of metal/alloy ingots in a furnace in a protective gas atmosphere, mixing nanoparticles with molten metal using an impeller followed by solidification. In this process, both the incorporation of nanoparticles into the melt and pouring of the composite slurry into the mould are carried out in a fully liquid state. Figure 8.1 is a schematic diagram showing a typical setup commonly used for manufacturing MMNCs using melt stirring [16]. A graphite impeller stirs a melt mixture vigorously, generating a vortex in the melt for dispersing nanoparticles. The main drawbacks of stir casting are poor wettability between molten metal and ceramic nanoparticles, and high-porosity content of the composite products. Moreover, reinforcing particles tend to float or sink depending on their density relative to the liquid metal. These issues become especially significant as the reinforcement size decreases due to greater agglomeration tendency and reduced wettability of the particles with the melt (Figure 8.2) [17].

The wetting of a solid surface with molten metal plays an important role in the dispersion of ceramic nanoparticle in the metal matrix. Wetting relates to the contact between a liquid and a solid surface, describing the ability of a liquid to spread

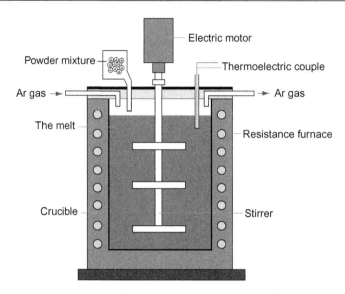

Figure 8.1 Setup for stir casting.
Source: Reprinted from Ref. [16] with permission of Elsevier.

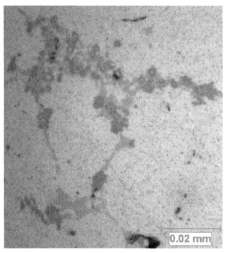

Figure 8.2 Agglomeration of alumina nanoparticles (47 nm) at the grain boundaries of 5 wt% Al_2O_{3p}/A206 nanocomposite.
Source: Reprinted from Ref. [17] with permission of Elsevier.

over a solid surface by minimizing surface free energy. For a liquid droplet on a solid surface (Figure 8.3), the surface energy (tension) of different components can be expressed by

$$\gamma_{SV} = \gamma_{SL} + \gamma_{LV} \cos \phi \qquad (8.1)$$

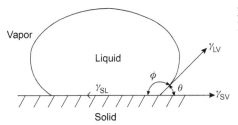

Figure 8.3 Schematic diagram showing a liquid droplet on a solid.

$$\cos \phi = \frac{\gamma_{SV} - \gamma_{SL}}{\gamma_{LV}} \tag{8.2}$$

where ϕ is the contact angle, γ_{SV}, γ_{SL}, and γ_{LV} are the surface tensions of solid−vapor, solid−liquid, and liquid−vapor, respectively. For $\phi = 0°$, the liquid droplet spreads over entire solid surface. At $\phi < 90°$, the liquid droplet wets the solid. The liquid does not wet solid for $\phi > 90°$, especially when $\phi = 180°$. The contact angle can be determined using sessile drop measurements. In general, ceramic nanoparticle has poor wettability with molten metal. This issue can be addressed partly by mixing ceramic nanoparticles with reactive metals, such as Mg and Li, using ball milling or heat treatment [17,18].

Alternatively, better dispersion of reinforcing particles can be realized by reducing the casting temperature via compocasting or rheocasting process, in which the reinforcements are added to metal in a semisolid state. For conventional MMCs, compocasting enables the attainment of improved wettability and better distribution of ceramic microparticles compared to stir casting [19,20]. Accordingly, this process has been employed increasingly by the researchers to fabricate cast MMNCs [16,21]. In some cases, compocasting-assisted ultrasonic cavitation can achieve even more uniform dispersion of ceramic nanoparticles [13−16]. The ultrasonic probe system can generate intense transient cavitation at a temperature of $\sim 5000°$C and pressure of ~ 1000 atm. The probe induces a violent collapse of micro gas bubbles around nanoparticle clusters, thereby causing breakdown of the clusters in the melt.

8.2.1.1 Al-Based Nanocomposites

Tahamtan et al. [18] employed mechanical stir mixing to fabricate 5 vol% Al_2O_3p/ A206 nanocomposite. The A206 alloy consists of 4.2−5.0% Cu, 0.2−0.5% Mg, 0.15−0.35% Mn, 0.15−0.3% Ti, <0.05% and Al balance. Below 1000°C, the contact angle between aluminum and Al_2O_3 is >90°, resulting in poor wetting by the liquid metal. This poor wetting behavior favors clustering of the alumina particles and their floating on the surface of the melt. To improve wettability of alumina nanoparticles (100 nm), alumina nanoparticles were ball-milled with Mg and Al powders followed by compression into disc specimens. Ball-milled discs were introduced into molten A206 alloy for forming nanocomposite. As a result, wettability of ball-milled alumina with molten metal improves considerably, leading to better distribution of nanoparticles in the melt. Mazahery et al. [22] also ball milled alumina nanoparticles (50 nm) with Al particles (16 μm) prior to introduction to molten A356 aluminum alloy. A356 is a hypoeutectic Al−Si alloy with a nominal composition of 7.5 wt% Si, 0.38 wt% Mg, 0.02 wt% Zn, 0.107 wt% Fe, and Al balance.

Very recently, Sajjadi et al. [19] fabricated Al_2O_3/A356 nanocomposites using both stir casting and compocasting processes. They reported that the alumina nanoparticles act as effective nucleation sites for the Al grains, producing nanocomposites with fine-grained microstructure. Due to the improved wettability of particles with the melt during compocasting, the grain size of compocast nanocomposite is finer than that of stir-cast composite (Figure 8.4A−C). Moreover, compocast nanocomposites have lower porosity content than stir-cast materials. Similarly, El-Mahallawi et al. [23] also reported that compocast Al_2O_3/A356 nanocomposites show good distribution and low agglomeration of alumina nanoparticles in the alloy matrix.

Li et al. [13] fabricated SiC/A356 nanocomposite by means of ultrasonic vibration processing for enhancing dispersion of ceramic nanoparticles. Figure 8.5 is a schematic diagram showing the setup of a typical ultrasonic processing system. The optical images of the A356 alloy and its nanocomposite with 2 wt% SiC are

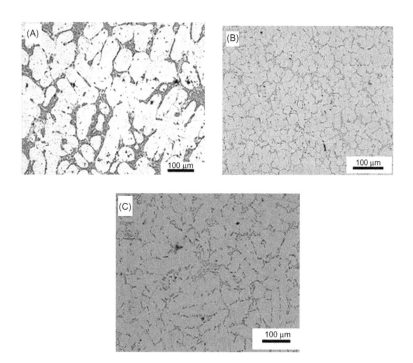

Figure 8.4 Optical images of (A) cast A356 alloy, (B) compocast 1 wt% Al_2O_3/A356 nanocomposite, and (C) stir-cast 1 wt% Al_2O_3/A356 nanocomposite.
Source: Reprinted from Ref. [19] with permission of Elsevier.

shown in Figure 8.6A−C, respectively. The microstructure reveals the formation of aluminum dendrites surrounded by a network of eutectic Al−Si lamellae. In a recent study, they also introduced 1.5 vol% TiCN nanoparticles into molten Al−9Mg alloy under ultrasonic mixing at 715°C [15]. The resulting composite showed significant grain refinement with TiCN nanoparticles dispersed mainly at the grain boundaries of matrix alloy (Figure 8.7A and B).

Su et al. [16] fabricated Al₂O₃/AA2024 nanocomposites using compocasting-assisted ultrasonic vibration. Figure 8.8A shows the microstructure of cast 1 wt% Al₂O₃/AA2024 nanocomposite. The nanocomposite exhibits fine-grained morphology. Alumina nanoparticles are well dispersed within the grains and at the grain boundaries of the matrix alloy (Figure 8.8B and C). The microstructural evolution of the composite melt during the solidification under an ultrasonic field is shown in Figure 8.9. During the solidification, primary α-Al dendrites first nucleate in the

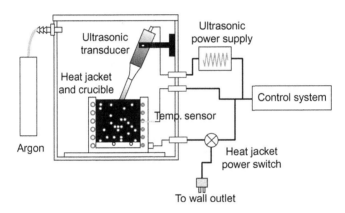

Figure 8.5 Experimental setup for ultrasonic processing unit.
Source: Reprinted from Ref. [13] with permission of Elsevier.

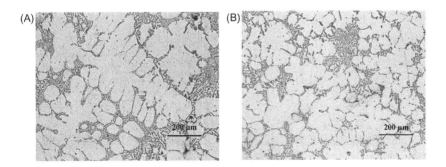

Figure 8.6 Microstructures of as-cast (A) A356 alloy without ultrasonic treatment and (B) 2 wt% SiC/A356 nanocomposite with ultrasonic processing.
Source: Reprinted from Ref. [12] with permission of Elsevier.

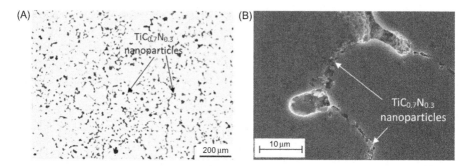

Figure 8.7 (A) Optical micrograph of unetched 1.5 vol% TiCN/Al−9Mg composite.
(B) SEM micrograph of etched 1.5 vol% TiCN/Al−9Mg composite showing nanoparticles
distributed along the grain boundaries.
Source: Reprinted from Ref. [15] with permission of Elsevier.

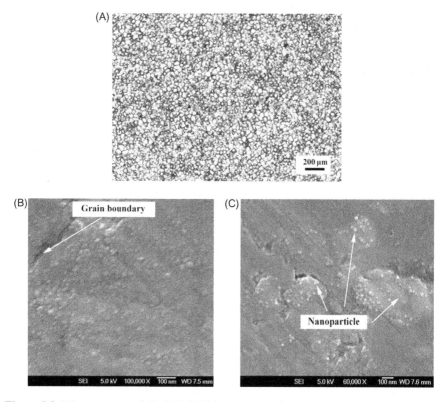

Figure 8.8 Microstructure of Al_2O_3/AA2024 nanocomposite prepared under ultrasonic
vibration. (A) Optical micrograph. SEM images showing (B) grain interior and (C) grain
boundary.
Source: Reprinted from Ref. [16] with permission of Elsevier.

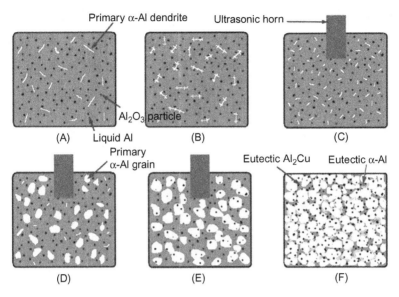

Figure 8.9 Scheme of evolution of Al₂O₃/AA2024 nanocomposite melt to final microstructure under an ultrasonic field: (A) the formation of primary α-Al dendrites, (B) the formation of dendritic arms, (C) the breakage of dendrites by ultrasonic cavitation, (D and E) the growth of α-Al grains, and (F) the completion of the solidification. *Source*: Reprinted from Ref. [16] with permission of Elsevier.

melt, while alumina nanoparticles are pushed into the melt by the dendrites (Figure 8.9A and B). When an ultrasonic probe is introduced into the melt, the dendrites are broken by intense ultrasonic cavitation, producing more nucleation sites for the α-Al grains (Figure 8.9C and D). Some alumina nanoparticles are trapped inside the grains under intensive ultrasonic cavitation (Figure 8.9E). As the temperature of the composite slurry drops to the solidus temperature, the eutectic reaction takes place eventually. Those alumina nanoparticles pushed by the growing grains are trapped within the eutectic phase (Figure 8.9F).

8.2.1.2   Mg-Based Nanocomposites

Magnesium has a density of 1.74 g/cm³, that is, about two-thirds of the density of aluminum with a value of 2.70 g/cm³. Magnesium offers several advantages for structural engineering applications, including good damping capacity, excellent castability, and large abundance. Comparing with Al, magnesium possesses low mechanical strength, poor creep, and corrosion resistance. Magnesium alloys with hexagonal close-packed (HCP) lattice are difficult to deform mechanically at room temperature due to their limited numbers of slip system. Despite these shortcomings, light-weight magnesium alloys have attracted increasing attention for use as structural materials in the automotive and aerospace industries recently because of the weight-reduction consideration [24]. Typical examples are the AZ series having Al contents ≤ 9 wt%, including AZ31 (Mg$-$3Al$-$1Zn$-$0.2Mn) and AZ91 (Mg$-$9Al$-$1Zn$-$0.3Mn) alloys. Zinc is added to improve the corrosion resistance and strength of magnesium alloys. Generally, the low mechanical strength of Mg-based alloys can be improved greatly by adding ceramic nanoparticles.

Li and coworkers fabricated Mg-based MMCs reinforced with SiC or AlN nanoparticles using ultrasonic cavitation method [13,14,25,26]. The additions of SiC nanoparticles to Mg under ultrasonic-assisted casting process result in a significant reduction of the matrix grain size. However, some SiC microclusters still exist in the microstructures, especially at the grain boundaries [25,26]. High-resolution TEM image shows a clean interface between the Mg matrix and SiC nanoparticles (Figure 8.10A and B). In another study, the dispersion of AlN nanoparticles in the alloy matrix is somewhat improved. Figure 8.11A and B shows low magnified SEM images of cast AZ91D alloy and 1 wt% AlN/AZ91D nanocomposite, respectively. The microstructure of AZ91D is mainly composed of α-Mg, massive β-phase and lamellar β-Mg$_{17}$Al$_{12}$ phase. The microstructure of 1 wt% AlN/AZ91D nanocomposite is similar to that of AZ91D, but with much finer β-phase, demonstrating that a small loading level of AlN nanoparticles affects the solidification process of AZ91D alloy. At high magnification image, AlN nanoparticles tend to disperse into individual particles in the alloy matrix (Figure 8.11C). Very recently, Nie et al. [27,28] employed ultrasonic vibration during stir casting or compocasting to fabricate 1 vol% SiC/AZ91 nanocomposite. In the latter process, AZ91 alloy was first heated to 700°C, followed by cooling to 590°C in which the alloy was in a semisolid state. 1 vol% SiC nanoparticles (60 nm) were quickly added into the semisolid alloy under mechanical stirring for 5, 10, or 15 min. The melt was then reheated to 700°C and an ultrasonic probe was dipped into the melt for dispersing SiC nanoparticles (Figure 8.12).

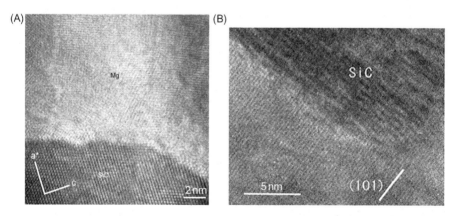

Figure 8.10 High-resolution TEM images showing a clean Mg$-$SiC interface in (A) 1 wt% SiC/Mg and (B) 2 wt% SiC/Mg$-$4Al$-$1Si nanocomposites.
Source: Reprinted from Refs. [25,26] with permission of Elsevier.

Gupta and coworkers systematically studied the microstructure and mechanical behavior of magnesium-based composites reinforced with alumina nanoparticles prepared by disintegrated melt deposition (DMD) technique [29–33]. The process involves mechanical stirring of Mg chips and reinforcing particles using an

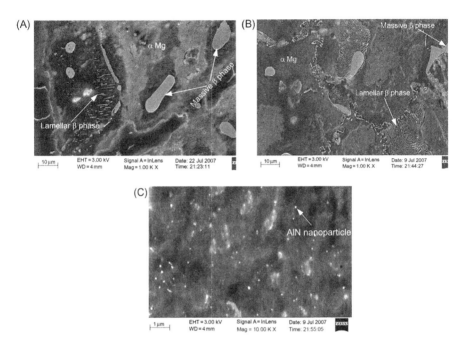

Figure 8.11 Low magnified SEM images of (A) AZ91D alloy and (B) 1 wt% AlN/AZ91D nanocomposite. (C) High magnification SEM image of 1 wt% AlN/AZ91D nanocomposite. *Source*: Reprinted from Ref. [14] with permission of Elsevier.

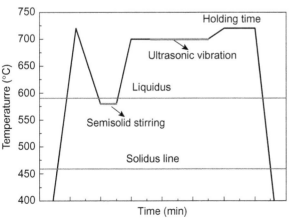

Figure 8.12 Schematic illustration of the temperature–time sequences for semisolid stirring assisted ultrasonic vibration. *Source*: Reprinted from Ref. [28] with permission of Elsevier.

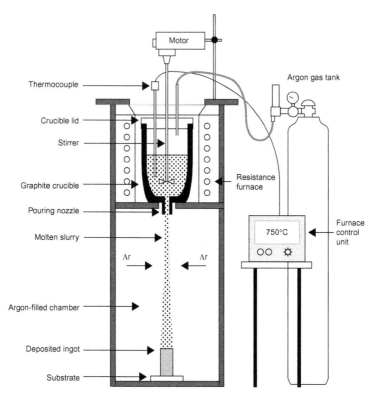

Figure 8.13 Schematic representation of DMD process.
Source: Reprinted from Ref. [34] with permission of Elsevier.

impeller under an argon atmosphere at a superheat temperature of 750°C (Figure 8.13) [34]. The melt was released through a pouring nozzle located at the base of crucible and disintegrated with argon gas jets. The ingot was hot extruded eventually. Microstructural examinations of the extruded composite samples revealed fairly uniform distribution of alumina nanoparticles.

8.2.2 Solid-State Processing

PM technique is a versatile process for manufacturing MMNCs due to its simplicity, flexibility, and near net-shape capability. The process involves mechanical blending of ceramic nanoparticles with metal/alloy powders in a rotary mill, followed by cold compaction and sintering. By simply mixing metal powders with the reinforcement material, homogeneous dispersion of nanoparticles in the metal matrix is difficult to achieve, especially at higher particle contents. The dispersion of ceramic nanoparticles can be somewhat improved through a wet mixing method in which the composite constituents are suspended in a solvent (e.g., ethanol),

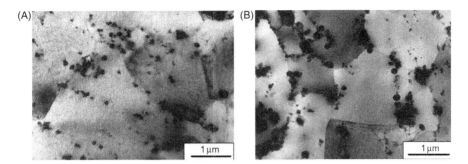

Figure 8.14 TEM micrographs of (A) 1 vol% Al_2O_3/Al and (B) 4 vol% Al_2O_3/Al nanocomposites.
Source: Reprinted from Ref. [35] with permission of Elsevier.

followed by the solvent evaporation, cold compaction, and sintering [35]. Figure 8.14A and B shows TEM micrographs of 1 vol% Al_2O_3/Al and 4 vol% Al_2O_3/Al nanocomposites prepared by wet powder mixing and sintering. Alumina nanoparticles together with few clusters are dispersed fairly in the Al matrix of 1 vol% Al_2O_3/Al nanocomposite. By increasing the particle content to 4 vol%, alumina nanoparticle clusters within the grains and at the matrix grain boundaries can be readily seen in the micrograph.

Mechanical alloying (MA) process is known to be effective for dispersing ceramic nanoparticles homogeneously in the metal matrix [36]. A uniform distribution of ceramic nanoparticles in the metal matrix is the crucial factor for attaining enhanced mechanical properties of MMNCs. MA is a solid-state processing that involves loading constituent powders into a high-energy ball mill containing grinding media, such as stainless steel or alumina balls. The powder mixture undergoes a series of repeating fracture, deform, and welding processes. This leads to intimate mixing of constituent powder particles on an atomic scale, producing a variety of supersaturated solid solutions, metastable crystallites, amorphous metal alloys, and grain size refinement down to nanometer scale. This process is commonly used to produce alloys and composites that are difficult to obtain from conventional melting and casting techniques. A process control agent (PCA), such as stearic acid or acrylic acid, is occasionally added. The PCA adsorbs on the surface of powder particles and minimizes cold welding between impacted particles, thereby preventing agglomeration. To minimize oxidation during high-energy milling, the operation can be carried out at cryogenic temperatures by introducing liquid nitrogen into the milling chamber. This process is termed as cryomilling. Several factors, such as the charge ratio (ratio of the weight of balls to the powder), ball mill design, milling atmosphere, time, speed, and temperature, can affect the dispersion of nanoparticles in metallic powders.

In recent years, FSP becomes quite popular for fabricating surface composites with ultrafine-grained microstructures via dynamic recrystallization [37]. In the process, a rotating tool pin is inserted to the substrate such that the friction and

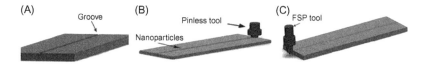

Figure 8.15 Schematic of the FSP process: (A) cutting a groove on the plate, (B) filling the groove with ceramic nanopowder and covering it with a pinless tool, and (C) performing FSP process with a pin tool.
Source: Reprinted from Ref. [38] with permission of Elsevier.

plastic deformation induced by the tool heats and softens the work piece. The tool pin then promotes intermixing of material in a local region. To manufacture MMNCs, a long groove cut on a metal surface in the path of the tool is filled with ceramic nanoparticles Then, the FSP is conducted along the groove to produce a thick surface composite (Figure 8.15) [38]. FSP can also serve as a secondary mechanical processing tool for achieving homogeneous dispersion of reinforcing nanoparticles and eliminating internal defects in the cast MMNCs [38,39].

8.2.2.1   Al-Based Nanocomposites

Razavi Hesabi et al. [40] studied morphological evolution and structural change of the Al_2O_3/Al composites by milling 5 vol% nano-Al_2O_3 (35 nm) and micro-Al_2O_3 (1 μm) to pure Al (48 μm) in a planetary ball mill under an argon atmosphere. They reported that the milling stages, such as plastic deformation, microwelding, and particle fragmentation, occur earlier in the composite with microalumina particles than in the powder mixture with nanoalumina particles. Furthermore, longer milling time is needed to achieve the steady-state condition in nanoalumina-reinforced composite. Zebarjad and Sajjadi [41,42] indicated that alumina micro-powders became finer and dispersed more uniformly in aluminum with increasing milling time.

Bathula et al. [43] employed high-energy ball milling and SPS to fabricate SiC/AA5083 composite. AA5083 alloy (Al−4.5Mg, 0.9Mn, 0.4Si, 0.2Cu, 0.15Ti, 0.25Zn) powder with a size of ~ 15 μm was ball milled with 10 wt% SiC nanoparticles (~ 20 nm) for 15 h. Figure 8.16A and B is SEM images of the as-received AA5083 alloy and milled SiC/AA5083 composite for 15 h. The milled composite mixture exhibits faceted feature with sizes ranging from 40 to 50 μm. A large increase in the size of milled composite mixture is due to the powder constituents undergoing a series of welding and fracture of particles process during the MA process. By consolidating with SPS, a dense nanocomposite with homogeneous dispersion of SiC nanoparticles in the metal matrix is obtained (Figure 8.17A and B). Kollo et al. [44] employed both planetary and attritor ball milling with heptane as a milling agent to fabricate SiC/Al nanocomposites. In spite of different milling techniques employed, effective dispersion of SiC nanoparticles in the aluminum matrix was achieved.

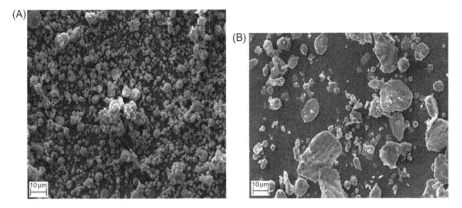

Figure 8.16 SEM images of (A) as-received AA5083 alloy and (B) SiC/AA5083 composite mixture after high-energy ball milling for 15 h.
Source: Reprinted from Ref. [43] with permission of Elsevier.

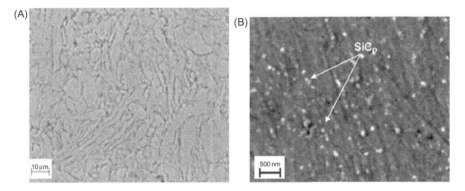

Figure 8.17 SEM images of spark plasma sintered SiC/AA5083 nanocomposite showing (A) dense feature and (B) dispersion of SiC nanoparticles.
Source: Reprinted from Ref. [43] with permission of Elsevier.

Recently, Shafiei-Zarghani et al. [45] employed FSP to fabricate Al_2O_3/Al composites. Their results showed alumina nanoparticles can disperse more uniformly in the metal matrix by increasing the number of FSP passes. Sharifitabar et al. [46] also reported a similar finding. They investigated the effects of processing parameters on microstructural evolution, deformation, and mechanical behavior of the nanocomposites. Bauri et al. [39] used FSP to homogenize the particle distribution in the cast 5 wt% TiC/Al composite. A single pass of FSP was enough to break the particle clusters. Two passes of FSP resulted in complete homogenization and elimination of casting defects. The grain size of the metal matrix was also refined substantially after each FSP pass.

8.2.2.2 Mg-Based Nanocomposites

Hassan and Gupta fabricated 1.1 vol% Al_2O_3/Mg nanocomposite by means of simple powder mixing and sintering [30]. Wong and Gupta also prepared SiC/Mg nanocomposites using dry powder mixing and microwave sintering [47]. Microwave sintering allows rapid densification of green products for short periods of time at lower temperatures. Ferkel and Mordike employed simple powder mixing or ball milling SiC nanoparticles (30 nm) with Mg powder (40 μm) followed by hot extrusion to fabricate 3 vol% SiC/Mg nanocomposite specimens [48]. For extruded composite prepared by powder mixing, SiC nanoparticles dispersed at the grain boundaries of coarse and elongated magnesium grains. For extruded composite fabricated by ball milling, SiC nanoparticles dispersed mainly at the shear bands of heavily deformed magnesium grains, with some nanoparticles located at fined magnesium grains of submicrometer scale.

FSP is an effective tool to eliminate casting defects of magnesium alloys. As aforementioned, the microstructure of cast AZ91 alloy consists of irregularly distributed β-$Mg_{17}Al_{12}$ phase at its grain boundaries (Figure 8.11A). The microstructure of FSPed AZ91-based nanocomposites is characterized by the homogeneous distribution of ceramic nanoparticles, the recrystallized grain structure, and the dissolution of β phase [38,49].

8.3 Mechanical Properties

The addition of second-phase particles in a crystal matrix is a common method for improving the mechanical properties of materials. In general, the additions of ceramic nanoparticles to metals increase the Young's modulus, yield stress, and tensile strength of MMNCs. The strengthening mechanisms responsible for the yield strength enhancement, include Orowan strengthening, grain size refinement due to the Hall—Petch effect, creation of dislocations to accommodate thermal expansion mismatch between the reinforcement and the matrix, and the load-bearing effect of the reinforcement. Among them, Orowan strengthening plays a decisive role in enhancing the yield strength of MMNCs due to the nanoscale dimension of reinforcing particles. These nanoparticles impede the movement of dislocations through dislocation bowing and looping effects. The bowing of dislocations between the nanoparticles builds up dislocation loops around each nanoparticle. On the other hand, Orowan strengthening is insignificant in the MMCs reinforced with microparticles, because the reinforcing particles are coarse and thus the interparticle spacing is large. The increment in yield stress due to Orowan strengthening can be described by the Orowan—Ashby equation [50]:

$$\Delta\sigma = \frac{0.13Gb}{\lambda}\ln\frac{r}{b} \tag{8.3}$$

where G is the shear modulus of the matrix material, b the burgers vector, r the radius of the particle, and λ the interparticle spacing defined by

$$\lambda = d_p \left[\left(\frac{1}{2V_p} \right)^{1/3} - 1 \right] \tag{8.4}$$

where V_p and d_p are the volume fraction and diameter of the nanoparticle, respectively. By decreasing the size of reinforcements to nanoscale level, the interparticle spacing is reduced, thereby restricting the activities of dislocations more effectively. The Orowan effect becomes less significant for the MMNCs with ceramic nanoparticles mainly located at the matrix grain boundaries.

The incorporation of ceramic nanoparticles into metals/alloys can lead to grain refining of the matrix phase. The yield strength enhancement of a composite due to the matrix grain refining can be described by the Hall–Petch equation:

$$\Delta\sigma_{HP} = \sigma_0 + kd^{-1/2} \tag{8.5}$$

where d is the grain size, σ_0 is friction stress, which includes contributions from the solutes and particles, and k is the Hall–Petch slope, that is, a measure of the resistance to dislocation motion caused by the presence of grain boundaries. The yield strength contribution by the dislocations generated from the difference in coefficient of thermal expansion between the reinforcement phase and the matrix can be written as [51]

$$\Delta\sigma_{CTE} = \sqrt{3}\alpha Gb\sqrt{\rho} \tag{8.6}$$

where ρ is the dislocation density and α the strengthening coefficient. This stress depends on the temperature from which the composite is cooled from its fabrication temperature. The load-bearing effect generally is insignificant in MMNCs with ceramic nanoparticles of a small aspect ratio, i.e., unity. The shear transfer of load from the soft matrix to the hard ceramic reinforcements during mechanical deformation can be expressed as [52]

$$\Delta\sigma_{load} = 0.5V_p\sigma_{ym} \tag{8.7}$$

where σ_{ym} is the matrix yield stress.

From these, the final yield strength of an MMNC is a linear summation of the strength imparted by each mechanism. Considering Orowan strengthening, dislocation generation, and load-bearing effects, Zhang and Chen derived the following equations for the yield stress of a composite (σ_{yc}) [52]:

$$\sigma_{yc} = (1 + 0.5V_p)\left(\sigma_{ym} + A + B\frac{AB}{\sigma_{ym}} \right) \tag{8.8}$$

$$A = 1.25 G_m b \sqrt{\frac{12(T_{process} - T_{test})(\alpha_m - \alpha_p)V_p}{bd_p(1 - V_p)}} \tag{8.9}$$

$$B = \frac{0.13 G_m b}{d_p\left[(1/2V_p)^{1/3} - 1\right]} \ln\frac{d_p}{2b} \tag{8.10}$$

where G_m is the shear modulus of the matrix, $T_{process}$ is the processing temperature, T_{test} is the testing temperature, α_m and α_p are the coefficient of thermal expansion of the matrix and particle, respectively.

8.3.1 Al-Based Nanocomposites

8.3.1.1 Nanocomposites Fabricated by Liquid-State Processing

The mechanical performance of MMNCs depends greatly on the attainment of homogeneous dispersion of ceramic nanoparticles in the metal matrix. Agglomeration of ceramic nanoparticles often occurs in the composites fabricated by liquid metallurgy route owing to the poor wettability of nanoparticles by molten metal. In this regard, special measures have been taken in the melt practices for enhancing wettability of ceramic nanoparticles. Mazahery et al. [22] fabricated Al_2O_3/A356 nanocomposites by stir casting at 800°C. Prior to the composite fabrication, alumina nanoparticles were ball-milled with Al particles to improve their wettability with liquid metal. Porosity level in the nanocomposites increased slightly with increasing particulate volume content due to the increase of contact surface area of alumina nanoparticles with molten aluminum alloy. Table 8.1 summarizes tensile behavior of stir-cast Al_2O_3/A356 nanocomposites. The yield strength, tensile strength, and ductility of the nanocomposites improved with the increase in alumina content. Furthermore, the tensile strength of 1.5 vol% Al_2O_3/A356 nanocomposite using Al particles is much higher than that of 1.5 vol%

Table 8.1 Tensile Properties of Stir-Cast Al_2O_3/A356 Nanocomposites

Materials	Casting at	Tensile Strength (MPa)	Ductility (%)	Grain Size (μm)
Al	800°C	115 ± 3	2.9	48
1.5 Al_2O_3/Al without using Al particles	800°C	126 ± 4	1.4	45
0.75 Al_2O_3/Al using Al particles	800°C	167 ± 2	2	16
1.5 Al_2O_3/Al using Al particles	800°C	182 ± 2	2.2	12
2.5 Al_2O_3/Al using Al particles	800°C	171 ± 2	2	15
3.5 Al_2O_3/Al using Al particles	800°C	163 ± 2	1.9	14
5 Al_2O_3/Al using Al particles	800°C	160 ± 3	1.7	15

Source: Reprinted from Ref. [22] with permission of Elsevier.

Al_2O_3/A356 nanocomposite without milling alumina with Al particles. More recently, Karbalaei Akbari et al. [53] also ball milled alumina nanoparticles with either Al or Cu powders. They fabricated 1.5 vol% Al_2O_3/A356 nanocomposite by casting at 850°C under mechanical stirring at different periods. They found that the porosity level increased with increasing stirring time. Thus, the yield stress and Young's modulus of the nanocomposite decrease with increasing stirring time (Figure 8.18A and B). It is noted that the as-cast A356 alloy exhibits low tensile ductility (only 0.7%). The stir-cast nanocomposite also exhibits low tensile ductility due to the formation of dendritic structure and porosity (Figure 8.19).

Sajjadi et al. [19] compared the microstructure and mechanical behavior of stir-cast and compocast Al_2O_3/A356 nanocomposites. They also fabricated Al_2O_3/A356 microcomposites using the same casting processes for the purposes of comparison. The mechanical results showed that the addition of alumina (micro and nano) to A356 led to the improvement in yield strength, ultimate tensile strength, compression strength, and hardness at the expense tensile ductility. They attributed this to the matrix grain size refinement and the creation of thermal mismatch dislocations due to the additions of reinforcing particles. Moreover, compocast nanocomposites exhibit higher yield stress and tensile strength than stir-cast nanocomposites. The porosity level and grain size of compocast nanocomposites were lower than those of stir-cast materials because the wettability of particles in compocasting was better than stir casting. Figure 8.20 shows typical stress−strain curves of A356 and its nanocomposites fabricated by compocasting. The variations of yield stress and

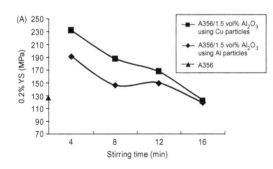

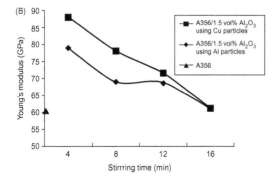

Figure 8.18 The variations of (A) yield strength and (B) Young's modulus with stirring time for stir-cast 1.5 vol% Al_2O_3/A356 nanocomposite materials.
Source: Reprinted from Ref. [53] with permission of Elsevier.

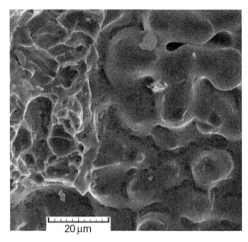

Figure 8.19 SEM fractograph of tensile-tested 1.5 vol% Al$_2$O$_3$/A356 nanocomposite showing dendritic feature and porosity.
Source: Reprinted from Ref. [53] with permission of Elsevier.

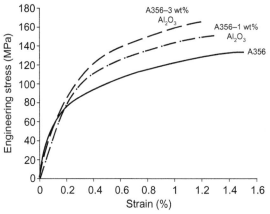

Figure 8.20 Tensile stress–strain curves of A356 and its nanocomposites fabricated by compocasting.
Source: Reprinted from Ref. [19] with permission of Elsevier.

ultimate tensile strength with alumina content of both micro- and nanocomposites are shown in Figure 8.21A and B, respectively.

Li and coworkers fabricated Al-based nanocomposites reinforced with SiC or TiCN nanoparticles using ultrasonic cavitation processing method [12,15]. Figure 8.22A shows the variations of yield stress and tensile strength with SiC nanoparticle content of SiC/A356 nanocomposites. The elongation versus nanoparticle content of the SiC/A356 nanocomposites is shown in Figure 8.22B. Apparently, SiC nanoparticle additions enhance the yield stress of A356 alloy markedly without sacrificing tensile ductility. The yield stress of A356 alloy is improved more than 50% by adding 2 wt% SiC nanoparticles. Su et al. [16] fabricated Al$_2$O$_3$/AA2024 nanocomposites using compocasting-assisted ultrasonic cavitation. Figure 8.23 shows tensile properties of as-cast AA2024 and its nanocomposites. Optimal yield stress and tensile strength of AA2024 can be achieved by adding 1 wt% alumina nanoparticles. Comparing with the matrix alloy,

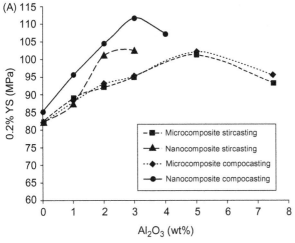

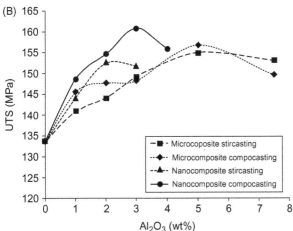

Figure 8.21 The variations of (A) 0.2% yield strength and (B) ultimate strength with alumina content of compocast and stir-cast composites. *Source*: Reprinted from Ref. [19] with permission of Elsevier.

the ultimate tensile strength and yield strength of 1 wt% Al_2O_3/AA2024 composite are enhanced by 37 and 81%, respectively. This is attributed to homogeneous distribution of alumina nanoparticles and grain refinement of the aluminum matrix (Figure 8.8). Representative tensile properties of cast MMNCs reported by several researchers are summarized in Table 8.2. In this table, the content of ceramic nanoparticles is given in volume percentage or weight percentage. The relation between the volume fraction (V_f) and weight fraction (W_f) of the reinforcing phase of a composite is given by

$$V_f = \frac{W_f/\rho_f}{(W_f/\rho_f) + (1 - W_f)/\rho_m} = \frac{W_f\rho_m}{W_f\rho_m + (1 - W_f)\rho_f} \tag{8.11}$$

where ρ_f and ρ_m are the reinforcing phase density and matrix density.

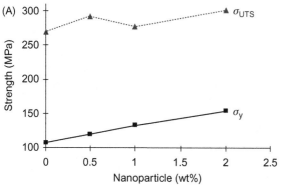

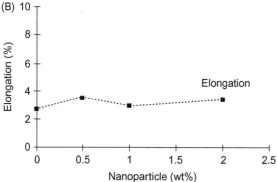

Figure 8.22 (A) Yield stress and tensile strength and (B) elongation versus SiC nanoparticle content of SiC/A356 nanocomposites. *Source*: Reprinted from Ref. [12] with permission of Elsevier.

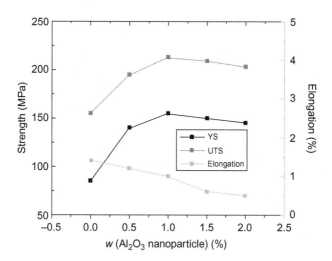

Figure 8.23 Tensile properties of as-cast AA2024 and Al_2O_3/AA2024 nanocomposites. *Source*: Reprinted from Ref. [16] with permission of Elsevier.

Table 8.2 Tensile Properties of Cast Aluminum Matrix Nanocomposites

Particle Content (vol%)	Particle Content (wt%)	Reinforcing Nanoparticle	Matrix Material	Fabrication Process	Yield Strength (MPa)	Tensile Strength (MPa)	Fracture Strain (%)	References
0	—	—	A356	Stir casting	94	135	—	[22]
1.5	—	Al_2O_3	A356	Stir casting at 800°C for 15 min	115	182	2.2	[22]
0	—	—	A356	Stir casting	127	137	0.7	[53]
1.5	—	Al_2O_3	A356	Stir casting at 850°C for 4 min	190	213	0.91	[53]
1.5	—	Al_2O_3	A356	Stir casting at 850°C for 12 min	150	168	0.58	[53]
	0	—	A356	Stir casting	85	134	1.55	[19]
	1	Al_2O_3	A356	Stir casting at 700°C for 30 min	88	143	1.31	[19]
	1	Al_2O_3	A356	Compocasting at 610°C	104	148	1.37	[19]
	3	Al_2O_3	A356	Stir casting at 700°C for 30 min	102	152	1.12	[19]
	3	Al_2O_3	A356	Compocasting at 610°C	112	160	1.19	[19]
0	—	—	Al–9Mg	Melting at 715°C assisted with ultrasonic vibration	137	167	1.4	[15]
0.5	—	TiCN	Al–9Mg	1.5 vol% TiCN/Al–9Mg nanocomposite was added to Al–9Mg melt for dilution	143	200	2.7	[15]
1.5	—	TiCN	Al–9Mg	Melting at 715°C assisted with ultrasonic vibration	146	209	2.9	[15]

8.3.1.2 Nanocomposites Fabricated by Solid-State Processing

Tjong et al. [10] were the first researchers to report the tensile behavior of aluminum reinforced with 1 vol% Si_3N_4 nanoparticles (15 nm). The nanocomposite was fabricated by dry powder mixing followed by hot pressing and hot extrusion. The results showed that the Si_3N_4 nanoparticles are very effective to enhance the yield stress and tensile strength of Al. The yield stress of Al increases significantly from 68 to 144 MPa by adding 1 vol% Si_3N_4. They attributed this to the Orowan strengthening mechanism. Kang and Chan [35] then also reported that low volume fractions of alumina nanoparticles are beneficial in improving yield stress and tensile strength of aluminum (Figure 8.24). The tensile strength of 1 vol% Al_2O_3/Al nanocomposite is similar to that of the 10 vol% SiCp (13 μm)/Al composite, while the yield strength of the former is higher than that of the latter. Furthermore, the yield stress and tensile strength of Al_2O_3/Al nanocomposites increase with increasing alumina content up to 4 vol% at the expense of tensile ductility. For the nanocomposites with Al_2O_3 content ≥ 4 vol%, the strengthening effect levels off due to the agglomeration of alumina nanoparticles (Figure 8.14B). The main strengthening mechanism in the Al_2O_3/Al nanocomposites derives from the Orowan effect. However, agglomeration of alumina nanoparticles reduces Orowan strengthening when the alumina content ≥ 4 vol%.

Kollo et al. [44] fabricated n-SiC/Al nanocomposites using planetary milling process followed by hot extrusion. Figure 8.25 shows tensile stress–strain curves of pure Al and n-SiC/Al nanocomposites. Apparently, an addition of 1 vol% n-SiC to pure Al enhances its ultimate tensile strength from 100 to 205 MPa, being a twofold increment. At 10 vol% n-SiC, the tensile strength reaches 405 MPa with a

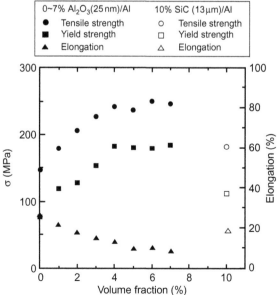

Figure 8.24 Tensile properties of PM-processed Al_2O_3/Al nanocomposites.
Source: Reprinted from Ref. [35] with permission of Elsevier.

drastic reduction in tensile ductility. More recently, Mobasherpour et al. [54] ball milled AA7075 alloy ((5.1−6.3 wt%)Zn−(2.1−2.9 wt%)Mg−(1.2−2 wt%)Cu and balance Al) with alumina nanoparticles. They reported that the hardness and ultimate tensile strength of Al_2O_3/AA7075 nanocomposites increase with increasing alumina content at the expense of tensile ductility.

For FSPed Al_2O_3/AA5052 composite specimens, alumina nanoparticles can disperse more uniformly in the alloy matrix by increasing the number of FSP passes as shown in Figure 8.26A and B[46]. The nominal composition of AA5052 alloy is 2.5 wt% Mg, 0.15 wt% Cr, 0.1 wt% Mn, 0.25 wt% Si, and 0.1 wt% Fe. The respective tensile strength, yield stress, and elongation of FSPed Al_2O_3/AA5052

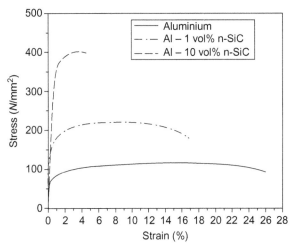

Figure 8.25 Tensile stress−strain curves of extruded Al and n-SiC/Al nanocomposites.
Source: Reprinted from Ref. [44] with permission of Elsevier.

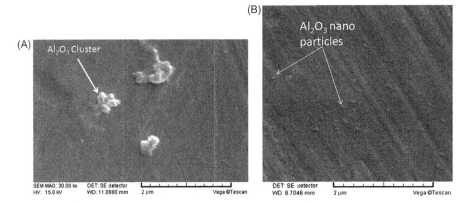

Figure 8.26 SEM micrograph of stir zone of the composite processed after (A) one pass and (B) four passes.
Source: Reprinted from Ref. [46] with permission of Elsevier.

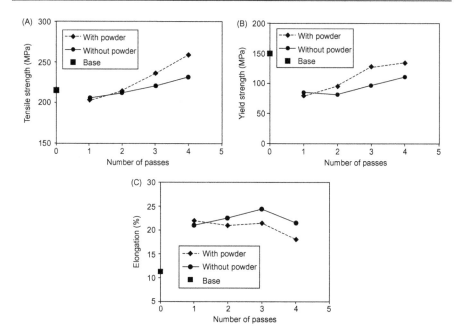

Figure 8.27 (A) Tensile strength, (B) yield strength, and (C) elongation as a function of number of passes for FSPed AA5052 alloy and Al_2O_3/AA5052 composite.
Source: Reprinted from Ref. [46] with permission of Elsevier.

specimens as a function of number of passes are shown in Figure 8.27A−C. Apparently, tensile and yield strengths of the composite specimens improve with increasing FSP passes because higher passes refine the matrix grain size in the stir processing zone effectively. Table 8.3 lists the tensile properties of PM aluminum-based nanocomposites.

8.3.2 Mg-Based Nanocomposites

8.3.2.1 Nanocomposites Fabricated by Liquid-State Processing

Li and coworkers fabricated Mg-based MMNCs reinforced with AlN or SiC nano-particles using ultrasonic cavitation method [13,14,25,26]. Figure 8.28A and B shows typical tensile properties of AZ91D and 1 wt% AlN/AZ91D nanocomposite at room temperature and 200°C, respectively. At room temperature, the yield stress of AZ91D alloy and 1 wt% AlN/AZ91D nanocomposite is 100 and 144 MPa, respectively. The yield stress and tensile strength of AZ91D are enhanced by 44% and 18%, respectively, due to the AlN addition. The mechanical properties of 1 wt % AlN/AZ91D nanocomposite at 200°C are also improved. The yield stress of AZ91D is 87 MPa but increases to 105 MPa by adding 1 wt% AlN. The levels of enhancement in yield stress and tensile strength at 200°C are ∼21% and 25%, respectively. The ductility of the 1 wt% AlN/AZ91D nanocomposite is improved

Table 8.3 Tensile Properties of PM Aluminum Matrix Nanocomposites

Particle Content (vol %)	Reinforcing Nanoparticle	Matrix Material	Fabrication Process	Yield Strength (MPa)	Tensile Strength (MPa)	Fracture Strain (%)	References
0	–	Al	PM process	68	103	–	[10]
1	Si_3N_4	Al	Dry powder mixing, hot pressing at 600°C, and hot extrusion at 420°C	144	180	17.4	[10]
0	–	Al	PM process	76	150	26	[35]
1.5	Al_2O_3	Al	Wet powder mixing in ethanol, cold isostatic pressing and sintering at 620°C for 2 h	124	176	21	[35]
0	–	Al	PM process	–	100	26	[44]
1	SiC	Al	MA followed by hot extrusion	–	205	17	[44]
10	SiC	Al	As above	–	405	4	[44]
0	–	AA7075	PM process	–	276	9	[54]
1	Al_2O_3	AA7075	MA followed by hot pressing at 450°C	–	317	3.2	[54]
3	Al_2O_3	AA7075	As above	–	365	3.1	[54]
5	Al_2O_3	AA7075	As above	–	443	2.1	[54]

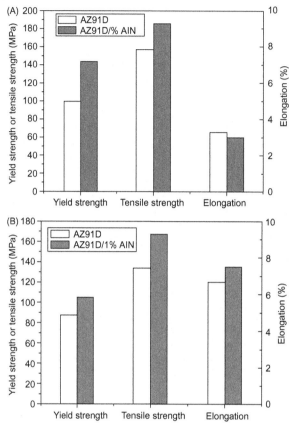

Figure 8.28 Tensile properties of AZ91D and 1 wt% AlN/AZ91D nanocomposite at (A) room temperature and (B) 200°C. *Source*: Reprinted from Ref. [14] with permission of Elsevier.

slightly at 200°C. They reported that ultrasonic cavitation-based solidification processing is very efficient in dispersing AlN nanoparticles in magnesium melt. These well-dispersed nanoparticles improve the mechanical properties of nanocomposites significantly through the Orowan strengthening mechanism.

In the case of SiC/Mg nanocomposites, the incorporation of SiC nanoparticles into the Mg matrix refines its grain size markedly. However, some SiC microclusters still exist in the microstructures, especially at the grain boundaries [25,26]. Nevertheless, the addition of only 1 wt% SiC to magnesium increases its yield strength from 47 to 67 MPa [26]. From Eqs. (8.8)−(8.10), the yield stress of the composite (σ_{yc}) due to the Orowan strengthening, dislocation generation, and load-bearing effects is estimated to be 65 MPa, assuming $\sigma_{ym} = 47$ MPa, $d_p = 66$ nm, $G_m = (E_m/[2(1 + \nu)]) = 16.5$ GPa, E_m (elastic modulus of Mg) $= 42.8$ GPa, ν (Poisson'ratio) $= 0.3$, $T_{process} = 750°C$, $T_{test} = 20°C$, $\alpha_m = 25 \times 10^{-6}$ $(°C)^{-1}$, $\alpha_p = 3.2 \times 10^{-6}$ $(°C)^{-1}$, and $b = 0.32$ nm. By further taking the Hall−Petch effect (Eq. (8.5)) into consideration, theoretical σ_{yc} value increases to 77 MPa. This predicted yield stress is somewhat larger than experimental yield stress of the nanocomposite, i.e., 67 MPa. This is due to the presence of SiC clusters in the nanocomposite. Agglomeration of nanoparticles reduces the effect of Orowan mechanism by changing the effective distance between nanoparticles and the nature of the obstacles.

Nie et al. [27] utilized ultrasonic vibration during stir casting to fabricate 1 vol% SiC/AZ91 nanocomposite. In the process, SiC nanoparticles (60 nm) were added to molten magnesium alloy at 700°C under ultrasonic vibration for 20 min followed by casting into a mold. Most SiC nanoparticles are well dispersed in the matrix, with the presence of some SiC microclusters. The ultimate tensile strength, yield strength, and elongation to fracture of the 1 vol% SiC/AZ91 nanocomposite are considerably enhanced compared with those of the AZ91 alloy (Table 8.4). In another study, Nie et al. [28] produced compocast 1 vol% SiC/AZ91 nanocomposite assisted with ultrasonic vibration as shown in Figure 8.12. The formation of SiC nanoparticle clusters at the alloy grain boundaries was also observed. However, SiC nanoparticles disperse more uniformly within the alloy grains (Figure 8.29). The ultimate tensile strength, yield strength, and elongation to fracture of 1 vol% SiC/AZ91 nanocomposite stirring for 5 min are markedly improved compared with those of cast AZ91 alloy. However, the ultimate tensile strength and elongation to fracture of the 1 vol% SiC/AZ91 nanocomposite decreases with increasing the stirring time (Figure 8.30).

Gupta and coworkers investigated mechanical behavior of magnesium-based composites reinforced with alumina nanoparticles fabricated by DMD and hot extrusion processes technique [29−33]. Representative results of their studies are listed in Table 8.4. It appears that small additions of alumina nanoparticles to pure magnesium and AZ31 alloy increase the yield stress, tensile strength, and tensile ductility considerably. It is noted that the tensile ductility of magnesium and AZ31 alloy improves greatly by incorporating alumina nanoparticles in their lattices. The improvement in tensile ductility of Mg-based nanocomposites is attributed to the activation of the basal slip system by ceramic nanoparticles. Magnesium generally exhibits low tensile ductility because of its HCP structure having three independent slip system only. In this regard, basal slip is difficult for pure Mg during mechanical deformation at room temperature. However, alumina nanoparticles in the magnesium matrix tends to activate basal slip during tension, leading to enhanced tensile ductility. Table 8.4 summarizes tensile properties of the Mg-based MMNCs prepared by liquid metallurgy route.

8.3.2.2 Nanocomposites Fabricated by Solid-State Processing

Gupta and coworkers also fabricated 1.1 vol% Al_2O_3/Mg nanocomposite using simple powder mixing and sintering [30]. Their results demonstrated that alumina nanoparticle addition increases the yield stress, tensile strength, and tensile ductility of magnesium. Similarly, low volume fractions of SiC nanoparticles are also beneficial in enhancing mechanical properties of magnesium nanocomposites fabricated by means of microwave sintering [47]. On the contrary, Ferkel and Mordike [48] indicated that the room temperature tensile strength of extruded 3 vol% SiC/Mg nanocomposite prepared by simple powder mixing is poorer than that of

Table 8.4 Tensile Properties of Cast Magnesium Matrix Nanocomposites

Particle Content (vol %)	Particle Content (wt %)	Nanoparticle	Matrix Material	Fabrication Process	Yield Strength (MPa)	Tensile Strength (MPa)	Fracture Strain (%)	References
–	0	–	AZ91D	Melt casting	100	160	3.3	[14]
–	1	AlN	AZ91D	Melting at 700°C assisted with ultrasonic vibration	144	187	3.0	[14]
–	0	–	Pure Mg	Melt casting	47	120	12.3	[26]
–	1	SiC	Pure Mg	Melting at 750°C assisted with ultrasonic vibration	67	133	6.3	[26]
0	–	–	Pure Mg	DMD and hot extrusion	97	173	7.4	[29]
0.22	–	Al_2O_3	Pure Mg	As above	146	207	8.0	[29]
0.66	–	Al_2O_3	Pure Mg	As above	170	229	12.4	[29]
1.11	–	Al_2O_3	Pure Mg	As above	175	246	14.0	[29]
–	–	Al_2O_3	AZ31	As above	172	263	10.4	[31]
–	1.5	Al_2O_3	AZ31	As above	204	317	22.2	[31]
–	0	–	Mg–2Al–1Si	Melt casting	48.6	137.9	9.2	[25]
–	2	SiC	Mg–2Al–1Si	Melting at 700°C assisted with ultrasonic vibration	73.8	157.2	7.5	[25]
–	0	–	Mg–4Al–1Si	As above	61.9	144.8	8.8	[25]
–	2	SiC	Mg–4Al–1Si	As above	82.1	177.9	9.5	[25]
0	–	–	AZ91	Casting without sonic probe	72	125	2	[27]
1	–	SiC	AZ91	Stir casting at 700°C assisted with ultrasonic vibration	90	220	8	[27]
1	–	SiC	AZ91	Compocasting assisted with ultrasonic vibration	88	188	5.5	[28]

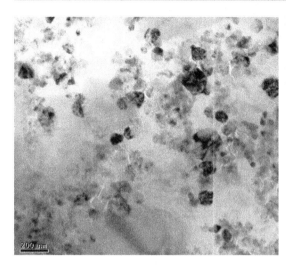

Figure 8.29 TEM image of 1 vol % SiC/AZ91 nanocomposite showing local dispersion of SiC nanoparticles.
Source: Reprinted from Ref. [28] with permission of Elsevier.

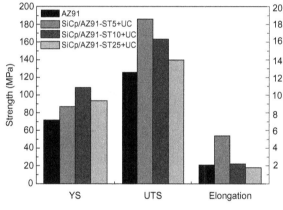

Figure 8.30 Tensile properties compocast AZ91 and 1 vol% SiC/AZ91 nanocomposite. ST5, ST10, and ST25 imply stirring liquid melt for 5, 10, and 25 min, respectively. UC implies ultrasonic cavitation.
Source: Reprinted from Ref. [28] with permission of Elsevier.

extruded Mg. This is due to nonuniform dispersion of SiC nanoparticles in the Mg matrix. However, the room temperature tensile strength of extruded 3 vol% SiC/Mg nanocomposite prepared by ball milling is much higher than that of extruded Mg (Figure 8.31A). At 100°C, SiC nanoparticles are beneficial to enhance its tensile strength due to the softening of the Mg matrix (Figure 8.31B).

Lee et al. [55] employed FSP to disperse SiO_2 nanoparticles (20 nm) in the AZ61 alloy having a nominal composition of $Mg-6.02\%$ $Al-1.01\%$ $Zn-0.30\%$ Mn. Silica nanoparticles are uniformly dispersed in surface alloy matrix after four FSP passes. The tensile properties of AZ61, FSP-processed AZ61 and its nanocomposites are tabulated in Table 8.5. The yield stress and tensile strength of FSPed SiO_2/AZ61 nanocomposites are considerably higher than those of pristine AZ61 and FSP-processed AZ61 as expected (Table 8.5). Furthermore, FSP treatment reduces the grain size of pristine AZ61 alloy markedly from 75 μm down to

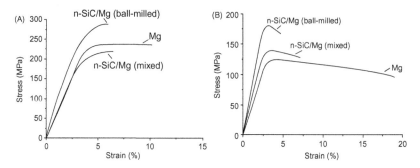

Figure 8.31 Stress–strain curves of extruded Mg and extruded 3 vol% SiC/Mg nanocomposite specimens prepared by powder mixing or ball milling at (A) room temperature and (B) 100°C.
Source: Reprinted from Ref. [48] with permission of Elsevier.

7−8 μm via dynamic recrystallization. This leads to improved yield stress and tensile strength of the alloy. Finally, FSP method is an effective secondary mechanical treatment to remove internal defects of cast magnesium alloy [38,49]. For example, FSP can dissolve hard β-$Mg_{17}Al_{12}$ phase in the grain boundaries of cast AZ91 alloy. This phase often acts as the favorable site for the crack nucleation. Moreover, porosities and voids that reduce tensile strength and elongation of the cast AZ91 alloy are also eliminated by the FSP treatment. Therefore, the resulting nanocomposites exhibit higher tensile ductility than the cast alloy (Table 8.5).

8.4 Conclusions

Ceramic nanoparticles are attractive reinforcing materials for metals to form MMNCs of high mechanical strength and stiffness for structural engineering applications. The mechanical performance of MMNCs depends greatly on the attainment of homogeneous dispersion of ceramic nanoparticles in the metal matrix. Ceramic nanoparticles of large surface-to-volume ratio are poorly wetted with metals, thus causing the formation of particle clusters in the metal matrix. As a result, the resulting nanocomposites exhibit poor mechanical properties.

Liquid-state processing route can manufacture MMNCs in large quantities at a relatively low cost using existing melting and casting equipments. However, ceramic nanoparticles pose serious difficulty in achieving a uniform dispersion because of their high viscosity in molten melt and poor wettability. In this regard, compocasting and ultrasonic vibration are effective techniques to disperse ceramic nanoparticles in molten metal or semisolid metal. PM route can manufacture dense

Table 8.5 Tensile Properties of Magnesium Matrix Nanocomposites Fabricated by Solid-State Processing Route

Particle Content (vol %)	Particle Content (wt %)	Nanoparticle	Matrix Material	Fabrication Process	Yield Strength (MPa)	Tensile Strength (MPa)	Fracture Strain (%)	References
–	0	–	Pure Mg	Dry powder mixing and sintering	132	193	4.2	[30]
0.66	1.5	Al$_2$O$_3$	Pure Mg	As above	194	250	6.9	[30]
–	0	–	Pure Mg	Dry powder mixing and microwave sintering	125	172	5.8	[47]
–	0.35	SiC	Pure Mg	As above	132	194	6.3	[47]
–	0.5	SiC	Pure Mg	As above	144	194	7.0	[47]
–	1.0	SiC	Pure Mg	As above	157	403	7.6	[47]
0	–	–	AZ61	Billet alloy	140	190	13	[55]
0	–	–	AZ61	Four FSP passes	147	242	11	[55]
5	–	SiO$_2$	AZ61	Two FSP passes	185	219	10	[55]
5	–	SiO$_2$	AZ61	Four FSP passes	214	233	8	[55]
10	–	SiO$_2$	AZ61	Two FSP passes	200	246	4	[55]
10	–	SiO$_2$	AZ61	Four FSP passes	225	251	4	[55]
0	–	–	AZ91	Cast alloy	–	139	6.74	[38]
8	–	SiO$_2$	AZ91	One FSP pass	–	193	12.56	[38]
8	–	SiO$_2$	AZ91	Two FSP passes	–	233	13.42	[38]
8	–	SiO$_2$	AZ91	Three FSP passes	–	279	14.67	[38]

MMNCs of higher tensile strength and ductility than the casting process. The high cost of powder materials restricts its use for manufacturing nanocomposite structural components in the industrial sectors.

References

[1] Dragone TL, Nix WD. Steady state and transient creep properties of an aluminum alloy reinforced with alumina fibers. Acta Metall Mater 1992;40(10):2781−91.

[2] Tjong SC, Mai YW. Processing structure−property aspects of particulate- and whisker-reinforced titanium matrix composites. Compos Sci Technol 2008;68 (3−4):583−601.

[3] Ma ZY, Mishra RS, Tjong SC. High-temperature creep behavior of TiC particulate reinforced Ti−6Al−4V alloy composite. Acta Mater 2002;50(17):4293−302.

[4] Ma ZY, Tjong SC. Microstructural and mechanical characteristics of *in situ* metal matrix composites. Mater Sci Eng R 2000;29(3−4):49−113.

[5] Tjong SC, Lau KC. Abrasive wear behavior of TiB_2 particle-reinforced copper matrix composites. Mater Sci Eng A 2000;282(1−2):183−6.

[6] Tjong SC, Lau KC. Properties and abrasive wear of TiB2/Al−4% Cu composites produced by hot isostatic pressing. Compos Sci Technol 1999;59(13):2005−13.

[7] Rawal J. Metal−matrix composites for space applications. JOM 2001;53(4):14−7.

[8] Micracle DB. Metal matrix composites—from science to technological significance. Compos Sci Technol 2005;65:2526−40.

[9] Tjong SC, Chen H. Nanocrystalline materials and coatings. Mater Sci Eng R 2004;45 (1−2):1−88.

[10] Ma ZY, Li YL, Liang Y, Zheng F, Bi J, Tjong SC. Nanometric Si_3N_4 particulate-reinforced aluminum composite. Mater Sci Eng A 1996;219(1−2):229−31.

[11] Ma ZY, Tjong SC, Li YL, Liang Y. High temperature creep behavior of nanometric Si_3N_4 particulate reinforced aluminium composite. Mater Sci Eng A 1997;225 (1−2):125−34.

[12] Yang Y, Lan J, Li XC. Study of bulk aluminum matrix nanocomposite fabricated by ultrasonic dispersion of nanosized SiC particles in molten aluminum alloy. Mater Sci Eng A 2004;380(1−2):378−83.

[13] Lan J, Yang Y, Li XC. Microstructure and microhardness of SiC nanoparticles reinforced magnesium composites fabricated by ultrasonic method. Mater Sci Eng A 2004;386(1−2):284−90.

[14] Cao G, Choi H, Oportus J, Konishi H, Li XC. Study on tensile properties and microstructure of cast AZ91D/AlN nanocomposites. Mater Sci Eng A 2008;494 (1−2):127−31.

[15] Wang D, De Cicco MP, Li XC. Using diluted master nanocomposites to achieve grain refinement and mechanical property enhancement in as cast Al−9Mg alloy. Mater Sci Eng A 2012;532:396−400.

[16] Su H, Gao W, Feng Z, Lu Z. Processing, microstructure and tensile properties of nano-sized Al_2O_3 particle reinforced aluminum matrix composites. Mater Des 2012;36:590−6.

[17] Schultz BF, Ferguson JB, Rohatgi PK. Microstructure and hardness of Al_2O_3 nanoparticle reinforced Al−Mg composites fabricated by reactive wetting and stir mixing. Mater Sci Eng A 2011;530:87−97.

[18] Tahamtan S, Halvaee A, Emamy M, Zabihi MS. Fabrication of Al/A206−Al₂O₃ nano/ micro composite by combining ball milling and stir casting technology. Mater Des 2013;49:347−59.

[19] Sajjadi SA, Ezatpour HR, Parizi MT. Comparison of microstructure and mechanical properties of A356 aluminum alloy/Al₂O₃ composites fabricated by stir and compocasting processes. Mater Des 2012;34:106−11.

[20] Selvam JD, Smart DS, Dinaharan I. Microstructure and some mechanical properties of fly ash particulate reinforced AA6061 aluminum alloy composites prepared by compocasting. Mater Des 2013;49:28−34.

[21] Sajjadi SA, Parizi MT, Ezatpour HR, Sedghi A. Fabrication of A356 composite reinforced with micro and nanoAl₂O₃ particles by a developed compocasting method and study of its properties. J Alloys Compd 2012;511(1):226−31.

[22] Mazahery A, Abdizadeh H, Baharvandi HR. Development of high performance A356/ nano-Al₂O₃ composites. Mater Sci Eng A 2009;518(1−2):61−4.

[23] El-Mahallawi I, Abdelkader H, Yousef L, Amer A, Mayer J, Schwedt A. Influence of Al₂O₃ nano-dispersions on microstructure features and mechanical properties of cast and T6 heat-treated AlSi hypoeutectic Alloys. Mater Sci Eng A 2012;556:76−87.

[24] Mordike BL, Ebert T. Magnesium: properties−applications−potential. Mater Sci Eng A 2002;302(1):37−45.

[25] Cao G, Konishi H, Li XC. Mechanical properties and microstructure of SiC-reinforced Mg-(2,4)Al−1Si nanocomposites fabricated by ultrasonic cavitation based solidification processing. Mater Sci Eng A 2008;486(1−2):357−62.

[26] Erman A, Groza J, Li XC, Choi H, Cao G. Nanoparticle effects in cast Mg−1 wt% SiC nanocomposites. Mater Sci Eng A 2012;558:39−43.

[27] Nie KB, Wang XJ, Hu XS, Xu L, Wu K, Zheng MY. Microstructure and mechanical properties of SiC nanoparticles reinforced magnesium matrix composites by ultrasonic vibration. Mater Sci Eng A 2011;528(15):5278−82.

[28] Nie KB, Wang XJ, Wu K, Xu L, Zheng MY, Hu XS. Processing, microstructure and mechanical properties of magnesium matrix nanocomposites fabricated by semisolid stirring assisted ultrasonic vibration. J Alloys Compd 2011;509(35):8664−9.

[29] Hassan SF, Gupta M. Development of high performance magnesium nanocomposites using solidification processing route. Mater Sci Technol 2005;20(11):1383−8.

[30] Hassan SF, Gupta M. Effect of type of primary processing on the microstructure, CTE and mechanical properties of magnesium/alumina nanocomposites. Comp Struct 2006;72(1):19−26.

[31] Paramsothy M, Hassan SF, Srikanth N, Gupta M. Enhancing tensile/compressive response of magnesium alloy AZ31 by integrating with Al₂O₃ nanoparticles. Mater Sci Eng A 2009;527(1−2):162−8.

[32] Paramsothy M, Nguyen QB, Tun KS, Chan J, Kwok R, Kuma JV, et al. Mechanical property retention in remelted microparticle to nanoparticle AZ31/Al₂O₃ composites. J Alloys Compd 2010;506(2):600−6.

[33] Hassan SF, Paramsothy M, Patel F, Gupta M. High temperature tensile response of nano-Al₂O₂ reinforced AZ31 nanocomposites. Mater Sci Eng A 2012;558:278−84.

[34] Ho KF, Gupta M, Srivatsan TS. The mechanical behavior of magnesium alloy AZ91 reinforced with fine copper particulates. Mater Sci Eng A 2004;369(1−2):302−8.

[35] Kang YC, Chan SL. Tensile properties of nanometric Al₂O₃ particulate-reinforced aluminum matrix composites. Mater Chem Phys 2004;85(2−3):438−43.

[36] Suryanarayana C, Al-Aqeeli N. Mechanically alloyed nanocomposites. Prog Mater Sci 2013;58(4):383−502.

[37] Mishra RS, Ma ZY. Friction stir welding and processing. Mater Sci Eng R 2005;50 (1−2):1−78.

[38] Khayyamin D, Mostafapour A, Keshmiri R. The effect of process parameters on microstructural characteristics of AZ91/SiO$_2$ composite fabricated by FSP. Mater Sci Eng A 2013;559:217−21.

[39] Bauri R, Yadav D, Suhas G. Effect of friction stir processing on microstructure and properties of Al TiC *in situ* composite. Mater Sci Eng A 2011;528(13−14):4732−9.

[40] Razavi Hesabi Z, Simchi A, Seyed Reihani SM. Structural evolution during mechanical milling of nanometric and micrometric Al$_2$O$_3$ reinforced Al matrix composites. Mater Sci.Eng A 2006;428(1−2):159−68.

[41] Zebarjad SM, Sajjadi SA. Microstructure evaluation of Al−Al$_2$O$_3$ composite produced by mechanical alloying method. Mater Des 2006;27(8):684−8.

[42] Zebarjad SM, Sajjadi SA. Dependency of physical and mechanical properties of mechanical alloyed Al−Al$_2$O$_3$ composite on milling time. Mater Des 2007;28 (7):2113−20.

[43] Bathula S, Anandani RC, Dhar A, Srivastava AK. Microstructural features and mechanical properties of Al 5083/SiCp metal matrix nanocomposites produced by high energy ball milling and spark plasma sintering. Mater Sci Eng A 2012;545:97−102.

[44] Kollo L, Bradbury CR, Veinthal R, Jaggi C, Carreno-Morelli E, Leparoux M. Nanosilicon carbide reinforced aluminium produced by high-energy milling and hot consolidation. Mater Sci Eng A 2011;528(21):6606−15.

[45] Shafiei-Zarghani A, Kashani-Bozorg SF, Zarei-Hanzaki A. Microstructures and mechanical properties of Al/Al$_2$O$_3$ surface nanocomposite layer produced by friction stir processing. Mater Sci Eng A 2009;500(1−2):84−91.

[46] Sharifitabar M, Sarani A, Khorshahian S, Shafiee Afarani M. Fabrication of 5052 AlAl$_2$O$_3$ ceramic particle reinforced composite via friction stir processing route. Mater Des 2011;32(8−9):4164−72.

[47] Wong WL, Gupta M. Simultaneously improving strength and ductility of magnesium using nano-size SiC particulates and microwaves. Adv Eng Mater 2006;8(8):735−9.

[48] Ferkel H, Mordike BL. Magnesium strengthened by SiC nanoparticles. Mater Sci Eng A 2001;298(1−2):193−9.

[49] Faraji G, Asad P. Characterization of AZ91/alumina nanocomposite produced by FSP. Mater Sci Eng A 2011;528(6):2431−40.

[50] Dieter GE. Mechanical metallurgy. 3rd ed. New York: McGraw-Hill; 1988. p. 212−20.

[51] Clyne TW, Withers PJ. An introduction to metal matrix composites. Cambridge: University Press; 1993.

[52] Zhang Z, Chen DL. Consideration of Orowan strengthening effect in particulate reinforced metal matrix nanocomposites: a model for predicting their yield strength. Scripta Mater 2006;54(7):1321−6.

[53] Karbalaei Akbari M, Mirzaee O, Baharvandi HR. Fabrication and study on mechanical properties and fracture behavior of nanometric Al$_2$O$_3$ particle-reinforced A356 composites focusing on the parameters of vortex method. Mater Des 2013;46:199−205.

[54] Mobasherpour I, Tofigh AA, Ebrahimi M. Effect of nano-size Al$_2$O$_3$ reinforcement on the mechanical behavior of synthesis 7075 aluminum alloy composites by mechanical alloying. Mater Chem Phys 2013;138(2−3):535−41.

[55] Lee CJ, Huang JC, Hsieh PJ. Mg based nanocomposites fabricated by friction stir processing. Scripta Mater 2006;54(7):1415−20.

9 Polymer Nanocomposites with High Permittivity

Zhi-Min Dang

Laboratory of Dielectric Polymer Materials and Devices, Department of Polymer Science and Engineering, School of Chemistry and Biological Engineering, University of Science & Technology Beijing, Beijing, China

9.1 Introduction

Ferroelectric/piezoelectric ceramics, e.g., barium titanate and barium zirconium titanate, find useful applications as dielectric materials for multilayer ceramic capacitors, electronic and optoelectronic devices due to their high dielectric constant, low dielectric loss, and low cost [1−8]. They are generally quite brittle and require high processing or sintering temperatures. In the energy storage technologies, dielectric capacitors possess an intrinsic high-power density due to their very fast energy uptake and delivery, and thus show great promise for making high-performance power electronics used in hybrid electric vehicles, medical devices, and electrical weapon systems as shown in Figure 9.1 [9−12]. However, such potential applications are severely hindered by the low-energy density of capacitors using current commercially available polymer dielectrics [10,13−15]. To improve the high-energy density of capacitors, materials with high dielectric permittivity and high electrical breakdown are needed.

Comparing with conventional ceramics, dielectric polymer materials have relatively high electrical breakdown field [16−19] and importantly offer processing advantages including mechanical flexibility and the capability of being molded into various shapes for power electronic devices with reduced volume and weight. Most polymers still fall short of rising demands for high-energy density capacitors because of their intrinsic low permittivity (ca. $\varepsilon_r < 10$). This constrains high-energy density in dielectric polymers that strongly relies on the high applied electric field (i.e., $E > 500$ MV/m) [15,17,20]. Thus it is necessary to substantially increase dielectric permittivity of the polymers while retains other beneficial physical properties. In order to acquire high-energy density in the capacitors, polymer materials must possess high dielectric permittivity and high electric breakdown field [9,10,21].

The introduction of inorganic particles into polymer matrices to form dielectric composites represents one of the most promising routes for achieving this purpose. The composite approach combines impressive permittivity of inorganic particles with

Nanocrystalline Materials. DOI: http://dx.doi.org/10.1016/B978-0-12-407796-6.00009-9

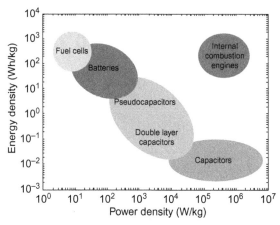

Figure 9.1 Comparison of the power density and energy density for batteries, capacitors, and fuel cells. (Energy is the capacity to do work; power is the rate at which work is done.)
Source: Reprinted from Ref. [9] with permission from Materials Research Society.

mechanical flexibility of polymers for producing materials with high-energy storage capacity. Earlier investigations on dielectric composites focus mainly on the polymer systems filled with particles of micrometer sizes. However, those microcomposites often show inferior dielectric properties due to the presence of surface defects and stress cracking after a long-serving time [22−24]. Furthermore, the thickness of microcomposite films is also often limited by the size of microparticles [25,26]. The miniaturization of modern electronic devices generally requires dielectric capacitors with a very small thickness.

The fast development in nanotechnology in recent years enables chemists and materials scientists to synthesize nanoparticles with high dielectric permittivity (or high-k). Such nanoparticles are effective to fill polymers to form functional nanocomposites with high permittivity. The advantages of polymer nanocomposites include low filler loading, low weight, and large internal surface areas [25,27−29]. Low nanofiller loadings permit the composite formation without sacrificing inherent polymeric properties, such as the density, flexibility, and facile processability. By overcoming the interfacial energy barriers due to the nanoparticle aggregation driven by the van der Wall forces, the fillers with interparticle distance of nanometer level would interact chemically and physically with the polymer matrices, resulting in the emerging of intermediate or mesoscopic properties. Such mesoscopic properties at the interfaces always bring about unexpectedly excellent macroscopic properties for the nanocomposites [21,25,29]. By decreasing the size of filler particles to the nanometer scale, large filler−matrix areas in the composites promote the exchange coupling effect through a dipolar interface layer, affording higher polarization level, dielectric response, and breakdown strength [21,25,30−37]. All these features offer new opportunities for designing different dielectric polymer nanocomposites (DPNCs) with nanometric interfaces, thereby offering higher energy storage or lower dielectric loss, and improved power dissipation [38]. The continuously increasing number of the scientific publications in this field indicates that DPNCs have attracted enormous research attention and gradually become a hot topic in the energy storage field, especially where high-power uptake or delivery is essential.

9.2 Routes of Permittivity Improvement in Polymeric Materials

In view of the basic concept of dielectric physics, polymer nanocomposites with high permittivity can be prepared by employing a polymer with suitable dielectric property as the matrix and nanoparticles with high permittivity/conductivity as the fillers [9−11].

9.2.1 Design and Adjustment of Molecular Structure of Polymers

Electrical polarization of polymer materials originates mainly from their own structures, including the configuration and conformation of polymeric chains and the morphology of polymer composites. The dynamic theory of the dielectric relaxation of dipolar materials is well documented, but is complicated by the local field factors [39,40]. The linear response theory illustrates that the complex permittivity is related to the molecular dipole moment quantities independent of any assumed model for relaxation. To design a suitable structure of polymer, in fact, a well-known Clausius−Mosotti equation relating the molecular weight of polymers, the polarizabilities of electrons and atoms, the dipole orientation of polymer chain structure, and the dipole moment of whole molecular, must be taken into consideration. Clausius−Mosotti equation is given as follows [41]:

$$\frac{k_r - 1}{k_r + 2}\frac{M}{\rho} = \frac{N}{3\varepsilon_0}\left(\alpha_e + \alpha_i \frac{\mu^2}{3KT}\right) \tag{9.1}$$

where k_r is the relative dielectric permittivity of material, M and ρ are the molecular weight and density, respectively, N is the Avogadro constant, α_e and α_i are, respectively, polarizabilities of electrons and the dipole orientation, and μ is the dipole moment. Therefore, the modification of polarizability and density of polymeric materials is also effective for acquiring new polymer materials with required dielectric properties [42]. Moreover, novel polymers with suitable molecular structure can be designed by considering their polarizability before they are synthesized.

9.2.2 A Simple Route by Introducing Nanoparticles into Polymers

Nanoparticles with high permittivity or high electrical conductivity are often incorporated into polymers to form polymer nanocomposite materials with high permittivity [9−11]. Their permittivity can be predicted theoretically based on the rule of mixtures and the percolation theory, respectively. These nanoparticles typically include insulating inorganic ferroelectric particles of high permittivity [43−47], semiconducting metal oxides [21,48−53], semiconducting organic nanoparticles [6,54−58], conducting metals [59−63], and carbon particles [64−74]. It is worth noting that the nanoparticle additions result in a remarkable increase in the filler−matrix interface such that the interfacial polarization contributes largely to the improved permittivity of resulting nanocomposites. From the report of Zhang and Cheng [75], all these polymer composites can be classified into two types, including dielectric—dielectric and conductor—dielectric composites. For dielectric—dielectric composites, it is very crucial to employ fillers with a high dielectric permittivity [43—47]. While for the conductor—dielectric composites, the high permittivity is based on the percolation theory [59—74]. There are many models and equations for simulating effective dielectric constant of a composite. However, each of these has their own limitations based on the assumptions and parameters used in the modeling.

9.2.3 Physical Parameters for Characterizing Dielectric Polymer Composites

There exist four parameters for characterizing dielectric properties of the polymer composites, including dielectric permittivity (dielectric constant), dielectric loss (loss tangent), leakage current, and breakdown strength [10]. For dielectric polymer composites used as the energy storage media in capacitors, low dielectric loss and low leakage current are necessary. Low dielectric loss often denotes a low leakage current in the polymer composites. In addition, it is beneficial to achieve high permittivity and high breakdown strength simultaneously in a polymer composite by adding nanoparticles. This is because high-permittivity polymer composites are commonly used to store electric energy. The maximum energy storage density (w_{max}, J/m^3) of a material can be acquired via the following equation:

$$w_{max} = \frac{1}{2} k_0 k_r E_{break}^2 \tag{9.2}$$

where k_0 is the dielectric permittivity in vacuum having a value of 8.854 pF/m, k_r and E_{break} are the relative permittivity and breakdown strength of materials, respectively. In order to possess maximum energy storage density, the material should have both high dielectric permittivity and high breakdown field. Nevertheless, in most practical applications, the electric field employed is less than the breakdown field. It is well known that one of the typical drawbacks of random composite approach lies in a substantial decrease in the material dielectric strength E_{break}. Higher values of k_r typically correspond to lower values of E_{break}. This experimental evidence is well recognized [43] and reasonably interpreted as the interfacial (Maxwell—Wagner) polarization phenomenon [76,77].

Herein are several approaches for describing E_{break}. The first model deals with a very rough schematization. Considering the polymer as a homogeneous material and a purely elastic body at low strains, and under Maxwell-stress activation condition, the nominal breakdown field of the material is given by the following expression:

$$E_{break} = \frac{V_{break}}{d_0} = e^{-1/2} \sqrt{\frac{Y}{2k_0 k'}} \cong 0.6 \sqrt{\frac{Y}{2k_0 k'}} \tag{9.3}$$

where k' is the dielectric permittivity of the material, Y its elastic modulus, V_{break} the breakdown voltage of the material, and d_0 its thickness at rest. This relation suggests that by increasing the dielectric permittivity of the material, its dielectric strength is expected to decrease as $1/\sqrt{k'}$. Interestingly, such a trend occurs commonly to give a good fit for the experimental data [78].

A delicate model is given by the local field theory of dielectric media subjected to an applied electric field E [79]. Although the space-averaged electric field is uniform, the actual internal electric field can locally vary from point to point, depending on interactions of local fields generated by the dipoles. Accordingly, the local electric field E_{local} is defined as the field that actually acts on an individual polarizable unit (such as a molecule or an atom). This is also known as the Lorentz local field and given by the following expression [79]:

$$E_{local} = \frac{k' + 2}{3} E \tag{9.4}$$

where k' is the dielectric permittivity of the material surrounding the local polarizable unit (excluding the unit itself). Equation (9.4) can be used to obtain an expression that relates the dielectric permittivity and the dielectric strength of the matrix $(k'_m, E_{break,m})$ with those of the resulting composite $(k'_c, E_{break,c})$. Assuming the value of local breakdown field of the matrix is the same as that of the composite, the following expression can be obtained as

$$\frac{k'_m + 2}{3} E_{break,m} = \frac{k'_c + 2}{3} E_{break,c} \tag{9.5}$$

Equation (9.5) can be further analyzed by using k'_c, determining from the mixing rules according to different models. It provides a useful tool to estimate the overall electrical behavior of the composite system.

9.3 Ceramic Nanoparticle/Polymer Nanocomposites with High Permittivity

A simple approach for acquiring high permittivity in the polymer composites is via the incorporation of ceramic nanoparticles with high permittivity into polymers. In this respect, the nanoparticle fillers contribute to the increment in effective dielectric permittivity of the composite in terms of effective medium concept, while the polymer matrices give a protection against dielectric breakdown. Numerous researches studied ceramic particle/polymer nanodielectric materials, and succeeded in improving their overall dielectric performances using ferroelectric metal oxides, such as $BaTiO_3$ [18,31,47,80−87] and $Pb(Zr,Ti)O_3$ (PZT) [88−90], and refractory metal oxides, such as TiO_2 [21,91−100] and ZrO_2 [101−103], in the past few years. However, from the energy storage point of view, there still exists several persistent issues that need to be addressed. The primary challenging issue is associated with the physical dispersion of the nanoparticles. The high surface energy of dielectric nanoparticles usually leads to agglomeration and phase separation from the polymer matrix, especially if they are forced into an incompatible matrix, yielding poor quality of films and weakened dielectric properties, such as high dielectric loss and low dielectric strength [32,47,80].

In order to improve the compatibility between the ceramic particles and polymer phases and thus avoid the agglomerated nanoparticles, voids and phase-separated mixture over large-length scales, surfactants such as phosphate esters and oligomers are usually employed, which allow for an increase in dispersion of the nanoparticles in the matrix, ultimately improving the overall film quality [43,44,104]. Unfortunately, residual free surfactant usually causes unexpectedly deteriorated dielectric properties at the applied fields. An alternative approach is thus to strongly bind surface modifiers to nanoparticles via robust chemical bonds. Ramesh et al. [26] have used trialkoxysilane surface modifiers for the dispersion of $BaTiO_3$ nanoparticles in an epoxy polymer host. The resulting composites exhibited a reasonably high permittivity value up to 45. Kim et al. [47] have also demonstrated a strong surface binding of phosphonic acid ligands onto $BaTiO_3$ nanoparticles and examined the use of such specific surface functionalities to control $BaTiO_3$−polymer interfacial interaction and, thus, facilitate the formation of homogenously dispersed high-energy density nanocomposites. As shown in Figure 9.2, the nanocomposites with unmodified $BaTiO_3$ nanoparticles exhibited large-scale aggregates and pinhole defects. In contrast, the ones with modified $BaTiO_3$ nanoparticles yielded comparatively uniform films with homogenous nanoparticle dispersions. The nanocomposites with 50 vol% modified $BaTiO_3$ particles afforded a large permittivity (~ 37), reasonably low loss (<0.007), and a high dielectric strength (over 200 MV/m). Thus, the modified $BaTiO_3$ nanocomposites featured a large calculated energy density of up to 6.1 J/cm^3. Being different from the Kim's surface modification, Wang et al. [30] successfully produced nanosized, ethylene diamine surface-modified $BaTiO_3$*in situ*. This surface modification afforded an enhancement in the dispersion into organic solvents and subsequently led to a uniform polymer nanocomposite films. As shown in Figure 9.3, Zhou et al. [46] have used hydrogen peroxide to prepare surface-hydroxylated $BaTiO_3$ and subsequently mixed them with PVDF to create hydrogen-bonded nanocomposites. Hydrogen bonds are formed between the electrophilic atoms on the PVDF and the −OH groups on the surface of the hydroxylated $BaTiO_3$ nanoparticle. As compared to the unmodified $BaTiO_3$ nanocomposite, the hydroxylated $BaTiO_3$ nanocomposite exhibited a relatively lower permittivity yet much higher dielectric strength.

An effective multicore interface is also realized to improve the dielectric properties of nanoparticles/polymer composites. For example, Huang and Jiang [105] successfully prepared core−shell structured $BaTiO_3$/poly(methyl methacrylate) (PMMA) nanocomposites via *in situ* atom transfer radical polymerization (ATRP) of methyl methacrylate (MMA) from the surface of $BaTiO_3$ nanoparticles. In this case, the $BaTiO_3$ core affords an increased permittivity (~ 16 at 1 kHz), while the flexible PMMA shell endows the material with good dispersability and inherent low

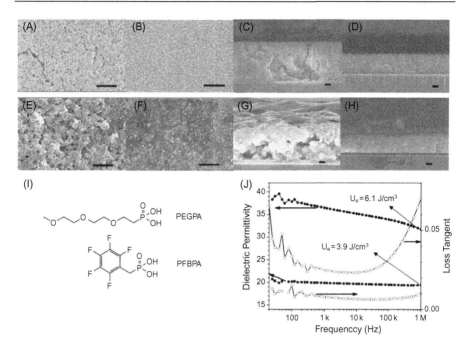

Figure 9.2 SEM images of surface and fractured cross section of spin-coated nanocomposite thin films. (A and C) BaTiO$_3$/polycarbonate (PC), (B and D) PEGPA-BaTiO$_3$/PC, (E and G) BaTiO$_3$/poly(vinylidenefluoride-*co*-hexafluoropropylene) [P(VDF-HFP)], and (F and H) PFBPA-BaTiO$_3$/P(VDF-HFP). All scale bars are 1 μm. (L) Structures of the phosphonic acid ligands used to modify BaTiO$_3$ for dispersion in PC {2-[-(2-methoxyethoxy)ethoxy]thyl} phosphonic acid (PEGPA) and P(VDF-HFP) pentafluorobenzyl phosphonic acid (PFBPA) host materials. (J) Frequency dependent of dielectric permittivity and loss tangent of PEGPA-BaTiO$_3$/PC (squares) and PFBPA-BaTiO$_3$/P(VDF-HFP) (circles).
Source: Reprinted from Ref. [47] with permission from Wiley.

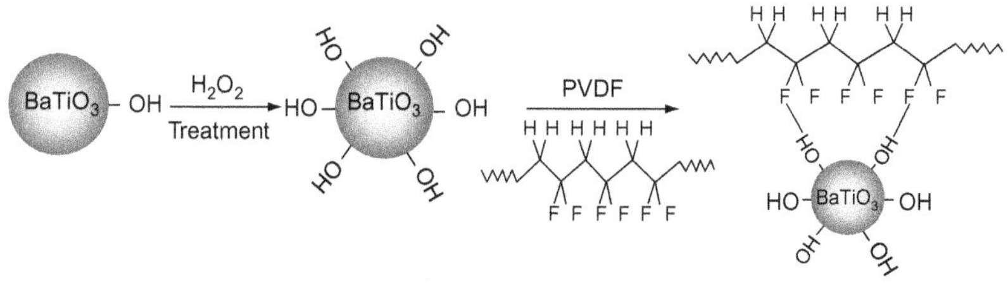

Figure 9.3 Schematic illustration of the hydroxylation of BaTiO$_3$ nanoparticles and formation of hydrogen bond in h-BaTiO$_3$/PVDF nanocomposites.
Source: Reprinted from Ref. [46] with permission from The American Chemical Society.

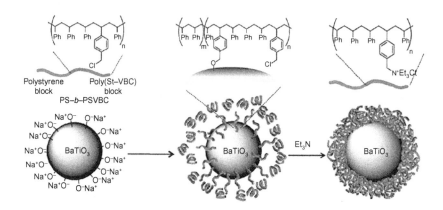

Figure 9.4 Synthesis route for BaTiO₃ particle with the polymer shell.
Source: Reprinted from Ref. [82] with permission from The American Chemical Society.

loss of the base polymer ($\sim$0.037 at 1 kHz). The insulating polymer shell around the nanoparticles is found to be capable of serving as effective barriers to retard the movement of the charge carriers in the composites. Recently, Jung et al. [82] investigated the effect of robust polymer shell on dielectric insulation and energy storage density in polystyrene (PS) nanocomposites having core–shell BaTiO₃ particles with diblock copolymer shielding layers. The adoption of diblock copolymer facilitated the formation of a well-defined layered polymer shell through a two-step processes, as illustrated in Figure 9.4. A weakly adsorbed layer was first formed by reacting chloromethyl group of polystyrene-block-poly(styrene-co-vinylbenzylchloride) (PS-*b*-PSVBC) with the oxy anion on the BaTiO₃ surface. This weakly adsorbed layer subsequently underwent a triethylamine treatment to transform nonpolar chloromethyl functional blocks to polar ammonium blocks, forming a self-assembled stable and dense layer that interacted strongly with the surface of BaTiO₃ particles. As a result, the layered polymer shell was self-assembled with nonpolar PS block in an outer layer while polar ammonium-containing block in an inner one. This interesting feature was beneficial for good dispersibility in a nonpolar medium, providing a high-energy storage capacity at high applied electric field. As shown in Figure 9.5, diblock copolymer-coated BaTiO₃ nanocomposite afforded a high dielectric permittivity ($\sim$44 at 10 kHz) with a significantly low leakage current density of 0.76 nA/cm², even the BaTiO₃ particle content reached as high as 75 wt%. The extremely large dielectric breakdown strength (up to 222 MV/m) at high permittivity consequently gave rise to a high calculated energy density of 9.7 J/cm³.

The introduction of anisotropy into the nanocomposites has been widely demonstrated to dramatically affect dielectric properties and thus energy storage capacity of the composites. To obtain high dielectric permittivity and breakdown strength materials with high-energy density, the orientation state or alignment degree of nanofillers must be well controlled. For instance, Sodano et al. prepared nanocomposites with aligned PZT nanowires using uniaxial strain assembly [89]. This method is effective to align

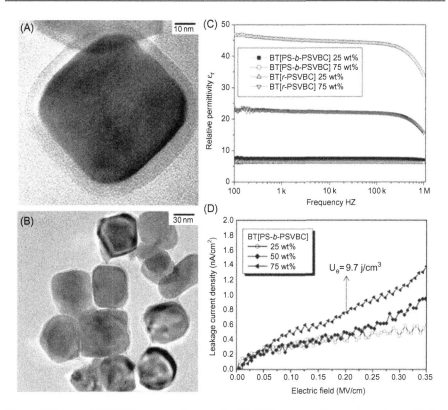

Figure 9.5 (A and B) TEM images of core−shell BaTiO₃ nanoparticles with a dense PS-*b*-PSVBC shell (after amine treatment). The thickness of the polymer shell is in the range of 5−7 nm. (C) Frequency dependence of dielectric permittivity of BaTiO₃[PS-*b*-PSVBC]PS and BaTiO₃[r-PSVBC]PS nanocomposites at different particle fractions. (D) Leakage current measurements for the BaTiO₃[r-PSVBC] nanocomposites.
Source: Reprinted from Ref. [82] with permission from The American Chemical Society.

nanowires in bulk nanocomposites even at high filler concentrations, in which dielectro-phoresis usually fails. As shown in Figure 9.6, the incorporation of PZT nanowires nota-bly increases electric displacement, especially in the case of 3-direction (*z*-aligned) nanocomposites. Therefore, the 3-direction aligned nanocomposites have the largest energy density. It should be also noted that the energy density increases with the PZT volume content. This demonstrates that the energy density of nanocomposites is strongly dependent upon the orientation state of PZT nanowires, regardless of the nano-wire concentration; the 3-direction aligned nanocomposites exhibit higher energy den-sity than randomly oriented nanowire composites. Herein we only cite the method of uniaxial strain assembly as an example to illustrate the impact of processing approaches on the energy storage performance of nanocomposites. Practically, any other methods that can align nanoparticles in the composites effectively can be employed to tailor the anisotropy and in turn, increase the energy density of the nanocomposites.

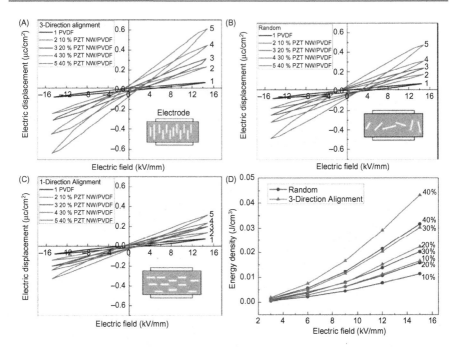

Figure 9.6 Electric displacement—electric field (D−E) loops for the polymer: (A) 3-direction (z-aligned), (B) randomly dispersed, and (C) 1-direction (x−y aligned) PZT/PVDF nanocomposites, measured under different applied fields at room temperature and 100 Hz. Insets in (A−C) show the schematic images of the orientation state of PZT nanowires in different nanocomposites. (D) Dependence of energy density on the PZT volume fraction in the PVDF matrix under 15 MV/m.
Source: Reprinted from Ref. [89] with permission from Wiley.

9.4   Metal Nanoparticle/Polymer Nanocomposites with High Permittivity

Inspite of great progress readily made in improving the permittivity of ceramic particle/polymer composites, the common drawback of limited permittivity enhancement remains a key challenging issue. By replacing ceramic nanoparticles with conductive nanomaterials, percolative polymer nanocomposites with high dielectric permittivity in the vicinity of the percolation threshold can be created [59−63]. Such percolative systems are alternatives to the ceramic/polymer nanocomposites acting as nanodielectrics for high-density energy storage capacitors due to their extremely high dielectric permittivity and low filler loading levels. Dang et al. [59] first proposed and demonstrated percolative Ni/PVDF composites as a dielectric material for the energy storage applications. The dielectric permittivity of the composites can reach 400 when the Ni volume content approaches the percolation threshold of 17 vol%. However, inevitable high dielectric loss and leakage current are often accompanied due to the metal particles contact each other in forming a conductive network. Comparing to conductive spherical particles, one-dimensional (1D) carbon nanotubes with large aspect ratios can give rise to even much lower percolation threshold.

During the past years, a few strategies have been proposed to prevent a direct contact between conductive particles near the percolation threshold (f_c), aiming at high dielectric loss suppression, maximizing dielectric strength, and also extending the compositional windows. One straightforward approach is to introduce an inter-particle barrier layer into percolative nanocomposites. As such, the tunneling current between neighboring conductive nanoparticles can be suppressed, leading to lower dielectric loss and higher breakdown strength. Additionally, the interfacial layers introduced can tailor sensitive microcapacitor structures near the f_c, which generally gives rise to a considerably expanded compositional window for the per-colative composites. By fine tuning the properties of barrier layers, a desirable energy density can be achieved in the percolative composites due to the largely increased permittivity and slightly reduced breakdown strength. The introduction of interfacial layer into percolative composites was first demonstrated by Wong and coworkers [106,107] in the metal/polymer composite system. Aluminum fillers (100 nm) were initially self-passivated to form a thin alumina (Al_2O_3) layer (2.8 nm), then compounded with epoxy to yield percolative composites. The incor-poration of such core−shell Al nanoparticles into the epoxy matrix causes a signifi-cant increase in dielectric permittivity without affecting inherent loss tangent of the matrix [106]. A high dielectric permittivity up to 60 can be achieved at 50 wt% filler content, while keeping the loss tangent as low as 0.02. The alumina interfacial layer effectively restricts the electron transfer between neighboring Al nanoparti-cles, ensuring the loss tangent at a low level. The slight decrease in dielectric per-mittivity at filler content above 50 wt% can be attributed to the void formation at high filler content as a result of the presence of fractal clusters. The resulting nano-composites are anticipated to show excellent low field dielectric properties. Because of the lacking of characterization technique for high field dielectric responses, it is difficult to assess the energy density of such composites.

Comparing to the self-passivated inorganic shell, coating organic interfacial layers on conductive nanoparticles is considered to be more versatile, allowing larger tunability in dielectric properties of percolative composites. Qi et al. [62] modulated Ag nanoparticles (40 nm) with a thin layer of mercaptosuccinic acid (MSA) molecules and then mixed with epoxy to form nanocomposites via *in situ* polymerization process. The dielectric permittivity of the composites can reach up to 300, while the loss tangent still remains <0.05 as shown in Figure 9.7A. The presence of organic coating renders the formation of conductive filler network very difficult, which thus significantly reduces the dielectric loss and affords a desirable high concentration tolerance for high permittivity. Such high permittivity and low loss tangent are attractive for the energy storage applications. However, the break-down strength decreases substantially with increasing Ag volume content (Figure 9.7B), indicating the organic surfactant coating fails to prevent tunneling current at high fields due to their loose morphology and possibly the presence of

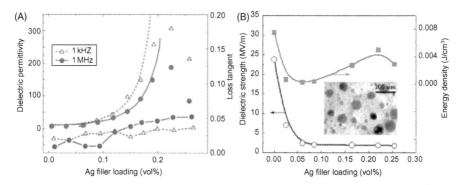

Figure 9.7 (A) Dependence of dielectric permittivity and loss tangent of MSA-modulated Ag/epoxy composites on the volume fraction of fillers, measured at different frequencies. The Ag nanoparticles used in these composites have a diameter of 40 nm. (B) Dielectric breakdown strength and maximum calculated energy density of MSA-modulated Ag/epoxy composites as a function of Ag volume fractions. The maximum energy density was calculated using the permittivity data at 1 kHz. Inset in (B) shows the morphology of surfactant-modified silver nanoparticles.
Source: Reprinted from Ref. [62] with permission from Wiley.

structural defects. Because of a significant reduction in dielectric strength, the energy density of the composites is smaller relative to the polymer matrix. Therefore, a denser organic layer with sound structural integrity is highly desirable to improve both low field and high field dielectric properties, thus, in turn, increases the energy density of percolative composites.

Nan and coworkers [108,109] prepared Ag@C core−shell nanoparticle/epoxy percolative composites recently. In the process, Ag nanoparticles were coated with carbonaceous organic shells via hydrothermal method, followed by compounding with the epoxy resin to create percolative nanocomposites. The morphologies of the Ag@C core−shell nanoparticles are clearly presented in Figure 9.8A and B. The thickness of shells can be fine tuned such that the dielectric properties are tailored to meet the specific requirements in practical applications.

Core−shell nanoparticles generally form a network near the percolation threshold. Each two neighboring particles can be treated as nanocapacitors, preventing the Ag cores from contacting each other directly even at high filler loading (30 vol%). As shown in Figure 9.8B and D, the Ag@C/epoxy nanocomposites show high permittivity ($\sim$450) and very low loss tangent (tan $\delta < 0.05$), and the permittivity does not change with high applied electric fields. However, the breakdown voltage drops with increasing filler concentration as expected, for example, from ca. 30 MV/m to 0.6−1 MV/m. So the energy density of the polymer matrix can be improved at certain optimal filler concentrations, and such enhancement becomes larger by increasing the thickness of the shells. Of particular note is that these optimal filler concentrations are far below the percolation threshold, suggesting that the percolative composites with high permittivity near f_c may not attain a high-energy density at high fields, though low field characteristics of the composites would be

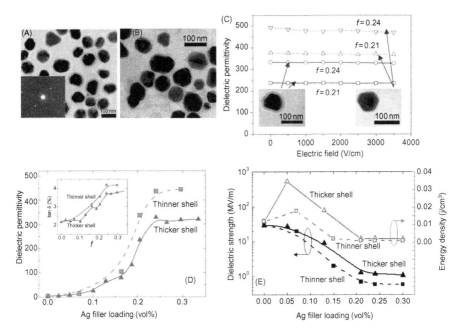

Figure 9.8 TEM images of the Ag@C core/shell nanoparticles with (A) thick shells and (B) thinner shells. The single Ag crystals can be readily seen from the selected area electron diffraction pattern of the particle. (C) High field dielectric permittivity for the composites with different volume fraction of thinner and thicker Ag@C core/shell nanoparticles. (D) Low field dielectric permittivty as a fucntion of Ag volume fraction in the Ag@C filled composites. (E) Dielectric breakdown strength and calculated energy density of the composites as a function of the filler contents. The energy density was calculated by using the permittivity data at 1 kHz.
Source: Reprinted from Refs. [108,109] with permission from Wiley.

significant. For the pulsed power applications, the high field behavior (applied electric voltage > 10 kV) is more important. Thus, the key topic for further development of high-energy-density percolative composites lies in the improvement of dielectric strength rather than the enhancement of permittivity near the percolation threshold. In this case, how to improve the high field dielectric responses of polymers with conductive nanofillers still remain an open question.

When metal particles are incorporated into polymers, dielectric permittivity at the percolation concentration often decreases dramatically with an increase in temperature [59], possibly attributed to the inevitable thermal expansion of the polymer matrix. Thus thermal expansion results in a breakdown of conducting network structure. A decrease in conductivity of the Ni/PVDF composites with temperature supports the thermal expansion effect. However, Qi et al. [62] reported that the dielectric permittivity of nanosized Ag particle/epoxy nanocomposites exhibits good temperature stability from 25 to 135°C. Nanosized Ag particles were formed directly in the epoxy resin via *in situ* method. As such, the Ag/epoxy composites

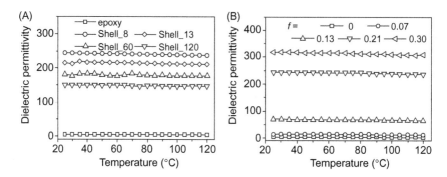

Figure 9.9 Temperature dependence of dielectric permittivity at 1 kHz: (A) the volume fraction is 0.21, 0.24, 0.19, and 0.18 for the shell_8, shell_13, shell_60, and shell_120 cases. (B) Temperature dependence of dielectric permittivity at 1 kHz of the composite films containing Ag@C particles with the thicker shells at different concentrations.
Source: Reprinted from Refs. [108,109] with permission from Wiley.

were more resistant to temperature-induced structural changes. In contrast, Nan's group reported an alternative method to acquire temperature stability of dielectric permittivity in the metal particle/polymer composites [108,109]. The use of Ag@C core/shell structure particles with a thin (nanosize) organic shell layer, as shown in Figure 9.9, resulted in remarkable temperature stability of dielectric permittivity in the Ag@C/epoxy composites. The organic nanoshells served as electrical barriers between the Ag cores to form a continuous interparticle-barrier-layer network and to retain a stable high dielectric permittivity and low loss. Obviously, the use of metal nanoparticles with a core−shell structure is effective at tuning dielectric properties of conducting filler/polymer composites.

9.5 Organic Nanoparticle/Polymer Nanocomposites with High Permittivity

Organic fillers, such as copper-phthalocyanine (CuPc) and polyaniline (PANI), are often used to prepare high-*k* polymer composites [6,20,56−58,110−114]. CuPc and PANI are organic materials, with their composites show excellent flexibility and typically employed as electrostriction materials [6,112]. The dielectric permittivity of those organic filler/polymer composites is also highly temperature and frequency dependent. For example, Zhang et al. [6] reported that the dielectric permittivity of the 40 wt% CuPc/poly(vinylidene fluoride-trifluoroethylene) [P (VDF-TrFE)] composites increases significantly with temperature and frequency when compared to the polymer matrix (Figure 9.10).

By irradiating CuPc/P(VDF-TrFE) composite, its permittivity also increased, especially at 50°C [92]. However the dielectric permittivity decreased dramatically

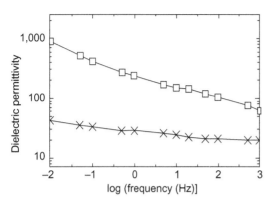

Figure 9.10 The weak field (100 V/cm) dielectric permittivity as a function of frequency for the composite (squares) and the polymer matrix (crosses). Data points are shown and solid curves are a guide to the eye.
Source: Reprinted from Ref. [6] with permission from Nature.

with temperature. From this view point, surface modification of the CuPc particles and chemical graft methods can be employed to realize composites with better permittivity. Wang et al. [115] prepared PVDF nanocomposite having poly(p-chloromethyl styrene) (PCMS) grafted with high dielectric constant CuPc (PCMS-g-CuPc) as the filler. Compared to neat PVDF and CuPc/PVDF materials, the (PCMS-g-CuPc)/PVDF composite displayed a marked increase in dielectric permittivity and a relatively weak frequency dependence of dielectric permittivity. The significant enhancement of the dielectric response can be attributed to an enhanced exchange coupling effect as well as the Maxwell–Wagner–Sillars (MWS) polarization mechanism. Wang et al. [56,57] also employed P(VDF-TrFE) and poly (vinylidene fluoride-trifluoroethylene-chlorofluoroethylene) [P(VDF-TrFE-CFE)] as the matrices, respectively, to study the effect of grafting and blending processes of CuPc/polymer composites on their dielectric properties. They found that compared with simple blend of the P(VDF-TrFE-CFE) and CuPc, a lower dielectric loss and a higher breakdown field were achieved in the grafted nanocomposite. They attributed this to a more uniform distribution of o-CuPc particles in the polymer matrix as well as a much smaller inclusion size. Moreover, the dielectric constant of grafted nanocomposite at frequencies above 1 kHz is much higher than that of the simple blend, indicating an increased interface effect like the exchange coupling that can enhance the dielectric response.

By employing polyurethane (PU) as the polymer matrix, Huang et al. [116] studied fully functionalized high-dielectric-constant nanophase polymers with high electromechanical responses. Three-component multifunctional PANI-*g*-PolyCuPc-*g*-PU composites exhibit relatively weak frequency dependence. Such a nanophase polymer has a strong interface effect in which the exchange coupling markedly increases dielectric response to an extent that exceeds the prediction of simple mixing rules for dielectric composites [36]. Consequently, these self-organized nanophase polymers exhibited high dielectric permittivity, reduced dielectric loss, and improved breakdown field and reliability, all of which are highly desirable for high-dielectric-permittivity materials for actuator and dielectric applications.

In the case where organic conducting particles (PANI) are employed as the fillers, the frequency dependence of dielectric properties is relatively strong [20,113,114,117].

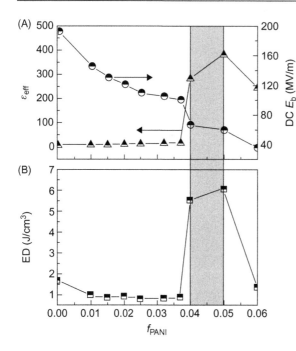

Figure 9.11 Dependence of (A) dielectric permittivity (ε_{eff}) measured at room temperature and 10^3 Hz, DC breakdown strength (DC E_b) and (B) energy density (ED) of the PANI/PVDF films on f_{PANI}. *Source*: Reprinted from Ref. [20] with permission from RCS.

Dang et al. [20] demonstrated that dielectric permittivity of the PANI/PVDF nanohybrid films was as high as 385 (at 10^3 Hz), with the breakdown strength of 60 MV/m, and the energy storage density of 6.1 J/cm³ (Figure 9.11), being three times higher than that of neat PVDF. Moreover, the dielectric permittivity of such nanohybrid films was frequency independent, even the PANI concentration approaches f_c. This route is demonstrated to be effective to prepare high-energy-density capacitor material used for wide frequency ranges. In addition, Bobnar et al. also reported the occurrence of enhanced dielectric responses in all-organic PANI/P(VDF-TrFE-CFE) composites [117]. Although the conductivity at the lowest frequencies approaches direct current conductivity, both dielectric permittivity and conductivity data are strongly remindful of the dielectric relaxation with the characteristic frequency of $\sim 10^4$ Hz. Thus, Maxwell–Wagner effect—the charging of interfaces that occurs in electrically heterogeneous materials, resulting in dielectric response with the frequency character being identical with that of the Debye dipolar absorption. Obviously, pure terpolymer exhibits dielectric properties typical of inorganic relaxors, yet its dielectric constant is much lower. On the other hand, an admixture of PANI, whose conductivity follows the Mott's variable range hopping behavior, strongly increases the dielectric constant of the terpolymer, especially when approaches the percolation threshold. The frequency dependence of dielectric response of the composite is tentatively described in terms of the extended Maxwell–Wagner mixture approach. Wang et al. observed an enhancement of the electrical properties of ferroelectric polymers by adding PANI nanofibers with controllable conductivities [113]. As shown in Figure 9.12, the dielectric constant of the PANI-HCl nanofiber (2.9 wt%)/P(VDF-TrFE) composites is two to four times

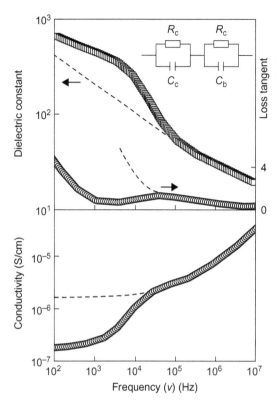

Figure 9.12 The dielectric constant, loss, and conductivity as a function of frequency measured for the composite with 2.9 wt% PANI-HCl nanofibers.
Source: Reprinted from Ref. [113] with permission from Wiley.

higher than that of the polymer matrix at frequencies above 0.1 MHz. The exchange coupling between the nanosized filler and P(VDF-TrFE) with a large difference in dielectric response or inhomogeneous field distribution in the matrix is likely responsible for the observed dielectric enhancement [118].

It is widely recognized that the difference in the electron work function of a semiconductor and the conductor yields a thin layer in the semiconductor close to the junction that depletes of mobile electron carriers. When the concentration of conductive PANI nanofibers randomly distributed in insulating ferroelectric polymer approaches the percolation threshold, the composites are semiconducting and the depletion regions are located at the interfaces between the composites and metallic electrodes. Another explanation for the occurrence of depletion layer is partial overlapping of nanofiber grain boundaries near the percolation threshold. By doping PANI with dodecylbenzene sulfonic acid (DBSA) and perfluorosulfonic acid (PFSA) resins, respectively, the Debye-like relaxation of the PANI-PFSA nanofiber/P(VDF-TrFE) composite takes place at a lower frequency than that of the PANI-HCl/P(VDF-TrFE) composite and effectively restrains the leakage current of the former. However, in comparison to the PANI-PFSA/PANI composite, the PANI-DBSA/PANI composite presents weak frequency dependence of dielectric permittivity.

9.6 Carbon Nanoparticle/Polymer Nanocomposites with High Permittivity

Carbon nanoparticles with high electrical conductivity are also used for improving dielectric permittivity and subsequently energy storage density in the carbon nanoparticle/polymer composites. In fact, this route is principally based on the percolation effect as mentioned above. These nanoparticles include electrical conducting carbon black (CB), carbon nanotube, carbon nanofiber, graphite nanoplatelet, and graphene sheet. By virtue of their spherical shape, the function of CBs is almost the same as that of metal nanoparticles. Namely the percolation concentration of the particle/polymer composites is relatively high. For example, Stoyanov et al. [119] prepared CB/styrene ethylene−butylene−styrene (SEBS) composites, and found that the increase in permittivity of the composites cannot counter adverse impact of a reduction in dielectric strength. The energy density of CB/SEBS composite continuously decreased with increasing CB concentration. However, the corresponding low field dielectric spectroscopy displayed an insulating behavior up to 3.5 vol% CB. So they emphasized that the low field dielectric spectra do not provide adequate information to judge the usefulness of percolation dielectrics at high field for the energy storage. In addition, both low field and high field dielectric properties of these percolative systems are extremely sensitive to the filler concentration, especially near the percolation threshold. It has been demonstrated that, in some cases, a small deviation of 0.05% from the percolation threshold can decrease dielectric permittivity by 100%. Such a narrow processing window indeed imposes a serious challenge in controlling percolative composites with the threshold composition for practical energy storage applications [120]. Thus, the optimal way is to adopt filler concentrations a little far below from the percolation threshold.

Geometric factors and/or conductivity of the carbon fillers play a key role in determining the dielectric properties of the nanocomposites mainly through affecting the value of percolation threshold [67,68,121,122]. For example, the additions of those nanoparticles with 1D or 2-dimension (2D) with high aspect ratios to polymers can yield a low percolation threshold in the polymer composites. Monte Carlo simulations predict the influence of the aspect ratio of ellipsoidal fillers on the percolation threshold as

$$f_c \propto \frac{1}{p} \tag{9.6}$$

where p is the ratio of length of long axis to short axis. This expression has been well verified by the experimental results [123−126]. For instance, nanocomposites filled with $p \sim 12$ carbon fiber [64], $p \sim 200$ exfoliated graphite nanoplatelet [72], and $p \sim 1000$ graphene sheet [127] display rapidly decreased f_c values of 0.074, 0.010, and 0.001, respectively, as a result of the increase in aspect ratios. Moreover, a rather high dielectric permittivity of 2700 at 100 Hz is observed at such low filler concentrations [72]. The low percolation threshold in such conductive composites renders high mechanical flexibility, reduced volume, and weight, which is thus highly desirable for satisfying the requirements of high-performance flexible energy storage devices.

From the literature, a higher dielectric permittivity of 600 was observed near the percolation threshold of 8 vol% in multiwalled carbon nanotube (MWNT)/PVDF nanocomposites, where MWNT fillers were chemically modified with 3,4,5-trifluorobromobenzene (TFBB) [67]. Such high dielectric permittivity originated from the MWS effect at the percolation threshold and associated with a thin insulating layer of PVDF covering functionalized MWNT. The mobile charge carriers were blocked at the interfaces between functionalized MWNT and the PVDF, and the large π-orbital of MWNT provided mobile electrons with large domains. Strongly electrophilic F groups on functionalized MWNTs further reinforced the MWS effect. Another promising candidate for creating high-permittivity percolative composites is the deployment of 2D conductive filler with a disk or platelet shape. For example, He et al. [72] fabricated graphite nanoplatelet/polymer nanocomposites using solvent casting process. The graphite nanoplatelets were aligned with their faces parallel to each other but separated by a thin host polymer layer, forming numerous microcapacitors. As a result, a large dielectric permittivity of 2700 was observed near the percolation threshold as low as 0.01.

In general, a large deviation of dielectric permittivity observed in different composite systems can be largely attributed to the size, shape, and spatial arrangement of nanoparticles as well as intrinsic properties of the polymer matrices and nanofillers [128−131]. Yao et al. [128] studied the effect of carbon nanotube aspect ratio on the percolation threshold and found that the threshold first increased with increasing aspect ratio, reached a maximum and subsequently followed a decrease with further increasing aspect ratio. A decrease in the percolation threshold at high aspect ratio can be explained in term of the nanotube entanglement.

By forming 1D or 2D nanoparticles with a core−shell structure, the dielectric properties of these particle/polymer composites can be improved as expected. The dense and integral organic shells would minimize the decrease in the dielectric strength and give rise to an increase in the energy density relative to the neat polymer matrix. Very recently, Yang et al. [132] successfully coated MWNTs with polypyrrole (PPy) organic shells by means of inverse microemulsion polymerization process. TEM images of MWNTs revealed that the thickness of PPy layer was in the order of 10 nm as shown in Figure 9.13. Such PPy-coated MWNT/PS composites displayed a high dielectric permittivity ($\sim$44), low loss (<0.07), and high-energy density (4.95 J/cm^3). This value is about two times higher than that of pure PS and also larger than that of state-of-the-art commercial biaxially oriented polypropylene (BOPP) capacitor dielectrics. The enhanced dielectric performance originated from the organic shell PPy, which not only enabled good dispersion of MWNTs in the polymer matrix but also screened the charge movement to shut off leakage current. Liu et al. [133] modified MWNTs with jeffamines T403 for establishing a strong interaction between functionalized MWNT-T403 and polysulfone (PSF) polymer matrix via the ester and ether groups on the MWNT-T403 surface. Consequently, electrospinning method was used to prepare MWNT-T403/PSF

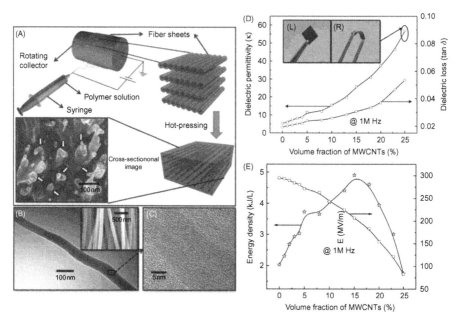

Figure 9.13 (A) Schematic for preparation of MWNT-T403/PSF composites. The inset shows the cross-sectional SEM image of a composite perpendicular to the fiber direction, MWNTs are indicated by the arrows. (B) TEM image for MWNT-T403/PSF core/shell nanostructure. SEM image in the inset shows the MWNT-T403/PSF nanofibrous sheet prepared by electrospinning. (C) High-resolution TEM image of an MWNT. (D) Dependence of dielectric permittivity and loss of the composites on MWNT-T403 loading, measured at room temperature and 1 MHz. The flexibility of the composite is demonstrated in the photographs in the insets. (E) Energy density and breakdown strength of the composites with different MWNT-T403 loadings.
Source: Reprinted from Ref. [133] with permission from Wiley.

composite nanofibers with MWNT-T403 coated by PSF as shown in Figure 9.13. Such electrospun composite nanofibers were collected in the form of a sheet on a rotating drum, with several plies of the sheets stacked along the same fiber axis direction. These sheets were finally molded by hot pressing near the softening temperature of PSF, and perpendicular to the fiber direction to form the MWNTs array (Figure 9.13A). The organic, dielectric PSF coating on MWNTs not only acts as a barrier to prevent MWNTs from direct contact but also as a polymer matrix after fusion by hot pressing at 175°C. The resulting MWNT-T403/PSF composites show a high permittivity (ca. 58) when the MWNT-T403 content reaches 25 vol%, while there is only a slight change in loss (0.02→0.05 at 1 MHz) (Figure 9.13D). And also, a maximum energy density of 4.98 kJ/L is observed when the MWNT-T403 content approaches 15 vol%, as shown in Figure 9.13E.

Apparently, the shells on conductive fillers with good integrity are desirable for attaining high dielectric permittivity and low loss tangent such that they provide a better insulating barrier for electron tunneling and suppress the leak current resulting from quantum tunneling. Although the low field characteristics of percolation nanocomposites are attractive, the high field dielectric behaviors are much more important for the power energy storage applications. The introduction of thickness monitoring and integral organic shells with high breakdown voltage can effectively optimize the high field tolerance, which is beneficial in maximizing the energy density of the percolative composites.

9.7 Current Issues and Conclusions

Though a great deal of works have been reported, it is still difficult to attain excellent polymer nanocomposites with high permittivity and other integrated properties. Proper understanding of intrinsic mechanisms responsible for dielectric properties of the nanoparticle/polymer composites is needed so that advanced dielectric composite materials can be fabricated. The general aim is to accomplish the universally desirable dielectric properties (large k_{eff} and high E_b) necessary for a large energy density in dielectric polymer materials.

To meet this goal, some processing methods are considered of technological importance. The first is to exploit the manipulation of anisotropy in high field dielectric properties. Designed anisotropy in dielectric properties provides an approach for developing ceramic particle/polymer nanocomposites with high permittivity and high breakdown and for understanding the mechanisms that control dielectric breakdown strength and nonlinear conduction at high fields. The second approach is more versatile and related the fillers with a multicore structure [134]. A properly engineered multilayered nanoparticle–polymer interface is essential for achieving favorable high field dielectric properties, thus attaining high-energy storage capacity. The insulating nonpolar outer layer not only effectively prevents the growth of a breakdown path and minimizes the leakage current but also increases the dispersibility of hybrid particles by matching the solubility parameters of the outer surface of particles with the polymer matrix. Thus, such features enable a large incorporation of nanoparticles to reach high enough permittivity while maintaining high dielectric strength in the composites. The third method is that the conductive fillers are well coated with a sound and integral insulating shell. This insulating barrier is capable of preventing electron tunneling and giving rise to a superior low field dielectric response, but unable to suppress the leakage current due to the tunneling at high fields. Because of this, an improved dielectric strength relative to the neat polymer matrix has not been reported so far despite higher density has been attained in certain conductive core–shell filler/polymer composites with percolative behavior.

As a newly emerging and highly interdisciplinary field, the progress in dielectric polymer materials for the capacitive energy storage is critically dependent on collaborative efforts from various disciplines, such as nanoscience and nanotechnology, materials science, and physical chemistry. This interesting subject deserves to receive further study in chemical and materials communities.

Acknowledgments

This work was financially supported by National Science Foundation (NSF) of China (grant no. 51073015), NSF of Beijing City (grant no. 2122040), State Key Laboratory of Power System (SKLD11KZ04), Key Laboratory of Engineering Dielectrics and Its Application (Ministry of Education), and the Special Fund of the National Priority Basic Research of China under Grant 2014CB239503.

References

[1] Saito Y, Takao H, Tani T, Nonoyama T, Takatori K, Homma T, et al. Lead-free piezo-ceramics. Nature 2004;432(7013):84−7.

[2] Junquera J, Ghosez P. Critical thickness for ferroelectricity in perovskite ultrathin films. Nature 2003;422(6931):506−9.

[3] Kingon AI, Maria JP, Streiffer SK. Alternative dielectrics to silicon dioxide for memory and logic devices. Nature 2000;405(6799):1032−8.

[4] Switzer JA, Shumsky MG, Bohannan EW. Electrodeposited ceramic single crystals. Science 1999;284(5412):293−6.

[5] Blackwood KM. Device applications of side-chain ferroelectric liquid crystalline polymer films. Science 1996;273(5277):909−12.

[6] Zhang QM, Li HF, Poh M, Xia F, Cheng ZY, Xu HS, et al. An all-organic composite actuator material with a high dielectric constant. Nature 2002;419(6904):284−7.

[7] Ma MM, Guo L, Anderson DG, Langer R. Bio-inspired polymer composite actuator and generator driven by water gradients. Science 2013;339(6116):186−9.

[8] Tavakkoli KGA, Gotrik KW, Hannon AF, Alexander-Katz A, Ross CA, Berggren KK. Templating three-dimensional self-assembled structures in bilayer block copolymer films. Science 2012;336(6086):1294−8.

[9] Whittingham MS. Materials challenges facing electrical energy storage. MRS Bull 2008;33(4):411−9.

[10] Dang ZM, Yuan JK, Zha JW, Zhou T, Li ST, Hu GH. Fundamentals, processes and applications of high-permittivity polymer matrix composites. Prog Mater Sci 2012;57 (4):660−723.

[11] Barber P, Balasubramanian S, Anguchamy Y, Gong S, Wibowo A, Gao H, et al. Polymer composite and nanocomposite dielectric materials for pulse power energy storage. Materials 2009;2(4):1697−733.

[12] Wang Q, Zhu L. Polymer nanocomposites for electrical energy storage. J Polym Sci Part B 2011;49(20):1421−9.

[13] Rabuffi M, Picci G. Status quo and future prospects for metallized polypropylene energy storage capacitors. IEEE Trans Plasm Sci 2002;30(5):1939−42.

[14] Zebouchi N, Malec D. Combination of thermal and electromechanical breakdown mechanisms to analyze the dielectric breakdown in polyethylene terephthalate. J Appl Phys 1998;83(11):6190−2.

[15] Wang Y, Zhou X, Chen Q, Chu B, Zhang Q. Recent development of high energy density polymers for dielectric capacitors. IEEE Trans Dielect Elect Insul 2010;17(4):1036−42.

[16] Job AE, Alves N, Zanin M, Ueki MM. Increasing the dielectric breakdown strength of poly(ethylene terephthalate) films using a coated polyaniline layer. J Phys D Appl Phys 2003;36(12):1414−7.

[17] Chu BJ, Zhou X, Ren KL, Neese B, Lin MR, Wang Q, et al. A dielectric polymer with high electric energy density and fast discharge speed. Science 2006;313(5785):334−6.

[18] Tuncer E, Sauers I, Randy James D, Ellis AR, Parans Paranthaman MP. Enhancement of dielectric strength in nanocomposites. Nanotechnology 2007;18(32):325704.

[19] Rahimabady M, Chen S, Yao K, Tay FEH, Lu L. High electric breakdown strength and energy density in vinylidene fluoride oligomer/poly(vinylidene fluoride) blend thin films. Appl Phys Lett 2011;99(14):142901.

[20] Yuan JK, Dang ZM, Yao SH, Zha JW, Zhou T, Li ST, et al. Fabrication and dielectric properties of advanced high permittivity polyaniline/poly(vinylidene fluoride) nanohybrid films with high energy storage density. J Mater Chem 2010;20(12):2441−7.

[21] Li JJ, Seok SI, Chu BJ, Dogan F, Zhang QM, Wang Q. Nanocomposites of ferroelectric polymers with TiO_2 nanoparticles exhibiting significantly enhanced electrical energy density. Adv Mater 2009;21(2):217−22.

[22] Nelson JK, Hu Y. Nanocomposite dielectrics—properties and implications. J Phys D Appl Phys 2005;38(2):213−22.

[23] Coppard RW, Bowman J, Dissado LA, Rowland SM, Rakowski RT. The effect of aluminum inclusions on the dielectric-breakdown of polyethylene. J Phys D Appl Phys 1990;23(12):1554−61.

[24] Sun YY, Zhang ZQ, Wong CP. Influence of interphase and moisture on the dielectric spectroscopy of epoxy/silica composites. Polymer 2005;46(7):2297−305.

[25] Dang ZM, Xu HP, Wang HY. Significantly enhanced low-frequency dielectric permittivity in the $BaTiO_3$/poly(vinylidene fluoride) nanocomposite. Appl Phys Lett 2007;90 (1):012901.

[26] Ramesh S, Shutzberg BA, Huang C, Gao J, Giannelis EP. Dielectric nanocomposites for integral thin film capacitors: materials design, fabrication, and integration issues. IEEE Trans Adv Pack 2003;26(1):17−24.

[27] Tanaka T, Montanari GC, Mulhaupt R. Polymer nanocomposites as dielectrics and electrical insulation-perspectives for processing technologies, material characterization and future applications. IEEE Trans Dielect Elect Insul 2004;11(5):763−84.

[28] Lewis TJ. Nanometric dielectrics. IEEE Trans Dielect Elect Insul 1994;1(5):812−25.

[29] Fan BH, Zha JW, Wang D, Zhao J, Dang ZM. Size-dependent low-frequency dielectric properties in the $BaTiO_3$/poly(vinylidene fluoride) nanocomposite films. Appl Phys Lett 2012;100(1):012903.

[30] Li J, Claude J, Norena-Franco L, Seok IS, Wang Q. Electrical energy storage in ferroelectric polymer nanocomposites containing surface-functionalized $BaTiO_3$ nanoparticles. Chem Mater 2008;20(20):6304−6.

[31] Kim P, Doss NM, Tillotson JP, Hotchkiss PJ, Pan MJ, Marder SR, et al. High energy density nanocomposites based on surface-modified $BaTiO_3$ and a ferroelectric polymer. ACS Nano 2009;3(9):2581−92.

[32] Lewis TJ. Interfaces are the dominant feature of dielectrics at the nanometric level. IEEE Trans Dielect Elect Insul 2004;11(5):739−54.

[33] Thomas P, Varughese KT, Dwarakanath K, Varma KB. Dielectric properties of poly (vinylidene fluoride)/$CaCu_3Ti_4O_{12}$ composites. Compos Sci Technol 2010;70(3):539−45.

[34] Murugaraj P, Mainwaring D, Mora-Huertas N. Dielectric enhancement in polymer−nanoparticle composites through interphase polarizability. J Appl Phys 2005;98 (5):054304.

[35] Guo N, DiBenedetto SA, Kwon DK, Wang L, Russell MT, Lanagan MT, et al. Supported metallocene catalysis for *in situ* synthesis of high energy density metal oxide nanocomposites. J Am Chem Soc 2007;129(4):766−7.

[36] Li JY. Exchange coupling in P(VDF-TrFE) copolymer based all-organic composites with giant electrostriction. Phys Rev Lett 2003;90(21):217601.

[37] Hanemann T, Szabo DV. Polymer—nanoparticle composites: from synthesis to modern applications. Materials 2010;3(6):3468−517.

[38] Chattopadhyay S, Ayyub P, Palkar VR, Gurjar AV, Wankar RM, Multani M. Finite-size effects in antiferroelectric PbZrO$_3$ nanoparticles. J Phys Cond Matter 1997;9 (38):8135−45.

[39] Sullivan DE, Deutch JM. Molecular theory of dielectric relaxation. J Chem Phys 1975;62(6):2130−8.

[40] Titulaer UM, Deutch JM. Analysis of conflicting theories of dielectric relaxation. J Chem Phys 1974;60(4):1502−13.

[41] Zha JW, Jia HJ, Wang HY, Dang ZM. Tailored ultra-low dielectric permittivity in high-performance fluorinated polyimide films by adjusting nanoporous characteristics. J Phys Chem C 2012;116(44):23676−81.

[42] Simpson JO, Clair AK. Fundamental insight on developing low dielectric constant polyimides. Thin Solid Films 1997;308:480−5.

[43] Rao Y, Wong CP. Material characterization of a high-dielectric-constant polymer—ceramic composite for embedded capacitor for RF applications. J Appl Polym Sci 2004;92(4):2228−31.

[44] Ogitani S, Bidstrup-Allen SA, Kohl PA. Factors influencing the permittivity of polymer/ceramic composites for embedded capacitors. IEEE Trans Adv Pack 2000;23 (2):313−22.

[45] Kim SW, Kim T, Kim YS, Choi HS, Lim HJ, Yang SJ, et al. Surface modifications for the effective dispersion of carbon nanotubes in solvents and polymers. Carbon 2012;50 (1):3−33.

[46] Zhou T, Zha JW, Cui RY, Fan B-H, Yuan JK, Dang ZM. Improving dielectric properties of BaTiO$_3$/ferroelectric polymer composites by employing surface hydroxylated BaTiO$_3$ nanoparticles. ACS Appl Mater Interf 2011;3(7):2184−8.

[47] Kim P, Jones SC, Hotchkiss PJ, Haddock JN, Kippelen B, Marder SR, et al. Phosphonic acid-modified barium titanate polymer nanocomposites with high permittivity and dielectric strength. Adv Mater 2007;19(7):1001−5.

[48] Wang GS, Deng Y, Guo L. Single-crystalline ZnO nanowire bundles: synthesis, mechanism and their application in dielectric composites. Chem Eur J 2010;16 (33):10220−5.

[49] Wang GS, Deng Y, Xiang Y, Guo L. Fabrication of radial ZnO nanowire clusters and radial ZnO/PVDF composites with enhanced dielectric properties. Adv Funct Mater 2008;18(17):2584−92.

[50] Dang ZM, Nan CW, Xie D, Zhang YH, Tjong SC. Dielectric behavior and dependence of percolation threshold on the conductivity of fillers in polymer-semiconductor composites. Appl Phys Lett 2004;85(1):97−9.

[51] Dang ZM, Fan LZ, Zhao SJ, Nan CW. Dielectric property and morphology of composites filled with whisker and nanosized zinc oxide. Mater Res Bull 2003;38 (3):499−507.

[52] Dang ZM, Fan ZL, Zhao SJ, Nan CW. Preparation of nanosized ZnO and dielectric propertie of composites filled with nanosized ZnO. Mater Sci Eng B 2003;99 (1−3):386−9.

[53] Dang Z, Fan L, Shen Y, Nan CW, Zhao S. Study of thermal and dielectric behavior of low-density polyethylene composites reinforced with zinc oxide whisker. J Therm Anal Cal 2003;71(2):635−41.

[54] Bobnar V, Levstik A, Huang C, Zhang QM. Dielectric properties and charge transport in all-organic relaxorlike CuPc−P(VDF-TrFE-CFE) composite and its constituents. Ferroelectrics 2006;338:107−16.

[55] Huang C, Zhang QM, Li JY, Rabeony M. Colossal dielectric and electromechanical responses in self-assembled polymeric nanocomposites. Appl Phys Lett 2005;87 (18):182901.

[56] Wang JW, Shen QD, Bao HM, Yang CZ, Zhang QM. Microstructure and dielectric properties of P(VDF-TrFE-CFE) with partially grafted copper phthalocyanine oligomer. Macromolecules 2005;38(6):2247−52.

[57] Wang JW, Shen QD, Yang CZ, Zhang QM. High dielectric constant composite of P (VDF-TrFE) with grafted copper phthalocyanine oligomer. Macromolecules 2004;37 (6):2296−8.

[58] Bobnar V, Levstik A, Huang C, Zhang QM. Distinctive contributions from organic filler and relaxorlike polymer matrix to dielectric response of CuPc−P(VDF-TrFE-CFE) composite. Phys Rev Lett 2004;92(4):047604.

[59] Dang ZM, Lin YH, Nan CW. Novel ferroelectric polymer−matrix composites with a high dielectric constant at the percolation threshold. Adv Mater 2003;15(19):1625−9.

[60] Panda M, Srinivas V, Thakur AK. On the question of percolation threshold in polyvinylidene fluoride/nanocrystalline nickel composites. Appl Phys Lett 2008;92(13):132905.

[61] Huang XY, Jiang PK, Xie LY. Ferroelectric polymer/silver nanocomposites with high dielectric constant and high thermal conductivity. Appl Phys Lett 2009;95(24):242901.

[62] Qi L, Lee BI, Chen SH, Samuels WD, Exarhos GJ. High-dielectric-constant silver−epoxy composites as embedded dielectrics. Adv Mater 2005;17(14):1777−80.

[63] Li YJ, Xu M, Feng JQ, Dang ZM. Dielectric behavior of a metal−polymer composite with low percolation threshold. Appl Phys Lett 2006;89(7):072902.

[64] Dang ZM, Wu JP, Xu HP, Yao SH, Jiang MJ, Bai J. Dielectric properties of upright carbon fiber filled poly(vinylidene fluoride) composite with low percolation threshold and weak temperature dependence. Appl Phys Lett 2007;91(7):072912.

[65] Yao SH, Dang ZM, Xu HP, Jiang MJ, Bai J. Exploration of dielectric constant dependence on evolution of microstructure in nanotube/ferroelectric polymer nanocomposites. Appl Phys Lett 2008;92(8):082902.

[66] Yao SH, Dang ZM, Jiang MJ, Xu HP, Bai JB. Influence of aspect ratio of carbon nanotube on percolation threshold in ferroelectric polymer nanocomposite. Appl Phys Lett 2007;91(21):212901.

[67] Dang ZM, Wang L, Yin Y, Zhang Q, Lei QQ. Giant dielectric permittivities in functionalized carbon-nanotube/electroactive-polymer nanocomposites. Adv Mater 2007;19 (6):852−7.

[68] Wang L, Dang ZM. Carbon nanotube composites with high dielectric constant at low percolation threshold. Appl Phys Lett 2005;87(4):042903.

[69] Li Q, Xue QZ, Hao LZ, Gao XL, Zheng QB. Large dielectric constant of the chemically functionalized carbon nanotube/polymer composites. Compos Sci Technol 2008;68(10−11):2290−6.

[70] Yuan JK, Yao SH, Sylvestre A, Bai J. Biphasic polymer blends containing carbon nanotubes: heterogeneous nanotube distribution and its influence on the dielectric properties. J Phys Chem C 2012;116(2):2051−8.

[71] Yuan JK, Yao SH, Dang ZM, Sylvestre A, Genestoux M, Bai J. Giant dielectric permittivity nanocomposites: realizing true potential of pristine carbon nanotubes in polyvinylidene fluoride matrix through an enhanced interfacial interaction. J Phys Chem C 2011;115(13):5515−21.

[72] He F, Lau S, Chan HL, Fan JT. High dielectric permittivity and low percolation threshold in nanocomposites based on poly(vinylidene fluoride) and exfoliated graphite nanoplates. Adv Mater 2009;21(6):710−5.

[73] Fan P, Wang L, Yang J, Chen F, Zhong M. Graphene/poly(vinylidene fluoride) composites with high dielectric constant and low percolation threshold. Nanotechnology 2012;23(36):365702.

[74] Wu C, Huang XY, Wang GL, Wu XF, Yang K, Li ST, et al. Hyperbranched polymer functionalization of graphene sheets for enhanced mechanical and dielectric properties of polyurethane composites. J Mater Chem 2012;22(14):7010−9.

[75] Zhang L, Cheng ZY. Development of polymer-based 0−3 composites with high dielectric constant. J Adv Dielect 2011;1:389−406.

[76] Gallone G, Carpi F, De Rossi D, Rossi DD, Levita G, Marchetti A. Dielectric constant enhancement in a silicone elastomer filled with lead magnesium niobate-lead titanate. Mater Sci Eng C 2007;27(1):110−6.

[77] Carpi F, De Rossi D. Improvement of electromechanical actuating performances of a silicone dielectric elastomer by dispersion of titanium dioxide powder. IEEE Trans Dielect Elect Insul 2005;12(4):835−43.

[78] McPherson J, Kim JK, Shanware A, Mogul H. Thermochemical description of dielectric breakdown in high dielectric constant materials. Appl Phys Lett 2003;82(13):2121−3.

[79] Blythe AR. Electrical properties of polymers. New York: Cambridge University Press; 1979.

[80] Dang ZM, Lin YQ, Xu HP, Shi CY, Li ST, Bai J. Fabrication and dielectric characterization of advanced BaTiO$_3$/polyimide nanocomposite films. Adv Funct Mater 2008;18 (10):1509−17.

[81] Cho SD, Lee JY, Hyun JG, Paik KW. Study on epoxy/BaTiO$_3$ composite embedded capacitor films (ECFs) for organic substrate applications. Mater Sci Eng B 2004;110 (3):233−9.

[82] Jung HM, Kang J-H, Yang SY, Won JC, Kim YS. Barium titanate nanoparticles with diblock copolymer shielding layers for high-energy density nanocomposites. Chem Mater 2010;22(2):450−6.

[83] Dang ZM, Wang HY, Xu HP. Influence of silane coupling agent on morphology and dielectric property in BaTiO$_3$/polyvinylidene fluoride composites. Appl Phys Lett 2006;89(11):112902.

[84] Song Y, Shen Y, Liu H, Lin Y, Li M, Nan CW. Enhanced dielectric and ferroelectric properties induced by dopamine-modified BaTiO$_3$ nanofibers in flexible poly(vinylidene fluoride-trifluoroethylene) nanocomposites. J Mater Chem 2012;22(16):8063−8.

[85] Song Y, Shen Y, Liu H, Lin Y, Li M, Nan CW. Improving the dielectric constants and breakdown strength of polymer composites: effects of the shape of the BaTiO$_3$ nanoinclusions, surface modification and polymer matrix. J Mater Chem 2012;22 (32):16491−8.

[86] Tuncer E, Sauers I, James DR, Ellis AR, Duckworth RC. Nanodielectric system for cryogenic applications: barium titanate filled polyvinyl alcohol. IEEE Trans Dielect Elect Insul 2008;15(1):236−42.

[87] Dou XL, Liu XL, Zhang Y, Feng H, Chen JF, Du S. Improved dielectric strength of barium titanate−polyvinylidene fluoride nanocomposite. Appl Phys Lett 2009;95 (13):132904.

[88] Banerjee S, Cook-Chennault KA. Influence of Al particles size and lead zirconate titanate (PZT) volume fraction on the dielectric properties of PZT-epoxy-aluminum composites. J Eng Mater Technol 2011;133(4):041016.

[89] Tang H, Lin Y, Sodano HA. Enhanced energy storage in nanocomposite capacitors through aligned PZT nanowires by uniaxial strain assembly. Adv Energ Mater 2012;2 (4):469−76.

[90] Tang H, Lin Y, Andrews C, Sodano HA. Nanocomposites with increased energy density through high aspect ratio PZT nanowires. Nanotechnology 2011;22(1):015702.

[91] Balasubramanian B, Kraemer KL, Reding NA, Skomski R, Ducharme S, Sellmyer DJ. Synthesis of monodisperse TiO_2-paraffin core−shell nanoparticles for improved dielectric properties. ACS Nano 2010;4(4):1893−900.

[92] Ma D, Hugener TA, Siegel RW, Christerson A, Martensson E, Onneby C, et al. Influence of nanoparticle surface modification on the electrical behaviour of polyethylene nanocomposites. Nanotechnology 2005;16(6):724−31.

[93] Lin S, Kuang X, Wang F, Zhu H. Effect of TiO_2 crystalline composition on the dielectric properties of TiO_2/P(VDF-TrFE) composites. Phys Stat Sol Rapid Res Lett 2012;6(8):352−4.

[94] Ouyang G, Wang K, Chen XY. TiO_2 nanoparticles modified polydimethylsiloxane with fast response time and increased dielectric constant. J Micromech Microeng 2012;22(7):074002.

[95] Dang ZM, Xia YJ, Zha JW, Yuan JK, Bai J. Preparation and dielectric properties of surface modified TiO_2/silicone rubber nanocomposites. Mater Lett 2011;65 (23−24):3430−2.

[96] Zha JW, Dang ZM, Zhou T, Song HT, Chen G. Electrical properties of TiO_2-filled polyimide nanocomposite films prepared via an *in situ* polymerization process. Synth Met 2010;160(23−24):2670−4.

[97] Zha JW, Fan BH, Dang ZM, Li ST, Chen G. Microstructure and electrical properties in three-component $(Al_2O_3-TiO_2)$/polyimide nanocomposite films. J Mater Res 2010;25(12):2384−91.

[98] Dey A, De S, De A, De SK. Characterization and dielectric properties of polyaniline−TiO_2 nanocomposites. Nanotechnology 2004;15(9):1277−83.

[99] McCarthy DN, Stoyanov H, Rychkov D, Ragusch H, Melzer M, Kofod G. Increased permittivity nanocomposite dielectrics by controlled interfacial interactions. Compos Sci Technol 2012;72(6):731−6.

[100] Nelson JK, Fothergill JC. Internal charge behaviour of nanocomposites. Nanotechnology 2004;15(5):586−95.

[101] Zou C, Kushner D, Zhang S. Wide temperature polyimide/ZrO_2 nanodielectric capacitor film with excellent electrical performance. Appl Phys Lett 2011;98(8):082905.

[102] Li J, Khanchaitit P, Han K, Wang Q. New route toward high-energy-density nanocomposites based on chain-end functionalized ferroelectric polymers. Chem Mater 2010;22(18):5350−7.

[103] Guo N, DiBenedetto SA, Tewari P, Lanagan MT, Ratner MA, Marks TJ. Nanoparticle, size, shape, and interfacial effects on leakage current density, permittivity, and breakdown strength of metal oxide−polyolefin nanocomposites: experiment and theory. Chem Mater 2010;22(4):5154−64.

[104] Chang SJ, Liao WS, Ciou CJ, Lee JT, Li CC. An efficient approach to derive hydroxyl groups on the surface of barium titanate nanoparticles to improve its chemical modification ability. J Colloid Interf Sci 2009;329(2):300−5.

[105] Xie L, Huang X, Wu C, Jiang P. Core−shell structured poly(methyl methacrylate)/ $BaTiO_3$ nanocomposites prepared by *in situ* atom transfer radical polymerization: a route to high dielectric constant materials with the inherent low loss of the base polymer. J Mater Chem 2011;21(16):5897−906.

[106] Xu JW, Wong CP. Low-loss percolative dielectric composite. Appl Phys Lett 2005;87(8):082907.

[107] Xu JW, Moon KS, Tison C, Wong CP. A novel aluminum-filled composite dielectric for embedded passive applications. IEEE Trans Adv Pack 2006;29(2):295–306.

[108] Shen Y, Lin Y, Li M, Nan CW. High dielectric performance of polymer composite films induced by a percolating interparticle barrier layer. AdvMater 2007;19 (10):1418–22.

[109] Shen Y, Lin Y, Nan CW. Interfacial effect on dielectric properties of polymer nano-composites filled with core/shell-structured particles. Adv Funct Mater 2007;17 (14):2405–10.

[110] Bobnar V, Levstik A, Huang C, Zhang QM. Intrinsic dielectric properties and charge transport in oligomers of organic semiconductor copper phthalocyanine. Phys Rev B 2005;71(4):041202.

[111] Huang C, Zhang QM, Su J. High-dielectric-constant all-polymer percolative composites. Appl Phys Lett 2003;82(20):3502–4.

[112] Li JY, Huang C, Zhang QM. Enhanced electromechanical properties in all-polymer percolative composites. Appl Phys Lett 2004;84(16):3124–6.

[113] Wang CC, Song JF, Bao HM, Shen QD, Yang CZ. Enhancement of electrical properties of ferroelectric polymers by polyaniline nanofibers with controllable conductivities. Adv Funct Mater 2008;18(8):1299–306.

[114] Yuan JK, Dang ZM, Bai J. Unique dielectric properties of ferroelectric polyaniline/poly(vinylidene fluoride) composites induced by temperature variation. Phys Stat Sol Rapid Res Lett 2008;2(5):233–5.

[115] Wang JW, Wang Y, Wang F, Li SQ, Xiao J, Shen QD. A large enhancement in dielectric properties of poly(vinylidene fluoride) based all-organic nanocomposite. Polymer 2009;50(2):679–84.

[116] Huang C, Zhang QM. Fully functionalized high-dielectric-constant nanophase polymers with high electromechanical response. Adv Mater 2005;17(9):1153–8.

[117] Bobnar V, Levstik A, Huang C, Zhang QM. Enhanced dielectric response in all-organic polyaniline–poly(vinylidene fluoride-trifluoroethylene-chlorotrifluoroethylene) composite. J Non-Cryst Sol 2007;353(2):205–9.

[118] Lunkenheimer P, Bobnar V, Pronin AV, Ritus AI, Volkov AA, Loidl A. Origin of apparent colossal dielectric constants. Phys Rev B 2002;66(5):052105.

[119] Stoyanov H, Mc Carthy D, Kollosche M, Kofod G. Dielectric properties and electric breakdown strength of a subpercolative composite of carbon black in thermoplastic copolymer. Appl Phys Lett 2009;94(23):232905.

[120] Pecharroman C, Moya JS. Experimental evidence of a giant capacitance in insulator–conductor composites at the percolation threshold. Adv Mater 2000;12 (4):294–7.

[121] Deepa KS, Kumari Nisha S, Parameswaran P, Sebastian MT, James J. Effect of conductivity of filler on the percolation threshold of composites. Appl Phys Lett 2009;94 (14):142902.

[122] Dang ZM, Zhang YH, Tjong SC. Dependence of dielectric behavior on the physical property of fillers in the polymer–matrix composites. Synth Met 2004;146(1):79–84.

[123] Neda Z, Florian R, Brechet Y. Reconsideration of continuum percolation of isotropically oriented sticks in three dimensions. Phys Rev E 1999;59(3):3717–9.

[124] Bug ALR, Safran SA, Webman I. Continuum percolation of permeable objects. Phys Rev B 1986;33(7):4716–24.

[125] Berhan L, Sastry AM. Modeling percolation in high-aspect-ratio fiber systems. I. Soft-core versus hard-core models. Phys Rev E 2007;75(4):041120.

[126] Garboczi EJ, Snyder KA, Douglas JF, Thorpe MF. Geometrical percolation-threshold of overlapping ellipsoids. Phys Rev E 1995;52(1):819−28.

[127] Stankovich S, Dikin DA, Dommett GHB, Kohlhaas KM, Zimney EJ, Stach EA, et al. Graphene-based composite materials. Nature 2006;442(7100):282−6.

[128] Yao SH, Yuan JK, Zhou T, Dang ZM, Bai J. Stretch-modulated carbon nanotube alignment in ferroelectric polymer composites: characterization of the orientation state and its influence on the dielectric properties. J Phys Chem C 2011;115 (40):20011−7.

[129] Kerr CJ, Huang YY, Marshall JE, Terentjev EM. Effect of filament aspect ratio on the dielectric response of multiwalled carbon nanotube composites. J Appl Phys 2011;109(9):094109.

[130] Dubnikova I, Kuvardina E, Krasheninnikov V, Lomakin S, Tchmutin I, Kuznetsov S. The effect of multiwalled carbon nanotube dimensions on the morphology, mechanical, and electrical properties of melt mixed polypropylene-based composites. J Appl Polym Sci 2010;117(1):259−72.

[131] Hou Y, Wang DR, Zhang XM, Zhao H, Zha JW, Dang ZM. Positive piezoresistive behavior of electrically conductive alkyl-functionalized grapheme/polydimethylsilicone nanocomposites. J Mater Chem C 2013;1(1):515−21.

[132] Yang C, Lin YH, Nan CW. Modified carbon nanotube composites with high dielectric constant, low dielectric loss and large energy density. Carbon 2009;47(4):1096−101.

[133] Liu H, Shen Y, Song Y, Nan CW, Lin Y, Yang X. Carbon nanotube array/polymer core/shell structured composites with high dielectric permittivity, low dielectric loss, and large energy density. Adv Mater 2011;23(43):5104−8.

[134] Tanaka T, Kozako M, Fuse N, Ohki Y. Proposal of a multi-core model for polymer nanocomposite dielectrics. IEEE Trans Dielect Elect Insul 2005;12(4):669−81.

10 Synthesis and Structural−Mechanical Property Characteristics of Graphene−Polymer Nanocomposites

Sie Chin Tjong[1,2]

[1]Department of Physics and Materials Science, City University of Hong Kong, Kowloon, Hong Kong, PR China, [2]Department of Physics, Faculty of Science, King Abdulaziz University, Jeddah, Saudi Arabia

10.1 Introduction

Carbon-based systems comprise an extensively broad and continually expanding range of materials, from the building blocks of graphene, to 0D buckyballs, 1D nanotubes, and 3D graphite. Graphene is one atom thick, two-dimensional (2D) layer of sp^2 hybridized carbon atoms covalently bonded in a honeycomb lattice with a carbon−carbon bond length of 0.142 nm. Graphene offers many remarkable properties, such as extraordinary high elastic modulus (1 TPa) and fracture strength (130 ± 10 GPa), and thermal conductivity (~ 5000 W/m K) [1]. The fracture strength of graphene is about 200 times higher than carbon steel (e.g., AISI 1040), making it the strongest material ever prepared. All these properties render graphene showing attractive potential for applications in several technological fields, including structural materials, electronics, optoelectronics, gas sensors, fuel cells, etc. [2]. To achieve these purposes, graphite must be fully exfoliated into a single graphene layer. However, it is difficult to obtain an isolated graphene since free-standing atomic layer is considered to be thermodynamically unstable in the past two decades.

Since the successful in mechanical exfoliation of graphite using a simple Scotch-tape technique, the properties of graphene have received enormous attention from global researchers recently. Mechanical exfoliation of graphite is simple, inexpensive, and no special equipment is needed. However, exfoliated graphene layer is uneven and the Scotch-tape technique has low production yield. In other words, exfoliated graphene is insufficient for large-scale technological applications.

Nanocrystalline Materials. DOI: http://dx.doi.org/10.1016/B978-0-12-407796-6.00010-5

Thus several methods have been explored by the researches to achieve large-scale exfoliation of graphene layer using chemical synthesis route, such as thermal evaporation of SiC substrate and chemical vapor deposition (CVD) of hydrocarbons on metal or carbide substrates [2−6]. Those processes generally require high vacuum facilities operating at high temperatures. Therefore, wet chemical oxidation of graphite in strong acids to yield graphene oxide (GO), followed by the chemical and thermal reduction is a cost-effective process for synthesizing graphene in large quantities.

Graphite oxide can be obtained by reacting graphite with strong oxidizers, for example, a mixture of sulfuric acid, sodium nitrate, and potassium permanganate [7]. Compared to pristine graphene, GO is heavily oxygenated and its basal plane carbon atoms are decorated with epoxide and hydroxyl groups, while its edge atoms with carbonyl and carboxyl groups as shown in Figure 10.1A and B [8−10]. Thus graphite oxide exhibits an increased interlayer spacing from original 0.34 nm

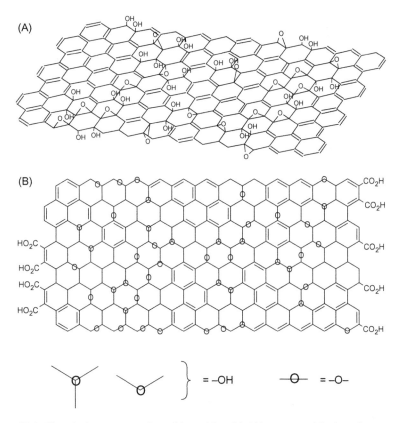

Figure 10.1 Chemical structures of graphite oxide with (A) epoxy and hydroxyl groups on its basal plane and (B) carboxylic groups at the edges.
Source: Reproduced from Refs. [8,9] with permission from Elsevier and The American Chemical Society, respectively.

of graphite to 0.6—1 nm, depending upon the amount of water intercalated within the stacked sheets [11]. Such functional groups render GO oxide hydrophilic, hence GO can be dispersed in aqueous media readily to form colloidal suspensions [12]. This facilitates exfoliation of graphite oxide into GO sheets via sonication [13]. Figure 10.2 is an atomic force microscopic (AFM) image of GO exfoliated in water via sonication showing the presence of graphene sheets with uniform thickness of ~1 nm. This thickness is somewhat larger than a theoretical value of 0.34 nm. This is attributed to the presence of covalently bound oxygen in GO.

GO is electrically insulating because the oxygenated functional groups distort intrinsic hybridized sp^2 C—C bonds in the graphene layers. To resume its electrical conductivity, chemical reducing agents, such as hydrazine, sodium borohydride, hydroquinone, and p-phenylene diamine, can react with GO for eliminating oxygen functionalities, producing the so-called reduced graphene oxide (RGO) or chemically modified graphene (CMG). [11,13—16]. Alternatively, RGO sheets can be obtained through solvothermal reduction of the GO colloidal dispersions in various solvents, e.g., ethanol, 1-butanol, ethylene glycol, N-methyl-2-pyrrolidone (NMP), etc. The reduction of GO generally occurs at relatively low temperatures of 100—200°C [17—19]. The Young's modulus of RGO single sheet is 0.25 TPa [20], being a quarter of that of pristine graphene.

Apart from the chemical reduction, thermal treatment at high temperatures can be used for the GO reduction, forming the so-called thermally reduced graphene oxide (TRG) or thermally expanded graphene oxide (TEGO). The thermal reduction process is performed by rapid heating GO (>2000°C/min) up to 1050°C in a vacuum or inert atmosphere, causing the evolution of carbon dioxide due to the

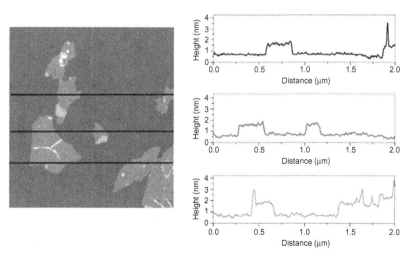

Figure 10.2 AFM image of exfoliated GO sheets. The sheets are ~1 nm thick. The horizontal lines in the image indicate the sections (in order from top to bottom) corresponding to the height profiles shown on the right.
Source: Reproduced from Ref. [13] with permission from Elsevier.

decomposition of hydroxyl and epoxide groups [21]. The evolved gas pressure then builds up, forcing the sheets apart and exfoliating GO sheets accordingly. TRG exhibits a wrinkled morphology (Figure 10.3A and B) [21−23]. It is noted that chemical or thermal reduction of GO is only partial, that is, some oxygenated groups still retain in GO [1,24]. Figure 10.4 shows the X-ray diffraction (XRD) patterns of graphite, GO, and TRG [25]. Graphite exhibits a sharp characteristic peak at $2\theta = 26.4°$. Upon chemical oxidation to form GO, this peaks shifts to $2\theta = {\sim}11.2°$, demonstrating an increase of the interlayer spacing from 0.34 to ${\sim}0.72$ nm due to the incorporation of the oxygen functional groups. However, TRG shows no reflections in the XRD pattern. The absence of the reflections implies the loss of periodic order in GO upon layer exfoliation. A weak reflection centered at $2\theta = 23{-}25°$ can be observed for densified TRG associated with the occurrence of restacked graphene.

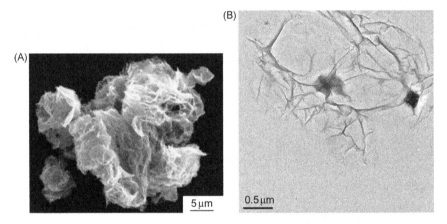

Figure 10.3 (A) SEM image of dry FGS powder. (B) TEM image of TRG showing wrinkled morphology.
Source: (A) Reproduced from Ref. [21] with permission from The American Chemical Society. (B) Reproduced from Ref. [22] with permission from The Royal Society of Chemistry.

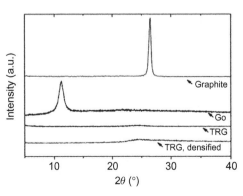

Figure 10.4 XRD patterns of graphite, GO, TRG, and densified TRG.
Source: Reproduced from Ref. [25] with permission from Elsevier.

Conventional polymer composites have been used as the structural components for industrial sectors due to their light weight, ease of fabrication, and low production cost compared with metal-based microcomposites [26−30]. However, polymer composites are generally reinforced with large volume fractions (ca. above 30%) of fillers of micrometer sizes for achieving desired mechanical and physical properties. Large filler volume contents are detrimental to the processing and mechanical properties of resulting polymer composites. Accordingly, inexpensive clay silicates and other ceramic nanoparticles (e.g., SiC, ZnO, Al_2O_3) are incorporated into polymers at low loadings to enhance their mechanical performance [31−38].

Compared with clay silicates and ceramic nanoparticles, graphene and its derivatives offer additional advantages for reinforcing polymers. Graphene exhibits excellent electrical and thermal conductivity in addition to their high mechanical stiffness and large aspect ratios. Furthermore, graphene has a higher surface-to-volume ratio than carbon nanotubes, rendering graphene potentially more favorable to form functional nanocomposites for industrial applications. Typical examples are the structural components of bipolar plates of fuel cells, orthopedic implants, electromagnetic interference shielding materials, chemical vapor sensors, and transducers [39−43]. Prior to the development of GO, graphene nanoplatelets (GnPs) derived from graphite intercalation compounds (GICs) are often used to reinforce polymers in producing conductive polymer nanocomposites [44−46]. GICs are produced by the insertion of selected chemical species between the layers of graphite, e.g., alkali metals, halogens, or acids [47]. Exfoliation of the GICs by rapid heating or microwave irradiation followed by sonication in solvents can produce large quantities of GnPs. However, GnP consists of multilayered sheets because of restacking of graphene layers after processing (Figure 10.5A and B) [47]. The XRD patterns of GIC, GnP, and functionalized GnP (m-GnP) are shown in Figure 10.5C [48]. GIC retains the stacked graphene structure, exhibiting a typical sharp and intense diffraction at 26.18°. This indicates the presence of a large amount of a crystalline phase in GIC with an interlayer spacing of 0.34 nm. Thus, GIC possesses a high structural integrity of graphene, in sharp contrast to the GO. The diffraction pattern of GnP displays the reflection at the same angle, with a large reduction in the peak intensity. GnP is directly prepared from the GIC via thermal treatment and ultrasonication. The thermal expansion and sonication cause a corrugation, a reduction in the lateral dimension, and an increase in the layer spacing; all of which contribute to the reduction in diffraction peak intensity. The intensity of m-GnPs shows a slightly increase in the peak intensity, since the washing and modification procedures cause realigning and stacking of wrinkled GnPs. In general, GnPs are commercially available at relatively low cost compared to TRGs [49,50].

10.2 Fabrication of Graphene/Polymer Nanocomposites

Polymer nanocomposites with good physical and mechanical properties require proper selection of a fabrication process such that graphene nanofillers can disperse

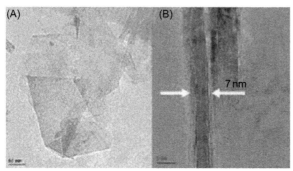

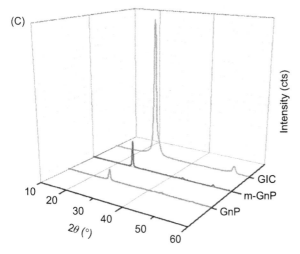

Figure 10.5 TEM images of (A) surface and (B) cross-sectional view of an individual graphite nanoplatelet particle. (C) XRD patterns of GIC, GnP, and surface-modified GnP. *Source*: (A and B) Reproduced from Ref. [45] with permission from Elsevier. (C) Reproduced from Ref. [48] with permission from The Royal Society of Chemistry.

uniformly in the polymer matrix. Graphene−polymer nanocomposites can be prepared by means of solution mixing, *in situ* polymerization, and melt blending. Generally, wet chemical processing route offers better dispersion of graphene nanofillers in the polymer matrix than the melt-mixing process. Carbonaceous nanofiller with large surface areas tends to agglomerate in the polymer matrix of nanocomposites, which may negate their properties greatly. As aforementioned, GO contains hydroxyl and epoxide functional groups on their basal planes, and carbonyl and carboxyl groups at its sheet edges. Such functional groups render GO hydrophilic, permitting its exfoliation in water readily. Therefore, GO can react with water-soluble polymers, such as poly(vinyl alcohol) (PVA) and poly(ethylene oxide), to form nanocomposites [51−54]. This strategy eliminates the use of organic solvents or chemical functionalization treatments that can be toxic and costly.

10.2.1 Functionalization of GO

Hydrophilic GO is poorly exfoliated in nonaqueous solvents, e.g., polar aprotic solvents such as *N,N*-dimethylformamide (DMF), dimethyl sulfoxide (DMSO), and

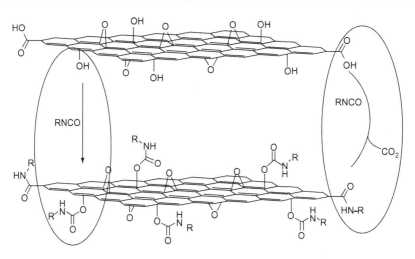

Figure 10.6 Organic isocyanates react with the hydroxyl (left oval) and carboxyl groups (right oval) of GO sheets to form carbamate and amide functionalities, respectively. *Source*: Reproduced from Ref. [57] with permission from Elsevier.

tetrahydrofuran (THF). Therefore, chemical functionalization of GOs is needed for exfoliating in such solvents and for reacting with nonpolar polymers, such as polyethylene (PE) and polypropylene (PP). Two main approaches are typically used for the surface modification of GO, i.e., covalent attachment of functional groups to GO and noncovalent attachment of surfactant molecules. In the formal route, GO can be chemically functionalized with long alkyl chain molecules (alkylation), isocyanates, and polyallylamine to form covalent linkages. As an example, acidic functional groups of GO react with long chain octadecylamine [$CH_3(CH_3)_{17}NH_2$] to yield graphene sheets that are soluble in nonpolar solvents [55,56].

GO can react with isocyanate to produce isocyanate-treated GO (iGO) [57]. During the reaction, edge carboxyl and surface hydroxyl groups of GO react with isocyanate to form amides and carbamate esters, respectively (Figure 10.6). Consequently, iGO can disperse in polar solvents, such as DMF, DMSO, and NMP. The iGO dispersion in DMF can be used as a medium for the synthesis of graphene/polyimide nanocomposites [58]. Generally, TRG disperses well in several solvents, such as NMP, DMF, 1,2-dichlorobenzene, nitromethane, and THF [21]. TRG with oxygenated groups can react with water-soluble poly(ethylene oxide) (PEO) and polar polymers, such as poly(methyl methacrylate) (PMMA), poly(acrylonitrile), and poly(acrylic acid) [59,60].

Noncovalent functionalization involves the wrapping of polymer chains onto graphene via the $\pi-\pi$ interaction. It can avoid defect generation and the disruption of graphene conjugation bond. Teng et al. [61] modified graphene through $\pi-\pi$ stacking of pyrene molecules with a functional segmented polymer chain, resulting in a remarkable improvement in thermal conductivity of epoxy resin.

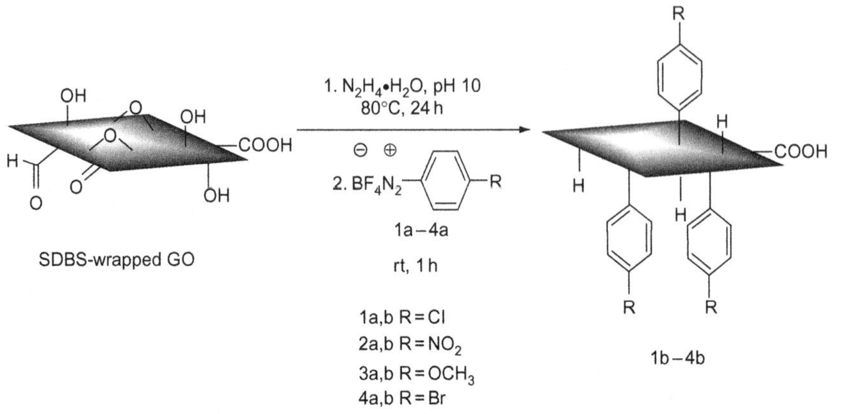

Figure 10.7 Wrapping GO with SDBS and functionalizaton of intermediate SDBS-wrapped RGO with diazonium salts.
Source: Reproduced from Ref. [62] with permission from The American Chemical Society.

Lomeda et al. [62] prepared surfactant-wrapped GO sheets by dispersing them in sodium dodecylbenzenesulfonate (SDBS) surfactant followed by the hydrazine reduction. The surfactant-wrapped RGO was then functionalized with diazonium salts (Figure 10.7).

10.2.2 Solution Mixing

Solution mixing is by far the most commonly used wet chemical blending process for fabricating polymer nanocomposites. The process involves the dissolution of polymer in a suitable solvent followed by mixing with graphene under vigorous stirring and/or sonication. As a result, polymer molecules coat graphene sheets effectively. By evaporating the solvent, the graphene sheets reassemble, sandwiching the polymers to form nanocomposites [1]. The advantages of solution mixing are its simplicity and effectiveness for dispersing nanofillers in the polymer matrix. The disadvantages are the necessities for solvent and sonication for mixing graphene fillers with the polymer suspension. Selected organic solvents are limited and the use of solvents in large quantities pollutes surrounding environment greatly. Furthermore, longtime sonication can induce structural defects and reduce the aspect ratios of graphene, resulting in poorer mechanical strength of resulting composites [63,64]. This strategy can be employed to synthesize graphene/polymer composites with a wide range of polymers, such as PMMA [60], polystyrene (PS) [65], polyvinyl chloride [66], PP [67], poly(vinylidene fluoride) [68], and epoxy [69,70]. Figure 10.8 is the SEM micrograph of 0.5 wt%TRG/epoxy nanocomposite showing the dispersion of wrinkled TRGs in the matrix [70].

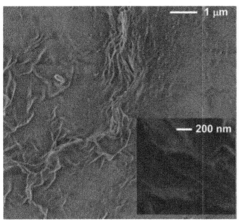

Figure 10.8 SEM image of cryofractured surface of 0.5 wt% TRG/epoxy nanocomposite, showing epoxy-coated TRG with wrinkled feature. The inset shows the wavy edges of the TRG sheets protruding out of the fracture surface. *Source*: Reproduced from Ref. [70] with permission from Elsevier.

10.2.3 In situ *Polymerization*

In situ polymerization requires the use of monomers, an initiator and a high-temperature reactor. In the process, GO or TRG is dispersed in a solvent, followed by adding a monomer and an appropriate initiator under agitation and thermal heating. This process enables the attainment of homogeneous dispersion of graphene nanofillers in the polymer matrix. The high costs of monomer and high-temperature vessel are the main limitations of this process. Further, the solvent removal issue is similar to that of the solvent casting technique. Potts et al. [71] synthesized *in situ* RGO/PMMA nanocomposites using MMA monomer, benzoyl peroxide initiator, GO, and hydrazine. Upon heating, the bond of benzoyl peroxide was cleaved producing free radicals. The radicals reacted with MMA to initiate polymerization. After the polymerization was completed and while the polymer was in solution, hydrazine was added for reducing GO into graphene and for preventing graphene from restacking. For comparison, GO/PMMA composites without using the reducing agent were also synthesized.

Polyamide (PA)-based composites can be synthesized via ring-opening polymerization of caprolactam in the presence of GO, and initiated by 6-aminocaproic acid at 250°C [72]. The PA6 chains are grafted to the graphene sheets via condensation reaction between the carboxylic acid of GO with the amide groups of PA6. During the polycondensation, GO is thermally reduced to graphene simultaneously. Yang et al. [22] synthesized graphene/poly(L-lactide) (PLLA) nanocomposite using *in situ* ring-opening polymerization of PLLA with graphene (Figure 10.9). TRGs are well dispersed in the PLLA matrix, leading to improved thermal stability and electrical conductivity of PLLA.

More recently, Wang et al. [73] synthesized graphene by dispersing graphite in NMP under sonication to form graphene. Thus graphene suspension was obtained by direct exfoliation of graphite in NMP. This novel process bypassed the GO synthesis route that required further reduction step for producing graphene. The polymerization was carried out by adding MMA monomer and 2,2'-azobisisobutyronitrile (AIBN) initiator

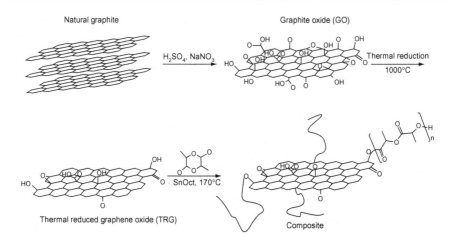

Figure 10.9 Synthesis scheme of TRG/PLLA nanocomposite via *in situ* polymerization.
Source: Reproduced from Ref. [22] with permission from Elsevier.

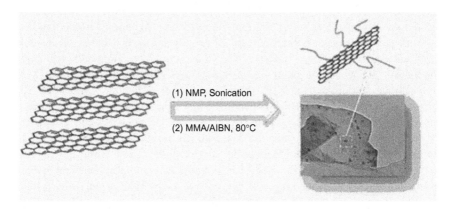

Figure 10.10 Synthesis scheme of graphene/PMMA nanocomposites via *in situ*
polymerization. The inset is a TEM micrograph showing good dispersion of graphene
nanofillers.
Source: Reproduced from Ref. [73] with permission from Elsevier.

to the graphene suspension at 80°C (Figure 10.10). Other graphene/polymer nanocomposites prepared by *in situ* polymerization with matrices based on PS [74,75], poly
(butylene terephthalate) [76], and polyimide [58] have been reported in the literature.

10.2.4   Melt Mixing

Melt mixing or blending is a cost-effective and versatile process for mass production of thermoplastics and their nanocomposites. The process employs a high

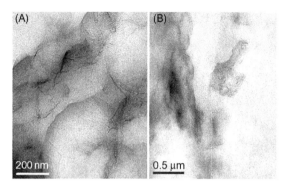

Figure 10.11 TEM micrographs of 1 wt% TRG/LLDPE nanocomposite specimen prepared by (A) solution mixing and (B) melt compounding.
Source: Reproduced from Ref. [25] with permission from Elsevier.

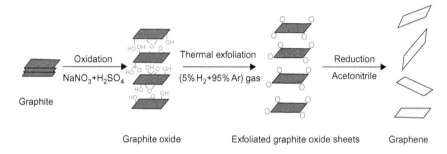

Figure 10.12 Reaction scheme for the preparation of graphene.
Source: Reproduced from Ref. [59] with permission from Elsevier.

temperature and shear force to disperse graphene nanofillers in the polymer matrix. Comparing with solution mixing and *in situ* polymerization processes, melt blending is less effective in dispersing graphene in the polymer matrix, particularly at higher filler loadings due to an increase of the melt viscosity during the processing [26]. Moreover, high shear forces during melt mixing can reduce the aspect ratio of graphene, leading to impaired mechanical properties. Melt blending has been employed by the researchers to fabricate graphene/polymer nanocomposites with matrices based on linear low-density PE (LLDPE) [25], PEO [59], PLA [77], polyurethane [78], PP [79], polycarbonate (PC) [80], and polyethylene terephthalate [81].

Kim et al. [25] investigated the effect of blending methods and PE functionalization on the dispersion of TRGs in the LLDPE matrix. Unlike fully separated, independent TRG sheets dispersed in the LLDPE matrix of the nanocomposites prepared by solution mixing, TRGs of melt-blended nanocomposites are mainly concentrated at local areas or islands. Such aggregated sheet islands are mostly isolated from one another, forming graphene-rich and graphene-poor phases (Figure 10.11A and B).

Mahmoud et al. [59] treated TRGs with acetonitrile agent to further reduce their residual oxygen groups to form graphene. The product is termed as foliated graphene sheets (FGSs) (Figure 10.12). AFM and TEM examinations revealed that the

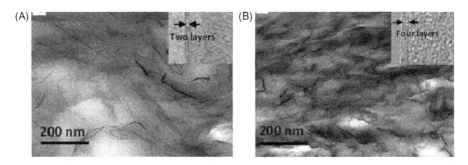

Figure 10.13 TEM micrographs of 0.3 vol% graphene/PEO nanocomposite specimens prepared by (A) solution mixing and (B) melt blending processes. Insets are HRTEM images showing exfoliation of graphene layers.
Source: Reproduced from Ref. [59] with permission from Elsevier.

FGSs consist of four to five graphene layers. They then prepared graphene/PEO nanocomposites using both solution mixing and melt blending processes. They also found that FGS sheets dispersed more uniformly in the matrix of solution-mixed nanocomposites than the melt-blended counterparts. Figure 10.13A and B shows TEM micrographs of the 0.3 vol% FGS/PEO nanocomposite prepared by solvent mixing and melt mixing, respectively. The sonication treatment during solution mixing can further exfoliate FGS into thin graphene layers. The polymer solution diffuses between the graphene sheets and eventually coats their surfaces. For melt-compounded composite, the polymer chains tend to intercalate into FGS sheets rather than diffuse between the graphene layers. This leads to less uniform dispersion of FGS sheets in the polymer matrix.

Graphene and its derivatives filled polymer nanocomposites have shown attractive applications in industrial sectors. Several fabrication processes described above show high potential for manufacturing industrial products, e.g., automotive panel body, automotive tire, gas storage container, supercapacitor, battery electrode, etc. based on novel graphene/polymer nanocomposites. Table 10.1 lists the recent patents granted by the United States Patent and Trademark Office for fabricating high-performance polymer nanocomposites reinforced with GO-, iGO-, and TRG fillers.

10.3   Mechanical Properties

Because of its relatively low cost, high aspect ratio, extraordinary high modulus and strength, graphene is regarded as an effective filler for fabricating high-performance polymer composites at very low loading content. The reinforcing efficiency of graphene/polymer nanocomposites depends exclusively on the effective stress transfer across the matrix—filler interface during the mechanical deformation.

Table 10.1 Patent Processes for Fabricating Polymer Nanocomposites Reinforced with Graphene Derivatives

Patent Number and Year	Type of Nanocomposites	Fabrication Processes	Inventor and Assignee	Patent Title
US 8192870 (2012)	TRG/PMMA	Solution mixing	Aksay IA, Yeh TC, Saville DA. Princeton University (USA)	Supercapacitor and battery electrode containing thermally exfoliated graphite oxide
US 8110524 (2012)	TRG/PMMA	Solution mixing	Prud'Homme RK, Aksay IA. Princeton University	Gas storage cylinder formed from a composition containing thermally exfoliated graphite
US 7935754 (2011)	TRG/PMMA	Solution mixing	Prud'Homme RK. Princeton University (USA)	Automotive body panel containing thermally exfoliated graphite oxide
US 7914844 (2011)	iGO/PS	Solution mixing	Stankovich S, Nguyen SB, Ruoff RS. Northwestern University (USA)	Stable dispersions of polymer-coated graphitic nanoplatelets
US 7923491 (2011)	GO/Elastomer, TRG/elastomer	Solution mixing, melt compounding	Weng W, Lohse D, Mota MO, Silva AS, Kresge EN. Exxon Mobil Chemical Patents Inc. (USA)	Graphite nanocomposites
US 7659350 (2010)	TRG/PMMA	*In situ* polymerization	Prud'Homme, RK. Princeton University (USA)	Polymerization method for formation of thermally exfoliated graphite oxide containing polymer
US 7745528 (2010)	TRG/Elastomer	Solution mixing, melt compounding	Prud'Homme RK. Princeton University (USA)	Functional graphene–rubber nanocomposites

Stress transfer from a relatively low modulus polymer matrix to a high modulus graphene leads to a marked increase of the composite modulus and strength. As expected, graphene/polymer nanocomposites offer large interfacial areas for the stress transfer due to the graphene nanofillers exhibit large aspect ratio and surface area. Furthermore, strong interfacial bonding between the graphene and polymer matrix is critical for effective reinforcement. Polymer nanocomposites with poor interfacial bonding exhibit inferior mechanical strength due to ineffective stress transfer across the matrix−filler interface. Covalent or noncovalent functionalization of graphene is effective for tailoring interfacial interaction such that the resulting nanocomposites exhibit high mechanical strength and stiffness. Generally, several factors such as the aspect ratio, dispersion state of graphene, graphene content, and the composite fabrication conditions affect mechanical performance of the graphene/polymer nanocomposites.

Recently, Koratkar et al. employed Raman spectroscopy to investigate the stress transfer effect in the 0.1% TRG/polydimethylsiloxane (PDMS) nanocomposite [82]. A single graphene layer possesses the well-known G-band around 1580 cm^{-1}. The shift of Raman G-band of this nanocomposite under applied strains was examined. For the purpose of comparison, the response of PDMS reinforced with 0.1 wt% single-walled carbon nanotube (SWNT) was also examined. Although SWNT is formed through the rolling of graphene sheet, the properties of the two are quite different. As recognized, the application of tensile or compressive stress to graphene can induce local strain, thus affecting carbon−carbon interatomic distance greatly. This in turn causes a change in the C−C bond vibrational frequency. Koratkar et al. reported that Raman G-mode peak shift in the 0.1% TRG/PDMS nanocomposite is at least on order of magnitude higher than that of the 0.1 wt% SWNT/PDMS material subjected to both tensile and compression deformations. The G-peak of 0.1% TRG/PDMS composite shifted by $\sim$2.4 cm^{-1}/composite strain %. In contrast, the peak shift was only $\sim$0.1 cm^{-1}/composite strain % in the 0.1% SWNT/PDMS nanocomposite. TRGs with rough and wrinkled surface features as shown in Figure 10.3A and B facilitate mechanical interlocking at the TRG−PDMS interfaces, resulting in effective load-transfer mechanism.

The mechanical properties of graphene/polymer nanocomposites are generally determined with tensile, flexural, microhardness, and nanoindentation tests. In particular, flexural test is more appropriate for measuring the strength and stiffness of graphene/epoxy nanocomposites with brittle polymer matrix [83]. The fracture behavior of polymer nanocomposites can be examined using Charpy and Izod impact tests [84]. These two techniques determine absorbed energy in the impact specimens fractured with a falling pendulum from a specified height. The impact energy values explain the failure behavior of notched composites in terms of brittle or ductile mode. The energy values are not the real material property data for designing structural components. In this regard, fracture mechanics approach is very useful to quantify the flaw tolerance of structural materials.

For the materials undergo elastic deformation or small scale yielding ahead the crack tip, linear elastic fracture mechanics (LEFM) is an effective method for characterizing the fracture toughness of graphene/epoxy nanocomposites [23,70].

The applied stress (σ) distribution in the vicinity of a crack tip is described by the stress intensity factor (K), defining as

$$K = \sigma(\pi a)^{1/2} \tag{10.1}$$

where a is the crack length. When K reaches a critical value, unstable crack growth causes the final fracture of a material. The critical stress intensity factor is referred to as the fracture toughness, K_C. The fracture toughness of brittle polymers and their composites can be measured using single edge notched bending (SENB) and compact tension (CT) specimens. Under a three-point bending mode, the stress intensity factor for the SENB specimen is expressed as [85]

$$K_{IQ} = \frac{P}{B\sqrt{W}} f(a/W) \tag{10.2}$$

where $f(a/W)$ is a geometrical shape factor. K_{IQ} can be treated as K_{IC} when the plane strain and linear elastic failure conditions are satisfied.

From the mechanical measurements and microstructural examinations, the strengthening and toughening mechanisms of the graphene/polymer nanocomposites can be systematically analyzed. Moreover, several micromechanical models of discontinuously fiber-reinforced composites, such as Halpin—Tsai (H—P) equations and modified rule of mixtures, can be used for predicting the mechanical properties of graphene/polymer nanocomposites [86]. According to the H—P model [87], the longitudinal modulus (E_{CL}) and transverse modulus (E_{CT}) of discontinuously short fiber-reinforced composite can be expressed as

$$\frac{E_{CL}}{E_m} = \frac{1 + \xi \eta_L V_f}{1 - \eta_L V_f} \tag{10.3}$$

$$\frac{E_{CT}}{E_m} = \frac{1 + 2\eta_T V_f}{1 - \eta_T V_f} \tag{10.4}$$

where E_m is the elastic modulus of matrix, V_f is volume fraction of the filler, η_L and η_T are given by the following expressions:

$$\eta_L = \frac{(E_f/E_m) - 1}{(E_f/E_m) + \xi} \tag{10.5}$$

$$\eta_T = \frac{(E_f/E_m) - 1}{(E_f/E_m) + 2} \tag{10.6}$$

where E_f is the elastic modulus of filler and $\xi = 2l/3t$ for sheet-like filler in which t and l are its thickness and length. For randomly oriented short fiber-reinforced composite, the elastic modulus of the composite is given by

$$E_{\text{random}} = \frac{3}{8}E_{\text{CL}} + \frac{5}{8}E_{\text{CT}} \qquad (10.7)$$

For the simulations of the stiffness of graphene/polymer nanocomposites, graphene nanofillers with large aspect ratios are considered analogous to conventional sheet-like fillers of micrometer dimensions. From the theoretical predictions, the reinforcing mechanisms of graphene nanofillers can be fundamentally elucidated by analyzing the discrepancies between the experimental results and predicted values.

10.3.1 Thermoplastic Matrix

Zhao et al. [51] prepared RGO/PVA nanocomposites reinforced with 0.5, 1, 3, and 5 wt% RGO (corresponding to 0.3, 0.6, 1.8, and 3 vol% RGO) using solution mixing process. The remnant oxygen functionalities of RGO react with water-soluble PVA easily. This facilitates the formation of strong interfacial bonding for effective stress transfer during the tensile tests. Figure 10.14A shows typical stress−strain curves of RGO/PVA nanocomposites. The tensile strength and elongation at break versus RGO content of the PVA nanocomposites are shown in Figure 10.14B. Apparently, the RGO additions improve the tensile strength of PVA at the expense

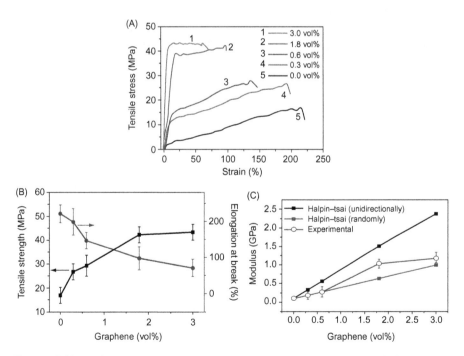

Figure 10.14 (A) Stress–strain curves, (B) tensile strength and elongation, and (C) Young's modulus versus RGO volume content of RGO/PVA nanocomposites.
Source: Reproduced from Ref. [51] with permission from The American Chemical Society.

of its ductility. The variation of Young's modulus with the RGO content of the RGO/PVA nanocomposites is shown in Figure 10.14C. The elastic modulus of PVA increases 10-fold by adding 1.8 vol% RGO. Simulated Young's moduli of the nanocomposites from the H—P model with aligned and randomly oriented RGO nanofillers are also shown in this figure. It is apparent that the measured elastic modulus data agree reasonably with those predicted from the H—P model with randomly dispersed RGO sheets when the filler contents ≤0.6 vol%.

It is noteworthy to mention that GO or RGO additions are beneficial in enhancing tensile elongation of PVA in addition to the improvements in tensile modulus and strength. This is because the oxygenated functionalities of GO or residual oxygenated groups of RGO can interact with water-soluble PVA (Figure 10.15) [88]. More recently, Layek et al. [89] indicated that sulfonated graphene (SG) is even more effective than GO in enhancing tensile stress, elastic modulus, and tensile ductility of PVA (Figure 10.16). The anchored $-SO_3H$ group in the SG is a stronger H-bonding group compared to hydroxyl and carboxyl groups of GO. Accordingly, SG can supramolecularly interact with the $-OH$ group of PVA through hydrogen bonding (Figure 10.17).

Mahmoud et al. [59] employed both solution mixing and melt blending processes to fabricate graphene/PEO nanocomposites. Figure 10.18A and B shows the stress—strain curves of solution-mixed and melt-blended graphene/PEO nanocomposites, respectively. Solution-mixed nanocomposites exhibit higher mechanical strength and stiffness than melt-blended counterparts due to homogeneous dispersion and further exfoliation of the graphene sheets (Figure 10.13A). In other words, the extent of the improvement is related to degree of uniform dispersion of the nanofillers in the polymer matrix. The variation of Young's modulus with graphene content of both nanocomposite systems is shown in Figure 10.19. The experimental elastic modulus of solution-mixed graphene/PEO nanocomposites matches well with the longitudinal modulus simulated from the H—P equation. In contrast, the experimental modulus of melt-blended nanocomposites agrees reasonably with that predicted from the H—P equation for the composite with randomly oriented fillers. Similarly, Kim et al. [25] also reported that the Young's modulus of solution-mixed TRG/LLDPE nanocomposites is higher than that of melt-blended TRG/LLDPE counterparts.

More recently, Achabi and Qaiss [90] studied the structure, thermal, and mechanical properties of melt-compounded RGO/PE nanocomposites. In the

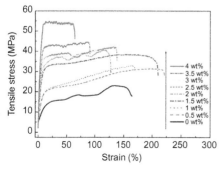

Figure 10.15 Stress—strain curves of solution-mixed RGO/PVA nanocomposites. *Source*: Reproduced from Ref. [88] with permission from Wiley.

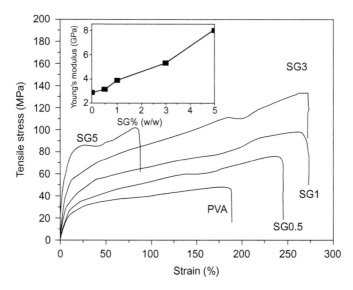

Figure 10.16 Mechanical properties of PVA and its nanocomposites with SG. The nanocomposites are designated as SG0.5, SG1, SG3, and SG5, respectively. The numbers indicate the weight percent of SG in the composites.
Source: Reproduced from Ref. [89] with permission from Elsevier.

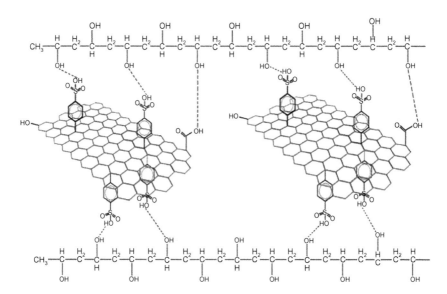

Figure 10.17 H-bonding interaction between PVA and SG.
Source: Reproduced from Ref. [89] with permission from Elsevier.

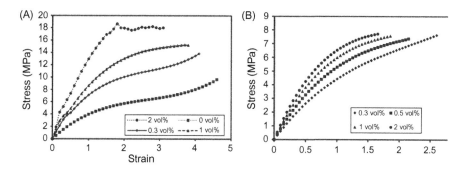

Figure 10.18 Stress–strain curves of (A) solution-mixed and (B) melt-blended graphene/PEO nanocomposites.
Source: Reproduced from Ref. [59] with permission from Elsevier.

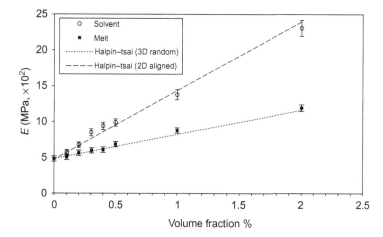

Figure 10.19 Young's modulus versus filler volume fraction for solution-mixed and melt-blended graphene/PEO nanocomposites.
Source: Reproduced from Ref. [59] with permission from Elsevier.

process, dry RGO powders were melt compounded with high-density PE (HDPE) in a Haake twin screw mixer followed by hot pressing. For comparison, HDPE reinforced with multiwalled carbon nanotubes (MWNTs) were also fabricated under the same processing conditions. Figure 10.20A and B shows the stress–strain curves of RGO/HDPE and MWNT/HDPE nanocomposites, respectively. The elastic modulus and tensile strength versus filler content of both nanocomposite systems are depicted in Figure 10.20C and D. Obviously, melt-compounded RGO/HDPE nanocomposites exhibit higher elastic modulus and tensile strength than the MWNT/HDPE nanocomposites at the same filler loading. This is because 2D graphene exhibits larger surface area than 1D-MWNT, resulting in a substantial increase of interfacial zone for effective stress transfer during tensile testing [82].

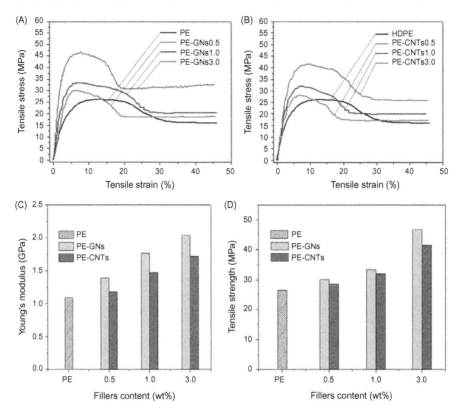

Figure 10.20 Stress—strain curves of (A) RGO/HDPE and (B) MWNT/HDPE nanocomposites with various fillers contents. (C) Young's modulus and (D) tensile strength as function of filler content for RGO/HDPE and MWNT/HDPE nanocomposites.
Source: Reproduced from Ref. [90] with permission from Elsevier.

In orthopedics, ultrahigh molecular weight polyethylene (UHMWPE) is widely used as the material for making acetabular cup of hip-joint prosthesis [91]. However, UHMWPE has low mechanical hardness, causing excessive wear and local inflammation after implantation into human body. Thus, graphene with good biocompatibility is an ideal nanofiller for reinforcing UHMWPE. More recently, Chen et al. [42] studied the structure, biocompatibility, thermal and mechanical properties of solution-mixed GO/UHMWPE nanocomposites. From the differential scanning calorimetric measurements, GO additions increase the crystallinity of UHMWPE, demonstrating that GO sheets act as effective sites for nucleating UHMWPE molecular chains (Figure 10.21). Moreover, GO sheets are dispersed homogeneously in the UHMWPE matrix (Figure 10.22A—B). The microhardness of UHMWPE and its nanocomposites is shown in Figure 10.23. The microhardness of UHMWPE improves markedly by adding GO of low loading levels. According to the Archard's law, the wear loss of a material is inversely proportional to its hardness [92]. Thus, the wear resistance of UHMWPE improves considerably by adding GOs.

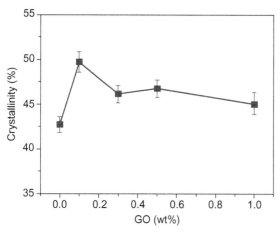

Figure 10.21 Crystallinity versus GO content of solution-mixed GO/UHMWPE nanocomposites. *Source*: Reproduced from Ref. [42] with permission from Elsevier.

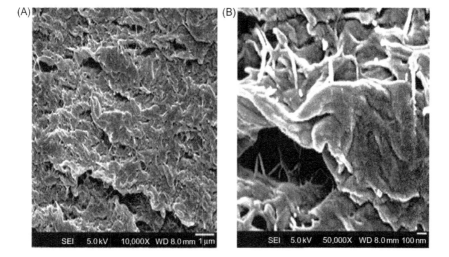

Figure 10.22 SEM images showing fracture surface of 0.5 wt% GO/UHMWPE nanocomposite at (A) low and (B) high magnification.
Source: Reproduced from Ref. [42] with permission from Elsevier.

As aforementioned, covalent functionalization of GOs with long alky chains like octadecylamine enables them to be dispersed in nonpolar xylene, a solvent for dissolving nonpolar PP [56]. Yun et al. [93] functionalized GO with a linear alkyl chain, that is, octylammonium by reacting with tetra-*n*-octylammonium bromide (TOAB) (Figure 10.24). Consequently, alkylated GO was dispersed in a xylene/PP solution under sonication followed by the solvent removal to form nanocomposites. They found that the elastic modulus of PP increases greatly (Figure 10.25), while the tensile strength shows no improvement by adding GO. In contrast, combined solution mixing and melt compounding has been reported to be effective for

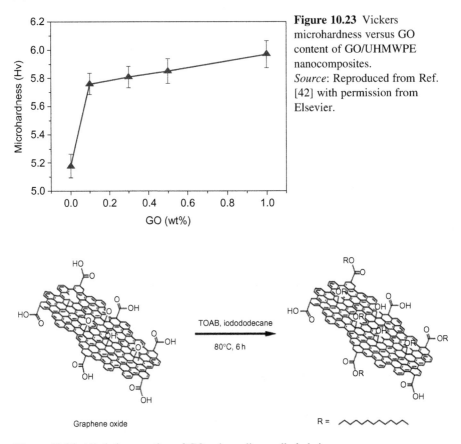

Figure 10.23 Vickers microhardness versus GO content of GO/UHMWPE nanocomposites. *Source*: Reproduced from Ref. [42] with permission from Elsevier.

Figure 10.24 Alkylation reaction of GO using a linear alkyl chain. *Source*: Reproduced from Ref. [93] with permission from Elsevier.

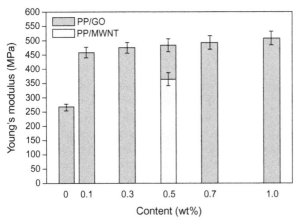

Figure 10.25 Young's modulus versus filler content for solution-mixed GO/PP and MWNT/PP nanocomposites. *Source*: Reproduced from Ref. [93] with permission from Elsevier.

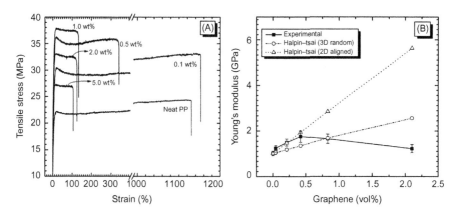

Figure 10.26 (A) Stress−strain curves and (B) Young's modulus versus filler content of RGO/PP nanocomposites prepared by combined solution mixing and melt blending processes.
Source: Reproduced from Ref. [79] with permission from Elsevier.

enhancing elastic modulus and tensile strength of the RGO/PP nanocomposites [79]. Figure 10.26A shows tensile stress−strain curves of the RGO/PP nanocomposites, respectively. The elastic modulus versus filler content of the RGO/PP nanocomposites is shown in Figure 10.26B. The modulus increases with RGO content up to 0.42 vol% (i.e., 1 wt%), thereafter decreases as the RGO content increases. Experimental moduli agree reasonably with those predicted from the H−P model (2D aligned) when the RGO contents ≤0.42 vol%. Above this content, experimental modulus deviates greatly from the H−P simulation due to the aggregation of RGO.

From Figure 10.10, *in situ* polymerization of MMA with solvent-exfoliated graphene can produce graphene/PMMA nanocomposites with strong interfacial bonding. As a result, the nanocomposites exhibit simultaneously improved elastic modulus, tensile strength, and elongation at break compared with pure PMMA [73]. Especially, the tensile stiffness and strength of PMMA are increased by 150.7% and 114%, respectively, by adding only 0.5 wt% nanofiller. In contrast, the additions of RGO to PMMA only result in a marginal enhancement of the elastic modulus [71]. Table 10.2 summarizes representative tensile properties of thermoplastics reinforced with graphene and its derivatives.

Recently, Ramanathan et al. [60] compared the reinforcing and stiffening effects of 1 wt% TRG, SWNT, and expanded graphite (EG) additions on the PMMA. They also studied the thermal properties of PMMA reinforced with those fillers (Figure 10.27). On the basis of dynamic mechanical analysis, TRG sheets were found to be more effective than EG and SWNT fillers for increasing elastic modulus, ultimate strength, glass transition temperature (T_g), and thermal stability of PMMA. They attributed this to the large surface area and wrinkled feature of TRGs, thereby favoring strong interfacial interaction (Figure 10.28A). In contrast, EG platelets exhibit poor bonding with the polymer matrix (Figure 10.28B).

Table 10.2 Mechanical Properties of Graphene/Thermoplastic Nanocomposites Prepared by Different Processes

Polymer	Filler	Filler content (wt%)	Young' modulus (GPa)	Tensile strength (MPa)	Elongation (%)	% Change, Elastic modulus	% Change, tensile strength	Processing method	Reference
PVA	GO	–	–	23 ± 2	165 ± 5	–	–	Solution mixing	[88]
	GO	0.5	–	31 ± 2	222 ± 10	–	34	As above	
	GO	1	–	33 ± 2	171 ± 8	–	43	As above	
	GO	1.5	–	38 ± 2	210 ± 9	–	65	As above	
	GO	2	–	40 ± 1	128 ± 7	–	74	As above	
	GO	2.5	–	42 ± 3	138 ± 5	–	83	As above	
	GO	3	–	43 ± 2	145 ± 6	–	87	As above	
	GO	3.5	–	45 ± 2	91 ± 7	–	96	As above	
	GO	4	–	55 ± 3	65 ± 8	–	139	As above	
PVA	SG	–	2.9	48	187	–	–	Solution mixing	[89]
	SG	1	3.9	97	266	36	102	As above	
	SG	3	5.3	133	272	86	177	As above	
	SG	5	8.0	101	84	180	111	As above	
PMMA	Graphene	–	2.09 ± 0.09	30.78 ± 3.32	1.67 ± 0.03	–	–	In situ polymerization	[73]
	Graphene	0.2	2.89 ± 0.48	58.52 ± 0.92	2.77 ± 0.47	38.3	90	As above	
	Graphene	0.5	5.24 ± 0.76	66.08 ± 1.00	3.32 ± 0.35	150.7	114	As above	
	Graphene	1	4.76 ± 0.93	49.15 ± 0.86	1.79 ± 0.18	127.8	59.7	As above	
PMMA	–	–	3.43 ± 0.06	58.1 ± 1.95	5.1	–	–	In situ polymerization and injection molding	[71]
	RGO	0.25	3.75 ± 0.57	59.5 ± 4.25	–	9.5	2.4	As above	
	RGO	0.5	4.04 ± 0.22	59.0 ± 4.75	–	17.8	1.5	As above	
	RGO	1	4.28 ± 0.17	53.0 ± 5.55	–	25.0	–8.8	As above	
	RGO	2	4.65 ± 0.35	57.0 ± 4.55	–	35.8	–1.9	As above	
	RGO	4	4.56 ± 0.18	48.9 ± 1.15	–	33.0	–15.8	As above	

PMMA	GO	0.25	3.65 ± 0.06	62.1 ± 0.60	4.3	6.4	6.9	In situ polymerization and injection molding	[71]
	GO	0.5	4.19 ± 0.03	66.4 ± 1.75	–	22.2	14.3	As above	
	GO	1	4.39 ± 0.18	70.9 ± 0.15	3.7	28.3	22.0	As above	
	GO	2	4.45 ± 0.00	66.8 ± 3.05	3.5	29.9	15.0	As above	
	GO	4	4.53 ± 0.32	65.4 ± 5.35	–	32.3	12.6	As above	
PP	–	–	1.02 ± 0.10	24 ± 0.8	1120 ± 100	–	–	Melt processing	[79]
	RGO	0.1	1.25 ± 0.12	33 ± 1.4	1150 ± 120	23	38	Solution mixing & melt blending	
	RGO	0.5	1.50 ± 0.15	36 ± 1.5	330 ± 30	47	50	As above	
	RGO	1	1.76 ± 0.25	37 ± 1.6	130 ± 20	74	54	As above	
	RGO	2	1.66 ± 0.29	32 ± 1.0	20 ± 8	63	33	As above	
	RGO	5	1.23 ± 0.13	27 ± 0.8	100 ± 15	21	12.5	As above	

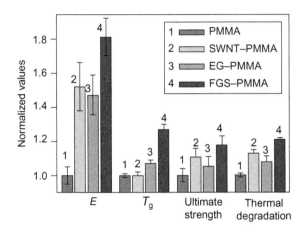

Figure 10.27 Summary of thermomechanical property improvements in solvent cast PMMA nanocomposites by reinforcing with 1 wt% EG, SWNT, or TRG. All property values are normalized to the values for pure PMMA and thus relative to unity on the scale above. Neat PMMA values are $E = 2.1$ GPa, $T_g = 105°C$, ultimate strength = 70 MPa, and thermal degradation temperature = 285°C.
Source: Reproduced from Ref. [60] with permission from Nature Publishing Group.

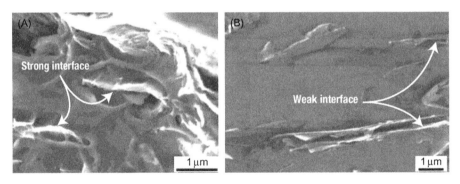

Figure 10.28 SEM fractographs of (A) 1 wt% TRG/PMMA and (B) 1 wt% EG/PMMA nanocomposites showing interfacial filler−matrix features.
Source: Reproduced from Ref. [60] with permission from Nature Publishing Group.

Moreover, SWNTs show high tendency to form clusters, hence their reinforcing effect in the PMMA matrix is reduced accordingly.

10.3.2 Thermosetting Matrix

Epoxy resin with excellent chemical resistance, high adhesion and electrical insulating behavior, low shrinkage and low cost is an important class of thermosetting

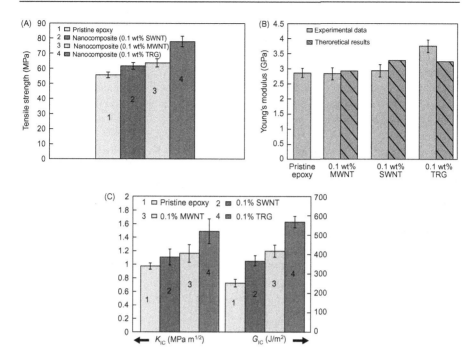

Figure 10.29 (A) Tensile strength, (B) Young's modulus, and (C) fracture toughness of pure epoxy, 0.1 wt% MWNT/epoxy, 0.1 wt% SWNT/epoxy, and 0.1 wt% TRG/epoxy nanocomposite samples. Theoretical simulation results from the H−P model with randomly oriented fillers are shown in (B).
Source: Reproduced from Ref. [23] with permission from The American Chemical Society.

polymers for use in various technological applications ranging from microelectronics to aerospace engineering. When cured, the highly cross-linked microstructure provides adequate mechanical strength but poor ductility or toughness. As a result, epoxy shows poor resistance to the crack propagation under static mechanical and cyclic fatigue deformations. In this respect, graphene and its derivatives with high aspect ratios and excellent mechanical strength are ideal nanofillers for epoxy resin for deflecting advanced crack front during static and cyclic deformation tests. However, pristine graphene cannot be used to blend with epoxy directly due to the lack of reactive sites for interacting with epoxide molecules. Thus, GO, RGO, or TRG is used instead in spite of their low structural integrity. For GO nanofillers, a reaction takes place between the carboxylic acid attached to GO surface and the epoxide ring of epoxy molecules during the composite fabrication, causing cross-linking of the epoxy chain with the GO layers [94].

Rafiee et al. [23] compared the reinforcing, stiffening, and toughening effects of the 1 wt% TRG, SWNT, and MWNT additions on the epoxy resin (Figure 10.29A−C). They also studied fatigue crack propagation rate (da/dN) of epoxy reinforced with those fillers. They reported that the tensile strength, Young's modulus, and fracture toughness of the 0.1 wt% TRG/epoxy nanocomposite are much higher than those of pure epoxy, 0.1 wt% MWNT/epoxy, and 0.1 wt% SWNT/epoxy nanocomposites. From Figure 10.29B, the critical strain energy release rate, G_c, of the above-mentioned samples is also shown. The G parameter refers to the energy available for a unit increase in crack length. Fracture occurs when G reaches a critical value G_c. The critical stress intensity factor under crack opening mode (K_{IC}) is related to G_{IC} by

$$K_{IC} = \sqrt{EG_{IC}}$$ (10.8)

Moreover, theoretical elastic modulus of all nanocomposite specimens determined from the H−P model with randomly oriented fillers is depicted in Figure 10.29B. The theoretical simulations underpredict elastic modulus of the 0.1 wt% TRG/epoxy nanocomposite by ~13%. TRGs with wrinkled and rough surface features promote mechanical interlocking at the TRG−matrix interface effectively. This mechanism explains the reinforcing and stiffening effects of epoxy by adding TRG. 2D TRGs with large aspect ratios can deflect propagating crack effectively, leading to enhanced fracture toughness and low fatigue crack growth rate of the 0.1 wt% TRG/epoxy nanocomposite.

Crack deflection is the process by which an initial crack tilts and twists its path when it encounters rigid fillers. This creates an increase in the total fracture surface areas, resulting in higher energy dissipation as compared to the unfilled polymers. Another toughening mechanism of polymer composites is the crack bridging, that is, the advanced crack opening is bridged by the pullout fillers. Crack bridging can be excluded from the toughening mechanism for the 0.1 wt%TRG/epoxy nanocomposite due to the strong mechanical locking effect. Because of the effective crack deflection by TRGs, the fatigue crack growth rate of the 0.1 wt% TRG/epoxy nanocomposite is considerably lower than that of pure epoxy, 0.1 wt% MWNT/epoxy, and 0.1 wt% SWNT/epoxy nanocomposites (Figure 10.30). In another work [70], Rafiee et al. investigated the effects of TRG loadings on the fracture

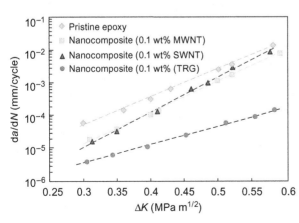

Figure 10.30 Fatigue crack growth rate versus stress intensity factor amplitude of pure epoxy, 0.1 wt% MWNT/epoxy, 0.1 wt% SWNT/epoxy, and 0.1 wt% TRG/epoxy nanocomposite samples. *Source*: Reproduced from Ref. [23] with permission from The American Chemical Society.

toughness and tensile properties of epoxy (Figure 10.31A and B). The K_{IC} and tensile strength of the TRG/epoxy nanocomposites attained maximum values at 0.125 wt% TRG.

Very recently, Tang et al. [83] studied the effect of RGO dispersion on the mechanical properties of RGO/epoxy nanocomposites prepared by solvent casting and curing. The van der Waals interaction among the reduced graphene sheets causes their high tendency to aggregate in solution. To ensure better dispersion of RGO in the epoxy matrix, the RGO/epoxy mixture was ball milled with zirconia balls in a planetary mill for 6 h. The high shear stress applied by the milling can break up the agglomerates of RGO sheets. Then the mixture was degassed at 80°C in vacuum for 6 days for the solvent (ethanol) removal, followed by adding a

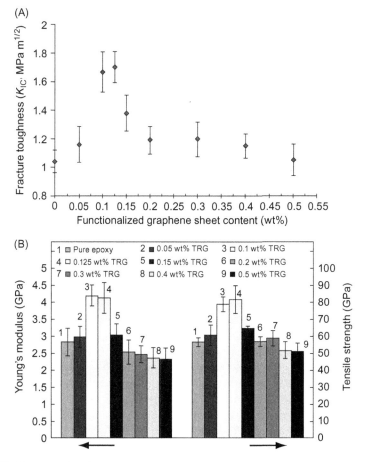

Figure 10.31 (A) Mode I fracture toughness (K_{IC}) versus filler content of TRG/epoxy nanocomposites. (B) Tensile properties of pure epoxy and TRG/epoxy nanocomposites reinforced with various TRG contents.
Source: Reproduced from Ref. [70] with permission from Wiley VCH.

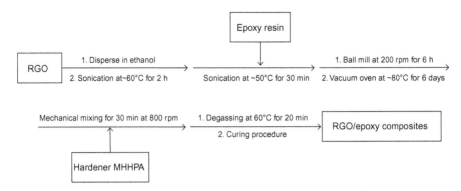

Figure 10.32 Schematic of the fabrication of RGO/epoxy nanocomposites.
Source: Reproduced from Ref. [83] with permission from Elsevier.

hardener under mechanical stirring. The last processing step was curing (Figure 10.32). For the comparison, RGO/epoxy nanocomposites without ball milling were also prepared under the same solution mixing and curing conditions, yielding poorly dispersed nanofillers as expected. Figure 10.33A and B shows the tensile properties of the highly- and poorly dispersed RGO/epoxy nanocomposites. The flexural properties of these nanocomposites are depicted in Figure 10.33C and D. The elastic and flexural moduli increase with the increase of RGO loading regardless of the dispersion level. Very low RGO additions up to 0.05 wt% lead to a small increase in both tensile and flexural strengths of highly dispersed RGO/epoxy nanocomposites. However, the tensile and flexural strengths decrease with increasing filler content for the poorly dispersed nanocomposites. The graphene agglomerates in poorly dispersed nanocomposites facilitate the initiation of cracks during mechanical deformations, resulting in reduced tensile and flexural strengths of the nanocomposites. The fracture toughness of those nanocomposites measured with the CT specimens is shown in Figure 10.34. Apparently, the fracture toughness of highly dispersed nanocomposites increases with increasing filler content. For poorly dispersed nanocomposites, the fracture toughness increases slowly with increasing RGO content up to 0.1 wt%, followed by a saturation as the filler content increases. Figure 10.35A is a SEM image showing the fracture surface of pure epoxy after fracture toughness testing. The fracture surface of pure epoxy is smooth and featureless, showing brittle failure characteristic with poor resistance to the crack propagation. For poorly dispersed 0.2 wt% RGO/epoxy nanocomposite, the fracture surface is rougher with the presence of few river flow patterns. The agglomeration of RGO in several local regions can be readily seen (Figure 10.35B and C). The 0.2 wt% RGO/epoxy nanocomposite with highly dispersed RGO also exhibits rough river patterns over entire surface. The dispersion of nanofillers is rather uniform due to the ball milling of RGO/epoxy mixture (Figure 10.35D).

Apart from the mechanical milling, functionalization of GO sheets plays a key role in enhancing interfacial filler—matrix interactions. Strong interfacial bonding enables

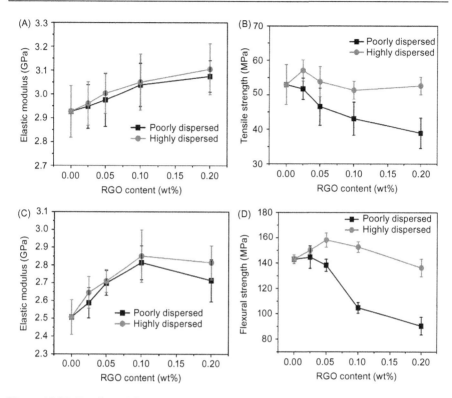

Figure 10.33 Tensile and flexural properties of cured RGO/epoxy nanocomposites: (A) and (C) modulus; (B) and (D) strength.
Source: Reproduced from Ref. [83] with permission from Elsevier.

effective stress transfer during mechanical deformations. As an example, Wang et al. [95] functionalized RGO suspension with 3-aminopropyltriethoxysilane (APTS). They reported that the addition of 1 wt% functionalized graphene sheets to epoxy increases its tensile modulus and strength markedly. Very recently, Wan et al. [96] functionalized TRG noncovalently via a surfactant-assisted process. They reported that the surface treatment of TRGs with a nonionic surfactant (Triton) is very effective to improve the compatibility of graphene with the epoxy resin. The pristine graphene sheets tend to reagglomerate during curing of resin, whereas the attachment of surfactant molecules effectively inhibits the reagglomeration of graphene by wrapping the sheets with the polymer chains. Therefore, the composites with noncovalently functionalized graphene exhibit much higher reinforcing and toughening effects compared to their counterparts with TRGs without surface modification (Figure 10.36A—C).

Ma et al. [48] functionalized GnPs with 4,4'-diaminophenylsulfone and then blended such m-GnPs with epoxy using solvent casting and curing processes. For comparison, epoxy composites reinforced with pristine GnPs were also fabricated

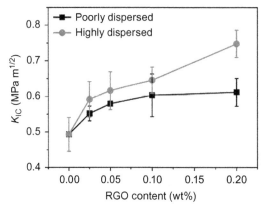

Figure 10.34 Fracture toughness versus RGO content of cured RGO/epoxy nanocomposites.
Source: Reproduced from Ref. [83] with permission from Elsevier.

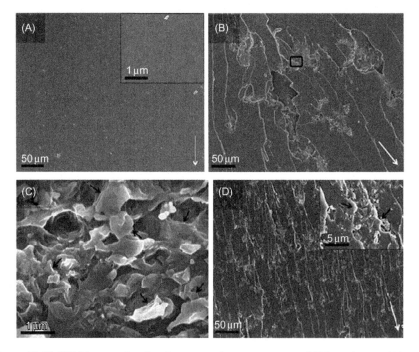

Figure 10.35 SEM images showing fracture surface appearances of CT specimens: (A) pure epoxy, (B) low magnification view of poorly dispersed 0.2 wt% RGO/epoxy composite, (C) high magnification view of a rectangle in (B), and (D) low magnification view of highly dispersed 0.2 wt% RGO/epoxy composite. The inset is a higher magnified image. White arrows indicate the crack growth direction. Black arrows show the RGO fillers.
Source: Reproduced from Ref. [83] with permission from Elsevier.

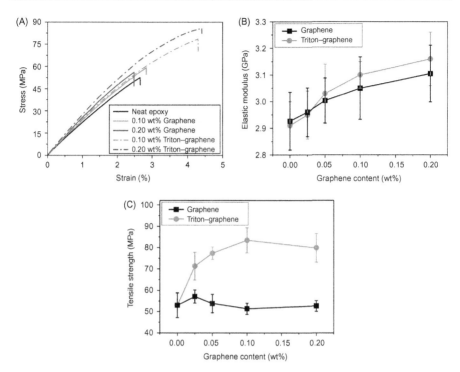

Figure 10.36 (A) Stress—strain curves of neat epoxy, TRG/epoxy, and Triton-functionalized TRG/epoxy nanocomposites. (B) Elastic modulus and (C) tensile strength versus TRG content for TRG/epoxy and Triton-functionalized TRG/epoxy nanocomposites. *Source*: Reproduced from Ref. [96] with permission from Elsevier.

using the same processing conditions. Figure 10.37 shows representative XRD patterns of the 0.244 vol% GnP/epoxy and 0.244 vol% m-GnP/epoxy nanocomposites. The patterns show the presence of a broad reflection extending from ~11 to 25°, resulting from the diffraction of cured epoxy molecules. Furthermore, there is a tiny, sharp reflection at $2\theta = 26°$ in the XRD pattern of unmodified 0.244 vol% GnP/epoxy nanocomposite, corresponding to the stacked GnPs with interlayer distance of 3.35 Å. However, this reflection is missing in the pattern of 0.244 vol% m-GnP/epoxy nanocomposite, demonstrating a lower degree of graphene stacking in m-GnPs. Figure 10.38A and B shows the respective plots of elastic modulus versus filler content and tensile strength versus filler content for GnP/epoxy and m-GnP/epoxy nanocomposites. For unmodified GnP/epoxy nanocomposites, elastic modulus increases slowly with increasing GnP content. The increase is more significant for functionalized m-GnP/epoxy system. At 0.489 vol% m-GnP, there is 57.45% improvement in modulus over pure epoxy. The tensile strength of the m-GnP/epoxy system peaks at 0.122 vol% m-GnP and then slightly decreases with increasing filler content. The improved tensile strength at 0.122 vol% m-GnP is

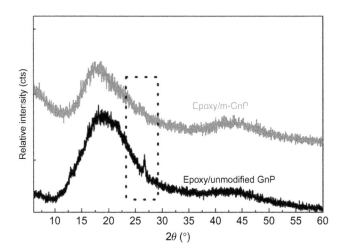

Figure 10.37 XRD patterns of 0.244 vol% GnP/epoxy and 0.244 vol% m-GnP/epoxy nanocomposites.
Source: Reproduced from Ref. [48] with permission from Elsevier.

due to the enhanced interfacial filler—matrix interaction. However, the tensile strength of unmodified GnP/epoxy nanocomposites is even poorer than that of pure epoxy and decreases markedly with increasing filler content. This is due to the ineffective stress transfer at the GnP—epoxy interface.

10.4 Conclusions

The demand for lightweight, strong, and tough materials for the composite industry has been growing continuously. Graphene and its derivatives are attractive nanofillers for reinforcing and stiffening polymers at very low loading levels. The resulting nanocomposites with high strength and light weight show great potential for use as the structural materials in aerospace and automotive industries. Wrinkled TRGs generally outperform MWNTs and SWNTs for reinforcing and toughening brittle PMMA and epoxy. Only 0.1 wt% TRG loading is needed to reinforce and toughen epoxy resin. This is because 2D TRGs with a higher surface-to-volume ratio than carbon nanotubes can lock mechanically at the filler—matrix interface, thereby promoting effective stress transfer at the interfacial region during the mechanical transformation. Further, TRGs can also deflect unstable crack propagation in the brittle polymer matrix, thus increasing energy dissipation and toughening the matrix accordingly. Similarly, melt-compounded RGO/HDPE nanocomposites exhibit higher elastic modulus and tensile strength than the MWNT/HDPE nanocomposites at the same filler loading.

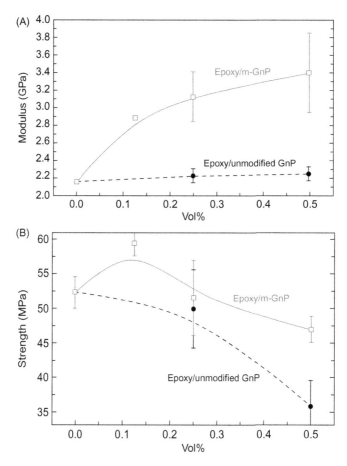

Figure 10.38 (A) Young's modulus and (B) tensile strength versus GnP volume content for GnP/epoxy and m-GnP/epoxy nanocomposites.
Source: Reproduced from Ref. [48] with permission from Elsevier.

The mechanical properties of polymer nanocomposites depend greatly on the interfacial adhesion and dispersion state of graphene nanofillers in the polymer matrix. These properties are mainly related to functionalization of nanofillers and the techniques employed for fabricating nanocomposites. GO is incompatible with hydrophobic polymers and does not form homogeneous composites. Thus, surface modification of GO is needed. Generally, solution mixing and *in situ* polymerization enable homogenous dispersion of graphene nanofillers in the polymer matrix compared with melt blending. Therefore, a higher reinforcing effect can be achieved in the polymer nanocomposites fabricated by means of solution mixing and *in situ* polymerization. Practically, melt-compounding process is preferred since it can manufacture graphene/polymer nanocomposites in large quantities.

Many attractive applications can be envisioned for novel graphene/polymer nano-composites, including automotive panel body, automotive tire, gas storage container, biomedical implant, supercapacitor, and battery electrode.

References

[1] Kuilla T, Bhadra S, Yao D, Kim NH, Bose S, Lee JH. Recent advances in graphene based polymer composites. Prog Polym Sci 2010;35(11):1350−75.

[2] Soldano C, Mahmood A, Dujardin E. Production, properties and potential of graphene. Carbon 2010;48(8):2127−50.

[3] Berger C, Song ZM, Li XB, Ogbazghi AY, Feng R, Dai Z, et al. Ultrathin epitaxial graphite: 2D electron gas properties and a route toward graphene-based nanoelectronics. J Phys Chem B 2004;108(52):19912−6.

[4] Reina F, Jia X, Ho H, Nezich D, Son H, Bulovich B, et al. Large area, few-layer graphene films on arbitrary substrates by chemical vapor deposition. Nano Lett 2009;9 (1):30−5.

[5] Kim SJ, Zhao Y, Jang H, Lee SY, Kim JM, Kim KS, et al. Large-scale pattern growth of graphene films for stretchable transparent electrodes. Nature 2009;457 (7230):706−10.

[6] Li X, Cai W, An J, Kim S, Nah J, Yang D, et al. Large-area synthesis of high-quality and uniform graphene films on Cu foils. Science 2009;324:1312−14.

[7] Hummers WS, Offeman RE. Preparation of graphitic oxide. J Am Chem Soc 1958;80 (6):1339−40.

[8] He HY, Klinowski J, Forster M, Lerf A. A new structural model for graphite oxide. Chem Phys Lett 1998;287(1−2):53−6.

[9] Lerf A, He HY, Forster M, Klinowski J. Structure of graphite oxide revisited. J Phys Chem B 1998;102(23):4477−82.

[10] Hontoria-Lucas C, Lopez-Peinado AJ, Lopez-Gonzalez J, Rojas-Cervantes ML, Martin-Aranda RM. Study of oxygen-containing groups in a series of graphite oxides: physical and chemical characterization. Carbon 1995;33(11):1585−92.

[11] Park S, Ruoff RS. Chemical methods for the production of graphenes. Nat Nanotechnol 2009;4(4):217−24.

[12] Paredes JI, Villar-Rodil S, Martinez-Alonso A, Tascon JM. Graphene oxide dispersions in organic solvents. Langmuir 2008;24(19):10560−4.

[13] Stankovich S, Dikin DA, Piner RD, Kohlhaas KA, Kleinhammes A, Jia Y, et al. Synthesis of graphene-based nanosheets via chemical reduction of exfoliated graphite oxide. Carbon 2007;45(7):1558−65.

[14] Shin HJ, Kim KK, Benayad A, Yoon SM, Park HK, Jung IS, et al. Graphite oxide by sodium borohydride and its effect on electrical conductance. Adv Funct Mater 2009;19 (12):1987−92.

[15] Chen C, Chen T, Wang H, Sun G, Yang X. A rapid, one-step, variable-valence metal ion assisted reduction method for graphene oxide. Nanotechnology 2011;22 (40):405602.

[16] Hu N, Wang Y, Chai J, Gao R, Yang Z, Kong E, et al. Gas sensor based on p-phenylenediamine reduced graphene oxide. Sens Actuat B Chem 2012;163(1):107−14.

[17] Nethravathi C, Rajamathi M. Chemically modified graphene sheets produced by the solvothermal reduction of colloidal dispersions of graphite oxide. Carbon 2008;46 (14):1994—8.

[18] Tien HN, Luan VH, Lee TK, Kong BS, Chung JS, Kim EJ, et al. Enhanced solvothermal reduction of graphene oxide in a mixed solution of sulfuric acid and organic solvent. Chem Eng J 2012;211—212(15):97—103.

[19] Dubin S, Gilje S, Wang K, Tung VS, Cha K, Hall AS, et al. A one-step, solvothermal reduction method for producing reduced graphene oxide dispersions in organic solvents. ACS Nano 2010;4(7):3845—52.

[20] Gomez-Navarro C, Burghard M, Kern K. Elastic properties of chemically derived single graphene shells. Nano Lett 2008;8(7):2045—9.

[21] (a)Schniepp HC, Li JL, McAllister MJ, Sai H, Herrera-Alonso M, Adamson DH, et al. Functionalized single graphene sheets derived from splitting graphite oxide. J Phys Chem B 2006;110(17):8535—9.(b)McAllister MJ, Li JL, Adamson DH, Schniepp HC, Abdala AA, Liu J, et al. Single sheet functionalized graphene by oxidation and thermal expansion of graphite. Chem Mater 2007;19(18):4396—404.

[22] Yang JH, Lin SH, Lee YD. Preparation and characterization of poly(L-lactide)—graphene composites using the *in situ* ring-opening polymerization of PLLA with graphene as the initiator. J Mater Chem 2012;22(21):10805—15.

[23] Rafiee MA, Rafiee J, Wang Z, Song H, Yu ZZ, Koratkar N. Enhanced mechanical properties of nanocomposites at low graphene content. ACS Nano 2009;3 (12):3884—90.

[24] Becerril HA, Mao J, Liu Z, Stoltenberg RM, Bao Z, Chen Y. Evaluation of solution-processed reduced graphene oxide films as transparent conductors. ACS Nano 2008;2 (3):463—70.

[25] Kim HW, Kobayashi S, Abdurrahim MA, Zhang ML, Khusainova A, Hillmyer MA, et al. Graphene/polyethylene nanocomposites: effect of polyethylene functionalization and blending methods. Polymer 2011;52(8):1837—46.

[26] Tjong SC, Xu SA, Li RK, Mai YW. Mechanical behavior and fracture toughness evaluation of maleic anhydride compatibilized short glass fiber/SEBS/polypropylene hybrid composites. Compos Sci Technol 2002;62(6):831—40.

[27] Tjong SC, Xu SA, Li RK, Mai YW. Short glass fiber-reinforced polyamide 6,6 composites toughened with maleated SEBS. Compos Sci Technol 2002;62 (15):2017—27.

[28] Tjong SC, Meng YZ. Microstructural and mechanical characteristics of compatibilized polypropylene hybrid composites containing potassium titanate whisker and liquid crystalline copolyester. Polymer 1999;40(26):7275—83.

[29] Tjong SC, Meng YZ. Performance of potassium titanate whisker reinforced polyamide-6 composites. Polymer 1998;39(22):5461—6.

[30] Tjong SC, Li RK, Cheung T. Mechanical behavior of $CaCO_3$ particulate-filled beta-crystalline phase polypropylene composites. Polym Eng Sci 1997;37(1):166—72.

[31] Liao CZ, Bao SP, Tjong SC. Microstructure and fracture behavior of maleated high-density polyethylene/silicon carbide nanocomposites toughened with poly (styrene—ethylene—butylene—styrene) triblock copolymer. Adv Polym Technol 2011;30(4):322—33.

[32] Bao SP, Liang GD, Tjong SC. Positive temperature coefficient effect of polypropylene/carbon nanotube/montmorillonite hybrid nanocomposites. IEEE Trans Nanotechnol 2009;8(6):729—36.

[33] Bao SP, Tjong SC. Impact essential work of fracture of polypropylene/montmorillonite nanocomposites toughened with SEBS-g-MA elastomer. Compos Part A 2007;38 (2):378−87.

[34] Tjong SC, Bao SP. Fracture toughness of high density polyethylene/SEBS-g-MA/montmorillonite nanocomposites. Compos Sci Technol 2007;67(2):314−23.

[35] Tjong SC. Structural and mechanical properties of polymer nanocomposites. Mater Sci Eng R 2006;53(3-4):73−197.

[36] Tjong SC, Liang GD. Electrical properties of low-density polyethylene/ZnO nanocomposites. Mater Chem Phys 2006;100(1):1−5.

[37] Tjong SC, Meng YZ, Hay AS. Novel preparation and novel properties of polypropylene−vermiculite nanocomposites. Chem Mater 2002;14(1):44−51.

[38] Chandra A, Turng LS, Li K, Huanf HX. Fracture behavior and optical properties of melt compounded semi-transparent polycarbonate (PC)/alumina nanocomposites. Compos Part A 2011;42(12):1903−9.

[39] Tjong SC. Polymer nanocomposite bipolar plates reinforced with carbon nanotubes and graphite nanosheets. Energy Env Sci 2011;4(3):605−26.

[40] Chiacchiarelli LM, Rallini M, Monti M, Puglia D. The role of irreversible and reversible phenomena in the piezoresistive behavior of graphene epoxy nanocomposites applied to structural health monitoring. Compos Sci Technol 2013;80(17):73−9.

[41] Ren L, Qiu J, Wang S. Thermo-adaptive functionality of graphene/polydimethylsiloxane nanocomposites. Smart Mater Struct 2012;21(10):105032.

[42] Chen Y, Qi Y, Tai Z, Yan X, Zhu F, Xue Q. Preparation, mechanical properties and biocompatibility of graphene oxide/ultrahigh molecular weight polyethylene composites. Eur Polym J 2012;48(6):1026−33.

[43] Yan DX, Ren PG, Pang H, Fu Q, Yang MB, Li ZM. Efficient electromagnetic interference shielding of lightweight graphene/polystyrene composite. J Mater Chem 2012;22 (36):18772−6.

[44] Li YC, Li RK, Tjong SC. Electrical transport properties of graphite sheets doped polyvinylidene fluoride nanocomposites. J Mater Res 2010;25(8):1645−8.

[45] Xiang JL, Drzal LT. Thermal conductivity of exfoliated graphite nanoplatelet. Carbon 2011;49(3):773−8.

[46] Chen GH, Weng W, Wu D, Wu C, Lu J, Wang P, et al. Preparation and characterization of graphite nanosheets from ultrasonic powdering technique. Carbon 2004;42 (4):753−9.

[47] Shioyama HH. The interactions of two chemical species in the interlayer spacing of graphite. Synth Met 2000;114(1):1−15.

[48] Ma J, Meng Q, Michelmore A, Kawashima N, Izzuddin Z, Bengtsson C, et al. Covalently bonded interfaces for graphene/polymer composites. J Mater Chem A 2013;1(13):4255−64.

[49] Available from: http://www.xgsciences.com/.

[50] Available from: http://www.vorbeck.com/.

[51] Zhao X, Zhang QH, Chen DJ, Lu P. Enhanced mechanical properties of graphene-based poly(vinyl alcohol) composites. Macromolecules 2010;43(5):2357−63.

[52] Yang X, Li L, Shang S, Tao X. Synthesis and characterization of layer-aligned poly (vinyl alcohol)/graphene nanocomposites. Polymer 2010;51(15):3431−5.

[53] Xu X, Hong WJ, Bai H, Li C, Shi GQ. Strong and ductile poly(vinyl alcohol)/graphene oxide composite films with a layered structure. Carbon 2009;47(15):3538−43.

[54] Matsuo Y, Tahara K, Sugie Y. Structure and thermal properties of poly(ethylene oxide)-intercalated graphite oxide. Carbon 1997;35(1):113−20.

[55] Niyoki S, Bekayarova E, Itkis ME, McWilliams JL, Hamon MA, Haddon RA. Solution properties of graphite and graphene. J Am Chem Soc 2006;128(124):7720—1.

[56] Cao YW, Feng JC, Wu PY. Alkyl-functionalized graphene nanosheets with improved lipophilicity. Carbon 2010;48(5):1683—5.

[57] Stankovich S, Piner RD, Nguyen SB, Ruoff RS. Synthesis and exfoliation of isocyanate-treated graphene oxide nanoplatelets. Carbon 2006;44(15):3342—7.

[58] Luong ND, Hippi U, Korhonen JT, Soininen AJ, Ruokolainen J, Johansson LS, et al. Enhanced mechanical and electrical properties of polyimide film by graphene sheets via in situ polymerization. Polymer 2011;52(23):5237—42.

[59] Mahmoud WE. Morphology and physical properties of poly(ethylene oxide) loaded graphene nanocomposites prepared by two different techniques. Eur Polym J 2011;47 (8):1534—40.

[60] Ramanathan T, Abdala AA, Stankovich S, Dikin DA, Herrera-Alonso M, Piner RD, et al. Functionalized graphene sheets for polymer nanocomposites. Nature Nanotechnol 2008;3(6):327—31.

[61] Teng CC, Ma CC, Lu CH, Yang SY, Lee SH, Hsiao MC, et al. Thermal conductivity and structure of non-covalent functionalized graphene/epoxy composites. Carbon 2011;49(15):5107—16.

[62] Lomeda JR, Doyle CD, Kosynkin DV, Hwang WF, Tour JM. Diazonium functionalization of surfactant-wrapped chemically converted graphene sheets. J Am Chem Soc 2008;130(48):16201—6.

[63] Vinodkopal K, Neppolian B, Salleh N, Lightcap IV, Grieser F, Ashokkumar M, et al. Dual-frequency ultrasound for designing two-dimensional catalyst surface: reduced graphene oxide—Pt composite. Colloids Surf A 2012;409:81—7.

[64] Singh V, et al. Graphene based materials: past, present and future. Prog Mater Sci 2011;56(8):1178—271.

[65] Stankovich S, Dikin DA, Dommett GH, Kohlhaas KM, Zimney EJ, Stach EA, et al. Graphene-based composite materials. Nature 2006;442(7100):282—6.

[66] Vadukumpully S, Paul J, Mahanta N, Valiyaveettil S. Flexible conductive graphene/poly(vinyl chloride) composite thin films with high mechanical strength and thermal stability. Carbon 2011;49(1):198—205.

[67] Yun YS, Bae YH, Kim D, Lee JY, Chin IJ, Jin HJ. Reinforcing effects of adding alkylated graphene oxide to polypropylene. Carbon 2011;49(11):3553—9.

[68] Jin Y, Huang X, Wu C, Jiang P. Permittivity, thermal conductivity and thermal stability of poly(vinylidene fluoride)/graphene nanocomposites. IEEE Trans Dielect Insul 2011;18(2):478—84.

[69] Gudarzi M, Sharif F. Enhancement of dispersion and bonding of graphene—polymer through wet transfer of functionalized graphene oxide. Express Polym Lett 2012;6 (12):1017—31.

[70] Rafiee MA, Rafiee J, Srivastava I, Wang Z, Song H, Yu ZZ, et al. Fracture and fatigue in graphene nanocomposites. Small 2010;6(2):179—83.

[71] Potts JR, Lee SH, Alam TM, An JH, Stoller MD, Piner RD, et al. Thermomechanical properties of chemically modified graphene/poly(methyl methacrylate) composites made by in situ polymerization. Carbon 2011;49(8):2615—23.

[72] Xu Z, Gao C. In situ polymerization approach to graphene-reinforced nylon-6 composites. Macromolecules 2010;43(16):6716—23.

[73] Wang J, Shi Z, Ge Y, Wang Y, Fan J, Yin J. Solvent exfoliated graphene for reinforcement of PMMA composites prepared by in situ polymerization. Mater Chem Phys 2012;136(1):43—50.

[74] Hu HT, Wang XB, Wang JC, Wan L, Liu FM, Zheng H, et al. Preparation and properties of graphene nanosheets—polystyrene nanocomposites via *in situ* polymerization. Chem Phys Lett 2010;484(4−6):247−53.

[75] Patrole AS, Patole SP, Kang H, Yoo JB, Kim TH, Ahn JH. A facile approach to the fabrication of graphene/polystyrene nanocomposite by *in situ* microemulsion polymerization. J Colloid Interf Sci 2010;350(2):530−7.

[76] Fabbri P, Bassoli E, Bon SB, Valentini L. Preparation and characterization of poly (butylene terephthalate)/graphene composites by *in situ* polymerization of cyclic butylene terephthalate. Polymer 2012;53(4):897−902.

[77] Bao C, Song L, Xing W, Yuan B, Wilkie CA, Huang J, et al. Preparation of graphene by pressurized oxidation and multiplex reduction and its polymer nanocomposites by masterbatch-based melt blending. J Mater Chem 2012;22(13):6088−96.

[78] Cai DY, Yusoh K, Song M. The mechanical properties and morphology of a graphite oxide nanoplatelet/polyurethane composite. Nanotechnology 2009;20(8):085712.

[79] Song PG, Cao ZT, Cai YZ, Zhao LP, Fang ZP, Fu SY. Fabricated of exfoliated graphene-based polypropylene nanocomposites with enhanced mechanical and thermal properties. Polymer 2011;52(18):4001−10.

[80] Kim HW, Macosko CW. Processing-property relationships of polycarbonate/graphene composites. Polymer 2009;50(15):3797−809.

[81] Zhang HB, Zheng WG, Yan Q, Yang Y, Wang J, Lu ZH, et al. Electrically conductive polyethylene terephthalate/graphene nanocomposites prepared by melt compounding. Polymer 2010;51(5):1191−6.

[82] Srivastava I, Mehta RJ, Yu ZZ, Schadler LS, Koratkar NN. Raman study of interfacial load transfer in graphene nanocomposites. Appl Phys Lett 2011;98(6):063102.

[83] Tang LC, Wan YJ, Yan D, Pei YB, Zhao L, Li YB. The effect of graphene dispersion on the mechanical properties of graphene/epoxy composites. Carbon 2013;60:16−27.

[84] Shen XJ, Liu Y, Xiao HM, Feng QP, Yu ZZ, Fu SY. The reinforcing effect of graphene nanosheets on the cryogenic mechanical properties of epoxy resins. Compos Sci Technol 2012;72(13):1581−7.

[85] ASTM Standard D 5045-99. Standard test methods for plane-strain fracture toughness and strain energy release rate of plastic materials. West Conshohocken: American Society for Testing and Materials; 1999.

[86] Longun J, Iroh JO. Nano-graphene/polyimide composites with extremely high rubbery plateau modulus. Carbon 2012;50(5):1823−32.

[87] Halpin JC, Kardos JL. The Halpin—Tsai equations: a review. Polym Eng Sci 1976;16 (5):344−52.

[88] Wang JC, Wang X, Xu CH, Zhang M, Shang XP. Preparation of graphene/poly(vinyl alcohol) nanocomposites with enhanced mechanical properties and water resistance. Polym Int 2011;60(5):816−22.

[89] Layek RK, Samanta S, Nandi AK. The physical properties of sulfonated graphene/poly (vinyl alcohol) composites. Carbon 2012;50(3):815−27.

[90] Achaby ME, Qaiss A. Processing and properties of polyethylene reinforced by graphene nanosheets and carbon nanotubes. Mater Des 2013;44:81−9.

[91] Laurent MP, Johnson TS, Crowninshield RD, Blanchard CR, Bhambri SK, Yao JQ. Characterization of a highly cross-linked ultrahigh molecular-weight polyethylene in clinical use in total hip arthroplasty. J Arthroplasty 2008;23(5):751−61.

[92] Archard JF. Contact and rubbing of flat surfaces. J Appl Phys 1953;24:981−8.

[93] Yun YS, Bae YH, Kim DH, Lee JY, Chin IJ, Jin HJ. Reinforcing effects of adding alkylated graphene oxide to polypropylene. Carbon 2011;49(11):3553−9.

[94] Shen B, Zhai W, Tao M, Lu D, Zheng W. Chemical functionalization of graphene oxide toward the tailoring of the interface in polymer composites. Compos Sci Technol 2013;77:87–94.

[95] Wang X, Xing W, Zhang P, Song L, Yang H, Hu Y. Covalent functionalization of graphene with organosilane and its use as a reinforcement in epoxy composites. Compos Sci Technol 2012;72(6):737–42.

[96] Wan WJ, Tang LC, Yan D, Zhao L, Li YB, Wu LB. Improved dispersion and interface in the graphene/epoxy composites via a facile surfactant-assisted process. Compos Sci Technol 2013;82:60–8.

[97] Novoselov KS, Geim AK, Morozov S, Jiang D, Zhang Y, Dubonos SV, et al. Electrically thin carbon films. Science 2004;306(5696):666–9.

11 Zener Tunneling in Polymer Nanocomposites with Carbonaceous Fillers

Linxiang He[1] and Sie Chin Tjong[1,2]

[1]Department of Physics and Materials Science, City University of Hong Kong, Hong Kong, PR China, [2]Department of Physics, Faculty of Science, King Abdulaziz University, Jeddah, Saudi Arabia

11.1 Background

Polymer composites reinforced with carbon fibers have been used widely as the structural components in aeronautical and space sectors because of their light weight, high mechanical strength and modulus as well as excellent corrosion resistance [1,2]. Carbon fibers are electrically conductive, rendering their polymer composites the ability to shield electromagnetic interference (EMI) effectively [3,4]. In addition, carbon fiber/polymer composites have successfully replaced heavy metallic components for structural engineering and EMI shielding applications in the aerospace industries. However, the limits of enhancing the properties of polymer have been reached due to the necessity of loading the polymers with high carbon fiber contents for achieving desired mechanical and physical properties. Large-volume fractions of carbon fibers generally lead to poor processability, inferior mechanical and physical properties of the resulting composite materials.

Recent advances in nanoscience and nanotechnology have provided tremendous opportunities for chemists and materials scientists to synthesize novel nanomaterials with functional properties. Among them, carbonaceous nanomaterials such as graphene, carbon nanotubes (CNTs), and carbon nanofibers (CNFs) are considered of technological importance. Graphene is one atom thick, two-dimensional (2D) layer of sp^2 hybridized carbon atoms covalently bonded in a honeycomb lattice. Graphene sheet serves as a building block to construct a variety of carbon-based nanostructures. It can wrap up to form zero-dimensional (0D) buckyball or roll into one-dimensional (1D) single-walled carbon nanotube (SWNT) (Figure 11.1). Multiple graphene sheets can roll into multiple concentric SWNT cylinders to yield multi-walled carbon nanotube (MWNT). Furthermore, graphene layers stack together along the *c*-axis to form three-dimensional (3D) graphite.

Nanocrystalline Materials. DOI: http://dx.doi.org/10.1016/B978-0-12-407796-6.00011-7

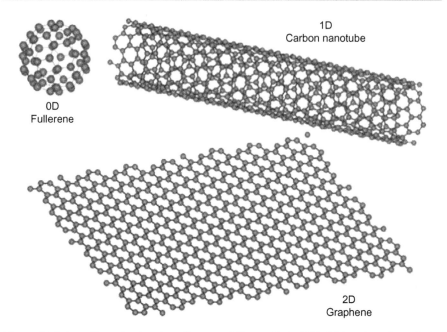

Figure 11.1 Low dimensional carbon allotropes: fullerene (0D), CNT (1D), and graphene (2D). *Source*: Reproduced from Ref. [5] with permission of Elsevier.

CNT and graphene have very large aspect ratios, extraordinary high mechanical strength and stiffness, and superior electrical and thermal conductivity. For example, SWNT and graphene exhibit exceptional high elastic modulus of about 1 TPa, and tensile strength of ∼50−500 MPa and 130 MPa, respectively. Graphene also exhibits excellent electrical conductivity of 10^6 S/m [6,7]. Therefore, the incorporation of such carbon nanomaterials into polymers at very low contents improves their physical, mechanical, and thermal properties markedly. Recently, miniaturization of electronic devices requires the use of conductive polymer nanocomposites with light weight and good electrical properties. The CNF/polymer, CNT/polymer, and graphene/polymer nanocomposites have been found to exhibit superior EMI shielding effectiveness and excellent gas sensitivity at low filler loading levels [8−12]. So these composite materials are very attractive for electrical and electronic applications.

The main drawbacks of using SWNTs and MWNTs to reinforce polymers are their high production cost and high tendency to form agglomerates. In contrast, inexpensive graphene has largely circumvented these issues [5]. Graphene sheets can be prepared using a variety of synthesis routes, as demonstrated in Figure 11.2 [13]. The reliable technique to produce high quality, pure, and single graphene layer is the micromechanical cleavage. However, this method is impractical for large-scale production of graphene sheets for reinforcing polymers. Large area graphene sheets can be grown by thermal evaporation of SiC substrates or by chemical

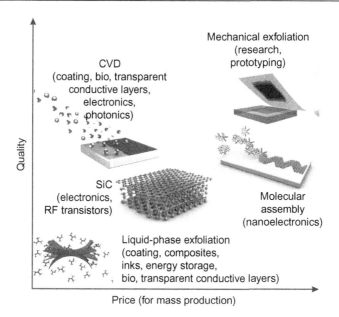

Quality

Price (for mass production)

Figure 11.2 Several preparation routes of graphene.
Source: Reproduced from Ref. [13] with permission of Nature Publishing group.

vapor deposition of hydrocarbons on specific substrates (e.g., transition metals, metal carbides) [14,15]. However, such graphene sheets are further transferred onto desirable substrates. This usually results in the formation of crystal defects, thereby degrading their physical properties. In this regard, wet chemical oxidation of graphite in strong oxidizing agents to yield graphene oxide (GO) is the cost-effective method to produce graphene-like sheets in large quantities. GO is an electrical insulator because its basal plane carbon atoms are decorated with epoxide and hydroxyl groups, while its edge atoms with carbonyl and carboxyl groups. Thus, its electrical conductivity can be restored either by using a chemical reducing agent or via high temperature treatment for removing those oxygenated groups. Recently, Hernandez et al. reported that high-yield production of graphene can be achieved by liquid phase exfoliation of graphite in an organic solvent [16]. Such a graphene suspension can be blended with polar polymers via solution mixing to form conductive graphene/polymer nanocomposites.

Prior to the successful mechanical and chemical exfoliation of graphite into graphene, graphite nanoplatelets (GNPs) were often used as the nanofillers to form conductive polymer composites [17]. GNPs consist of few stacked graphene sheets (ca. 10−30 sheets) resulting from thermal expansion or microwave irradiation of the graphite intercalation compound (GIC) (see Figure 10.5A and B). The use of GNP is desirable because they are cheaper and easier to produce than single layer graphene or CNTs [18]. In addition, CNFs or vapor grown carbon nanofibers (VGCNFs) synthesized from catalytic chemical vapor deposition of natural gas feedstock are available commercially at lower cost than CNTs [19,20]. The diameters of CNFs range typically from 50 to 500 nm. CNF consists of stacked graphene layers forming cones or cups such that stacking cone sheets yield a herringbone feature while stacking cups produce a bamboo-like morphology [21,22], as demonstrated in Figure 11.3. CNFs generally possess more crystalline defects than MWNTs.

Polymers are insulators and can be made electrically conductive by loading them with carbonaceous nanofillers. The high aspect ratio of carbon nanofillers enables the formation of conductive path network in the polymer matrix at a low filler concentration, commonly known as the percolation threshold. The main challenge of obtaining high electrical conductivity and low percolation threshold in the polymer nanocomposites is the attainment of uniform dispersion of carbon nanofillers in the polymer matrix. SWNT, MWNT, CNF, and pristine graphene are inert materials, hence having high tendency to form clusters in the polymer matrix. Covalent attachment of functional groups to carbon nanomaterials can improve their dispersion in organic solvents and the polymer matrix. In addition, a strong interfacial bonding is established at the filler–matrix interface due to the covalent functionalization. Thus covalent functionalization is beneficial in improving the mechanical properties of polymer nanocomposites. However, covalent functionalization is detrimental to electrical conduction of the polymer nanocomposites since the functional groups disrupt extended π-conjugation of the graphene sheets. Consequently, covalently functionalized polymer nanocomposites exhibit poorer electrical conductivity than unmodified composite counterparts [23,24]. As an example, silane-functionalized GNP/poly(methyl methacrylate) (PMMA) and MWNT/epoxy nanocomposites exhibit inferior conductivity than their unmodified counterparts. This is because silane molecules on the GNPs or MWNTs prevent electron tunneling between adjacent fillers, thereby degrading the electrical properties of treated composites [25,26]. In contrast, noncovalent functionalization of carbon nanofillers with nonionic surfactants improves electrical conductivity of resulting composites since the surfactant molecules do not disrupt π-conjugation of carbon nanomaterials [26].

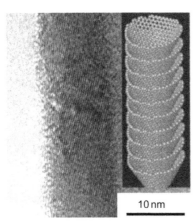

10 nm

Figure 11.3 Schematic diagram of the cup-stacked structure together with high-resolution transmission electron microscopic (HRTEM) image of a sidewall of a CNF showing such a structure.
Source: Reproduced from Ref. [22] with permission from the American Institute of Physics.

The dispersion of carbon nanofillers in the polymer matrix depends also greatly on the processing techniques employed for fabricating nanocomposites. Polymer composites with carbon nanofillers can be prepared via solution mixing, *in situ* polymerization of monomers, and melt compounding techniques. Solution processing and *in situ* polymerization generally lead to better dispersion of carbon nanofillers in the polymer matrix than melt compounding.

11.2 Electrical Conduction Behavior

11.2.1 Percolation

Percolation theory usually gives a phenomenological description of electrical properties of a conducting filler/insulator system. At low filler contents, conductive particles are isolated and dispersed independently in the insulating matrix. By increasing filler content, local clusters of filler particles are formed but the composite remains as an insulator. At the percolation threshold, these clusters connect each other to form a conductive path network through the matrix, resulting in a sharp rise in electrical conductivity. It has been established both theoretically and experimentally that the conductivity and the dielectric constant follow the power law relationship near the percolation threshold p_c [27−31]:

$$\sigma \sim (p_c - p)^{-s} \quad \text{for } p < p_c \tag{11.1}$$

$$\sigma \sim (p - p_c)^{t} \quad \text{for } p > p_c \tag{11.2}$$

$$\varepsilon \sim |p - p_c|^{-s'} \quad \text{for } p < p_c \text{ or } p > p_c \tag{11.3}$$

where σ is the conductivity, ε the permittivity (dielectric constant), and p the volume fraction of the conductive filler. The critical exponents s, s', and t are assumed to be universal. They depend only on the dimensions of the percolating system and independent on the geometry of conducting clusters. The currently accepted values of these exponents are: $s = t \approx 1.3$ for two dimensions and $s \approx 0.73$, $t \approx 2.0$ for three dimensions. Nevertheless, deviations in experimental values of s and t are frequently reported [32−34].

In addition, it is worth mentioning that the classical percolation theory applied to conductive-filler−polymer systems must meet certain conditions. The theoretically predicted value shows reasonable agreement with the experimental result provided the following criteria are fulfilled: the particles are spherical, monodisperse, and have an isotropic conductivity. If one or all these conditions are not fulfilled, the theoretical value can deviate significantly from the practical one. As an example, numerical calculations for the hard-core circles give $p_c \approx 16 \text{ vol}\%$ [35]. The percolation model based on the hard-core circles has its own limitation for predicting the percolation threshold of polymer composites with conducting nanofillers of large aspect ratios. Thermally- and chemically reduced GO fillers (2D) with large aspect ratios facilitate formation of conducting path network at very low percolation concentration of $\sim 0.07-0.1$ vol%. The percolation threshold of polymer nanocomposites is well below theoretical value of 16 vol%.

11.2.2 Nonlinear Electrical Conductivity

The percolation theory generally predicts static electrical conductivity of the conductor—insulator materials. By exposing a material system to an external electric field (E), its electrical current density (J) increases (Figure 11.4A). This is a very common feature of disordered conductor—insulator materials [36−44]. When the material no longer follows the ohmic law, the empirical relation $I = GV^\alpha$ ($\alpha \neq 1$) appears to fit reasonably for a variety of systems (Figure 11.4B). In this equation, G is a constant having the dimension of a conductance, V the applied voltage, and α the parameter relating to nonlinearity of the system. It should be noted that this equation is only an empirical formula, and not related to any specific, well-defined conduction mechanism.

Theoretically, nonlinear electrical transports can be categorized into two types. One is the electrical failures that take place irreversibly in extreme conditions either resulting from the application of high electric field (dielectric breakdown) or current density (burning of fuse). These failures can occur in both ordered and disordered materials, and considered as universal behavior of all materials, therefore are not discussed here. Another type corresponds to numerous cases where the electrical conduction becomes reversibly nonohmic (nonlinear) due to the application of a small electric field.

To obtain more insights into this issue, Gefen et al. proposed that such nonlinearity may arise from two situations: in one case, the conducting elements may themselves be nonlinear [45] and in another, the conducting elements are ohmic but the sample (macroscopic) conductivity become nonohmic due to possible

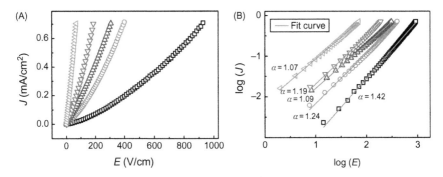

Figure 11.4 (A) $J-E$ relationships of GNP/epoxy nanocomposites. Other carbon-based polymer nanocomposites exhibit similar nonlinear behavior. (B) The experimental data are fitted with an empirical expression $J = E^\alpha$.

appearance of additional channels of conduction [45,46]. Accordingly, two theoretical models have been employed to study this behavior in the composite systems [45]. The first model involves a nonlinear random resistor network (NLRRN), assuming that the conducting backbone consists of microscopic circuit elements [47] The conduction behavior of each element is intrinsically nonlinear, and this nonlinearity is amplified due to the complicated structure of the infinite conducting network spanning the whole system. The second model is a dynamic random resistor network (DRRN). It is based on a network of bonds with ohmic characteristic, comprising a certain number of insulating channels. These originally insulating channels may become conductive upon exposure to a sufficiently strong local field. This transition is possibly caused by nonlinear hopping or tunneling of charge carriers. Recently, Zheng et al. [48] put forward a nonlinear random dynamic resistor network (NLRDRN) model, assuming that the nonlinearity originates from the amplification of each component coupled with the production of additional conducting channels subjected to an external field. However, no quantitative conclusions were made from this model.

From the $I-V$ characteristics of thin Au films, Gefen et al. [49] found that the DRRN model agreed well with their experimental results. Mathematically, the $I-V$ curves can be well described by the form:

$$I = GV + G'V^{\theta} \quad \text{where} \quad \theta = 3.04 \pm 0.08 \tag{11.4}$$

where G is the linear conductance and G' is the high order conductance. Since then, some authors demonstrated that the $I-V$ curves can also be well fitted with a linear plus a quadratic or cubic term [50], i.e.

$$I = GV + G'V^2 \tag{11.5}$$

$$I = GV + G'V^3 \tag{11.6}$$

The crossover current I_c, where the linear current is comparable to the nonlinear current, is supposed to scale with G, as $I_c \sim G^z$. By analyzing the value of the critical exponent z either NLRRN or DRRN model can be specified. Based on the results of Gefen et al. [45]

    for NLRRN: $z \leq (d-1)\varsigma/t$,   i.e., $z \leq 1.05$ in three dimensions and
    for DRRN: $z \leq 1 + \varsigma/t$, i.e., $z \leq 1.52$ in three dimensions

where ζ is the correlation length and t the critical exponent of conductivity, each of these exponents being applicable over a small range of volume concentrations above the percolation threshold. Aharony [51] later established stronger bound limits for the exponent z in the NLRRN $0.97 \leq z \leq 1.52$, in three dimensions.

Table 11.1 lists various carbon filler/polymer composite systems displaying non-linear conducting behavior. Figure 11.5A–E shows typical nonlinearity in

Table 11.1 Nonlinear Electrical Conduction Behavior of Various Carbon Filler/Polymer
Composite Systems

Composite System	Percolation Threshold	Nonlinear Curve Fitting Equation	Processing Route	References
MWNT/PDMS	< 0.1 wt%	$R = rV^{\alpha}I^{\alpha}$	In situ polymerization	[36]
GNP/epoxy	1.35 vol%	$J = \sigma_1 E + \sigma_3 E^3$	In situ polymerization	[39]
GNP/HDPE	11 vol%	$J = \sigma_1 E + \sigma_3 E^3$	Melt compounding	[41]
GNP/nylon-6	–	$J = \sigma_1 E + \sigma_b E^b$	In situ polymerization	[42]
CB/wax	0.76 vol%	$J = \sigma_1 E + \sigma_2 E^2$	Melt compounding	[50]
Graphite/epoxy	1.3 vol%	–	In situ polymerization	[52]
CB/HDPE	–	$J = \sigma_1 E + \sigma_2 E^2$	Melt compounding	[53]

electrical conduction of these systems. Although NLRRN and DRRN models can describe nonlinear electrical conduction of these composites qualitatively, none of them provides detailed information relating the nature of the microscopic conduction mechanism. Actually, Gupta and Sen found that the polynomial function cannot describe the whole nonlinear regime of an $I-V$ curve [54]. They developed a novel random resistor tunneling bond network (RRTN) model and attempted to fit the data using several exponential functions, but no satisfactory results were obtained. Till to present, the specific physical mechanism responsible for the nonlinear transports is still unavailable.

11.2.3 Electrical Transport Mechanisms

Several potential mechanisms have been proposed to describe electrical conduction behavior of the polymer composites with carbon fillers. For example, Pike and Seager [55] classified the conduction mechanisms into four major categories. Based on this classification, Celzard et al. [52] gave a very detailed analysis of the potential mechanisms involved in some graphite/polymer composites. The first scenario involves a uniform model (Figure 11.6A) in which very small conducting particles disperse homogeneously in the insulator. This leads to an impurity-conduction-type mechanism. A typical conduction process in this category is the charge carrier hopping between localized states. In such a process, the current density J is always exponentially dependent on the applied electric field E such that $J \sim \exp(E^{\alpha})$, where α is the power exponent [56–58]. Numerous researchers have investigated this behavior and proposed a variety of mathematical expressions [56–61]. However, it is found that no expressions can fit correctly the $J(E)$ curves as shown in Figure 11.4A for the whole range of electric fields. This is as expected

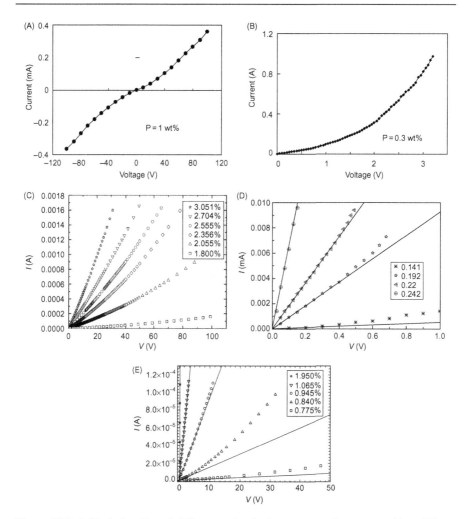

Figure 11.5 $I-V$ relationships of different carbon/polymer composite systems: (A and B) MWNT/polydimethylsiloxane (PDMS), (C) GNP/epoxy, (D) GNP/HDPE, and (E) GNP/nylon-6 composites.
(A and B) *Source*: Reproduced from Ref. [36] with permission from the American Institute of Physics. (C) Reproduced from Ref. [39] with permission from Elsevier. (D) Reproduced from Ref. [41] with permission of Springer. (E) Reproduced from Ref. [42] with permission from Wiley.

since conducting graphite particles are probably far too large to be considered as dilute impurities within the polymer.

The second type of conduction mechanism is based on a uniform channel model (Figure 11.6B): the particles give uniform-composition conducting paths that span the whole sample. In this case, the conductivity variations are mainly due to

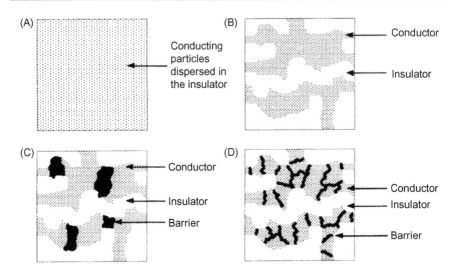

Figure 11.6 Major categories of conduction models in composite materials. (A) A uniform model. (B) A uniform channel model. (C) A nontunneling barrier model. (D) A tunneling barrier model.
Source: Reproduced from Ref. [55] with permission from the American Institute of Physics.

tortuosity and density modifications of the channels. This approach is most applicable to the composites with high particle concentrations.

The third is a nontunneling barrier model (Figure 11.6C): the conducting channels are randomly interrupted by weakly insulating material barriers. If the conduction process is not a tunneling effect, the mechanisms most frequently encountered are the Schottky emission, Poole–Frenkel effect, and space-charge-limited conduction. These three effects rely upon the possibility for an electrode to inject charge carriers into a dielectric. It has been shown that the potential barrier at a metal–insulator contact can be much lower than that of a metal–vacuum interface [60]. Consequently, there occurs electronic injection from the conductor into the insulator at ambient temperature and even below. However, Schottky emission and Poole–Frenkel effect generally yield the $J–E$ relationship in the form of $J \sim E^{1/2}$ [52], so they cannot be used to explain the behavior observed in the carbon nanofiller/polymer composites (Figure 11.4B). In the case of space-charge-limited conduction, the power exponent of the general $J–E$ relationship is often larger than 2 [62]. So this relation would yield $\log(J)$ versus $\log(E)$ curves with greater slopes than the observed α values as shown in Figure 11.4B. Moreover, some published results demonstrated that such a relationship does not necessarily hold for the conduction in amorphous materials [63].

The last configuration is based on a tunneling barrier model (Figure 11.6D): the channels are interrupted by sufficiently thin barriers to allow a tunneling process, either directly between particles or through one or several intermediate states within the insulator. Based on this, Sheng et al. [64] developed a thermal fluctuations induced tunneling effect, in which the conductivity increases when the temperature rises. They showed that the electrical behavior of a bond network within a composite material can be well described by the following $J-E$ relationship:

$$J \approx J_0 \exp\left[-a(T)(E/E_0 - 1)^2\right] \text{ and } a(T) = T_1/(T + T_0) \tag{11.7}$$

where T is the absolute temperature, and T_0 and T_1 are the parameters. This model can describe nonlinear conductivity of carbon/polyvinylchloride nanocomposites [64] and SWNTs network [65] near absolute temperature (0 K) very well. At room temperature, however, it is no longer able to fit the numerical data obtained.

11.3 Zener Tunneling in Polymer Nanocomposites with Carbonaceous Fillers

11.3.1 Zener Tunneling

In contrast to thermal fluctuation-induced tunneling, Zener proposed a conceptually simple model to explain the electrical breakdown in insulators [66,67]. This is referred to as Zener tunneling, or band-to-band tunneling/interband tunneling/internal field emission, as shown in Figure 11.7, illustrating an energy level diagram for an insulator or semiconductor with a bandgap between the valence and conduction bands. The electrons are effectively confined to the valence band with no electric field present. Upon the application of a field, the bands tilt in the space to an extent such that an electron from the valence band can tunnel into the conduction band, without the necessity of having an extra energy to cross the gap between the two bands. Now an electron with energy with respect to the valence band edge can make a transition from the valence to the conduction band not only vertically

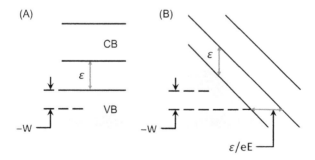

Figure 11.7 Model bands as a function of position of an insulator under (A) field free and (B) applied field configurations. The Zener tunneling of the valence band electron at energy $-W$ into the conduction band is analogous to vacuum field emission with ε playing the role of work function.

(requiring an amount of energy) but also horizontally. In other words, the valence band electron can tunnel into a current carrying conduction band state. There exist a number of methods to theoretically treat this phenomenon [68]. In all cases the rate limiting factor for electrons making transitions is an exponential and the probability (P) of electron tunneling from the valence to conduction band takes the following form:

$$P \sim \exp[a_0(m^*)^{1/2}\varepsilon^{3/2}/\hbar eE] \tag{11.8}$$

with the constant a_0 dependent upon the assumed model, m^* an appropriate effective mass, $\hbar$ the reduced Planck constant, and ε the bandgap. The similarity between Zener tunneling and field emission into vacuum is apparent, considering that ε plays the same role as the work function. Viewing as a Zener breakdown, vacuum field emission can be thought as a process in which electrons make time-dependent, field-induced transitions from the Fermi sea to the tilted conduction band. Therefore, the gap can be treated as an effective barrier and the tunneling process through such a barrier is analyzed. Hence, Zener tunneling can be regarded as a special form of Fowler–Nordheim tunneling. At present, Zener tunneling has been widely used to study the electrical characteristics of semiconductors [69–72], where it is known as the internal field emission, band-to-band tunneling, or interband tunneling. The J–E relationship caused by Zener tunneling has been generalized as [73]

$$J(E) = AE^n\exp(-B/E) \tag{11.9}$$

where A is a factor containing, among other quantities, the number of carriers available to attempt the tunneling transition and their frequency of attempting it. The field, E, appears to some power n, where n lies usually between 1 and 3, depending on the various corrections included in the theory such as the effects of image fields or Coulombic forces [74]. $B = a_0(m^*)^{1/2}\varepsilon^{3/2}/\gamma$, where γ is the field enhancement factor, which depends on the filler content and dispersion within the composite. It should be emphasized that the factor $\exp(-B/E)$ represents the probability of charge carrier transition from one site to another, it is expected to increase with the strength of external field.

11.3.2 The Emerging Model

As aforementioned, the percolation model can predict static electrical conductivity of the carbon nanofiller/polymer composites to a satisfactory level. This model can serve only as a first approximation for the description of electrical properties of the composites. In the presence of an electric field, however, carbon/polymer nanocomposites with the same filler type, content, and static electrical conductivity, their J–E ($\sigma(E)$) characteristics may vary greatly, owing to the difference in the shape, orientation, and dispersion of carbonaceous fillers. Obviously, the static conductivity $\sigma(p)$ alone cannot describe a comprehensive picture of the electrical phenomena in polymer nanocomposites with carbonaceous nanofillers. More parameters are therefore needed to give a precise description. These parameters should account for the geometrical aspects of the carbonaceous fillers, yielding dynamic electrical properties of the $J-E$ ($\sigma(E)$) characteristics. So in the presence of an electric field, a systematic modification to the static conductivity as given by the percolation theory is needed.

For a sufficient weak field, its influence on the electrical conduction in the composite can be neglected. In this case, percolation theory gives a current density $J = \sigma(p)E$. As the field strength increases, deviation from this linearity becomes pronounced and a nonlinear behavior is observed. Such nonlinearity depends greatly on the content and geometrical aspects (such as the shape and the arrangement) of carbonaceous fillers [75]. This can be explained in terms of the following aspects. Inside the composites, carbonaceous fillers connect one with other, forming conducting clusters (Figure 11.8); the size of which depends on the filler content and dispersion. By exposing the composites to an external electric field, local discontinuities in field strength can be expected, due to the presence of these conducting clusters. For narrow insulating gaps between the clusters, the field strength would be magnified by a field enhancement factor γ, equal to the ratio of the size of the conducting cluster to the width of the insulating gap. Around p_c this factor becomes very large therefore electrical breakdown may occur. The breakdown current would contribute to the electrical transport and result in a nonlinear conduction behavior. Therefore, the content and geometrical aspects of the fillers would influence the nonlinear conduction behavior seriously, through the field enhancement effect.

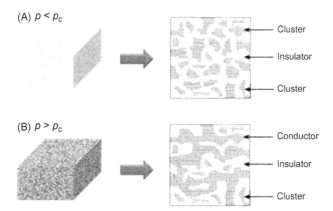

Figure 11.8 Schematic illustrations of carbon/polymer nanocomposite with filler content (A) below and (B) above p_c and their corresponding 2D models. For these two cases, the gaps between conducting region (the network conductor or isolated conducting clusters) and the insulating area compose a large number of tunneling junctions, which lead to a nonlinear electrical conduction behavior.

The electrical breakdown in insulators can be categorized into two main mechanisms. The most common is the avalanche breakdown, which takes place in extreme conditions under the application of very large electric field. The avalanche breakdown is caused by an impact ionization that produces a large amount of charge carriers. This is often associated with a large current increment. To gain enough energy for ionizing the atoms, electrons must move under a very strong electric field over a long distance. Avalanche breakdown usually refers to the permanent damage in the insulators caused by a large electric field. Thus even the field strength decreases, the current still maintains at a high level. In other words, the $J-E$ curve is irreversible. The second is the Zener breakdown, or Zener tunneling, as mentioned above. It causes no permanent damage to the material, since it relates to a field-induced generation of hole−electron pairs, leaving an undisturbed valence band structure. Therefore, Zener tunneling leads to a reversible nonlinear conduction behavior. The magnified internal field strength in combination with the reversibility in electrical conductivity lends full support for the occurrence of Zener tunneling [75,76]. Through Zener tunneling, large amounts of isolated clusters which are not connected to the whole conducting network may take part in the conduction process as well (Figure 11.9). In fact, van Beek and van Pul [76,77] applied the Zener tunneling approach to investigate nonohmic (nonlinear) electrical behavior of carbon black-filled polymer systems in early 1960s. However, they could not explain the electrical conduction behavior of the composites with a larger conductivity using Eq. (11.9). This is because they did not consider the percolation theory conceptually, since Eq. (11.9) predicts a zero static electrical conductivity. He and Tjong [78,79] modified this equation by taking $\sigma(p)$ (given by the percolation theory) into consideration and derived the $J-E$ relationship as follows:

$$J(E) = \sigma(p)E + A(p)E^{n(p)}\exp(-B(p)/E) \tag{11.10}$$

where $\sigma(p)E$ and $A(p)E^{n(p)}\exp(-B(p)/E)$ are defined as the linear part (J') and nonlinear part (J'') of the current density, respectively. They found that this equation gives a good explanation for the electrical conduction phenomena in a variety of the carbon/polymer composites [78−80]. In the next section, this equation will be employed to analyze the electrical conduction behavior of polymer composites with carbon nanofillers.

11.3.3   Zener Tunneling in CNF/CNT-Based Polymer Nanocomposites

Figure 11.9 shows the agreement of Eq. (11.10) with the experimental data for six CNF/high-density polyethylene (HDPE) composite samples with different filler contents. The fitted parameters are summarized in Table 11.2. It is found that this expression can well describe the electrical conduction behavior of CNF/HDPE composites. Based on the fitted parameters in Table 11.2, the two parts of current density, J' and J'', are calculated and also illustrated. It is very clear that the nonlinear conduction in percolated systems arises from the inclusion of J'' to J'.

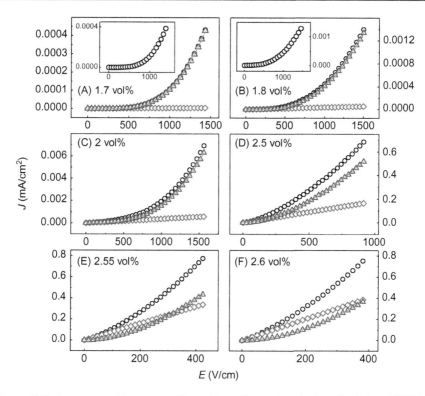

Figure 11.9 Agreement of Zener tunneling with nonlinear electrical conductivity of CNF/HDPE composites with $p < p_c$ (A−C) and $p > p_c$ (D−F). The open circles are data points and solid lines are fitting curves. The symbols "◇" and "△" denote J' and J'', respectively. Insets in (A) and (B) are the summation of J' and J'', showing good agreement between the simulation and experimental data.

Table 11.2 Parameters Characterizing $J-E$ Characteristics of The CNF/HDPE Composites

p (vol%)	σ (mS/cm)	$A(p)$	$B(p)$ (V/cm)	$n(p)$	$\gamma(p)$
1.7	3.15×10^{-9}	2.72×10^{-13}	1045.46	3.02	7.30×10^5
1.8	5.18×10^{-8}	5.68×10^{-13}	714.91	3.00	1.07×10^6
2.0	3.60×10^{-7}	2.09×10^{-12}	431.81	3.00	1.77×10^6
2.5	1.83×10^{-4}	4.56×10^{-6}	22.61	1.71	3.38×10^7
2.55	7.96×10^{-4}	6.03×10^{-6}	8.38	1.82	9.11×10^7
2.6	1.02×10^{-3}	7.00×10^{-6}	6.33	1.83	1.21×10^8

For comparison, we also tentatively tested the applicability of Eqs. (11.4)−(11.6), by converting $I(V)$ into $J(E)$ form and performing curve fitting. It was found that although these equations work well for all composite samples, at low field region the data points show some deviations from the fitted curve, as shown in

Figure 11.10. Furthermore, the fit was not very robust since fitting various data segments gave different results, as revealing by the correlation coefficient (Table 11.3). In contrast, Eq. (11.10) gives an excellent analytical fit as illustrated in Figure 11.9. To further explore the consistency of the fitted curve with the data points within the entire regime, we plot $E \ln(J/A - \sigma_0 E/A)$ against $E \ln E$ as shown in Figure 11.11. It is observed that all data points fall into a linear line for each composite. This implies that the fit of Eq. (11.10) to experimental data is very robust, and indirectly substantiate the presence of internal field emission in the CNF/HDPE composites. It is noted that the difference between Eqs. (11.4)−(11.6) and Eq. (11.10) is significant only at low field region. At a strong field strength, the exponential term in Eq. (11.10) approaches unity, so it can be simplified into Eq. (11.4), and further into Eqs. (11.5) and (11.6) for $n = 2$ or 3. This explains why Eqs. (11.4)−(11.6) cannot describe the electrical conductivity at low field region precisely.

For the composites with $p < p_c$ (Figure 11.9A−C), the whole electrical conduction is dominated by J''. With increasing filler content, the static electrical conductivity increases and gradually comparable to J'' (Figure 11.9D and E).

It is noted that Eq. (11.9) can fit the numerical data relatively well. However, for $p < p_c$ it always underestimates the practical current at low field strength. Figure 11.12A shows the plot of $E \ln(J/A)$ versus $E \ln E$ for the composite with 2 vol% CNF and the corresponding linear fit. Obviously, the data points gradually deviate from the fitting curve with decreasing field strength. In the case of $p > p_c$,

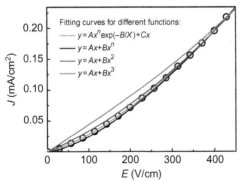

Figure 11.10 Fits of different formulas to the measured electrical conductivity of CNF/HDPE composite ($p = 2.5$ vol%).

Table 11.3 Correlation Coefficient (Adj. R^2) of Different Fitting Curves for The CNF/HDPE Composite ($p = 2.5$ vol%)

Equation	Adj. R^2
$y = Cx + Ax^n \exp(-B/x)$	0.99991
$y = Ax + Bx^n$	0.99979
$y = Ax + Bx^2$	0.99967
$y = Ax + Bx^3$	0.99701

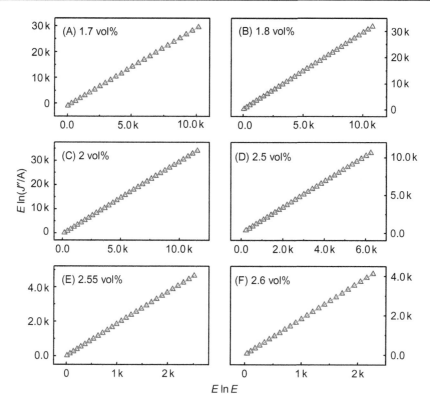

Figure 11.11 Relationship between $E \ln(J''/A)$ and $E \ln E$ for CNF/HDPE composites with $p < p_c$ (A–C) and $p > p_c$ (D–F). The "Δ" symbol represents data points and solid lines are linear fitting results. (The values of A are obtained from Table 11.2.)

very good fit results are achieved if all parameters are not constrained. However, a negative B is always obtained, as shown in Figure 11.12B. This implies that the transparency of the barrier is lowered with increasing field, contrary to what would be expected for the tunneling process. It is concluded that the apparent agreement has no physical meaning. As mentioned before, Zener tunneling describes a dynamic response without considering the static electrical conductivity (as predicted by the percolation theory). So Zener tunneling alone cannot describe the whole conduction process for the system, even it can fit the numerical data well. Similar to the CNF-based polymer composites, CNT-based polymer composites are also found to exhibit Zener tunneling when exposed to an electric field [79]. Figure 11.13 shows the consistence of our model with the experimental results.

11.3.4 Zener Tunneling in GNP-Based Polymer Nanocomposites

Similarly, GNP/polymer nanocomposites also exhibit nonlinear electrical conductivity [39, 41, 42]. Very recently, He and Tjong [80] made a detailed study of the

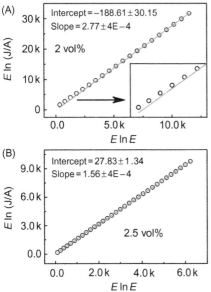

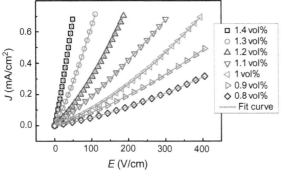

Figure 11.12 (A) Linear fit to $(E \ln E, E \ln (J/A))$ data of CNF/HDPE composite ($p = 2$ vol %) and the disagreement at low field regime (inset). (B) Linear fit to $(E \ln E, E \ln(J/A))$ data of CNF/HDPE composite ($p = 2.5$ vol%). The positive intercept obtained indicates an invalid negative energy gap. In each case, the open circles are data points and the solid lines are fitting curves; the values of A are obtained by fitting Eq. (11.9) to the (E, J) data.

Figure 11.13 Agreements between theory and experimental $J-E$ characteristics of CNT/HDPE nanocomposite. The open symbols are data points; the red solid lines are fitting curves. (For interpretation of the references to color in this figure legend, the reader is referred to the web version of this book.)

relationship between Zener tunneling parameters and filler content as shown in Figures 11.14 and 11.15 and Table 11.4. They found that at higher filler contents, larger amounts of conductive fillers act as the field emission sources inside, leading to larger A values. In addition, internal insulating gaps tend to become smaller at higher filler concentrations, if the fillers are dispersed uniformly in the insulating matrix. This makes the field emission easier to occur, thus producing a smaller B value. However, for the composites with high filler contents, the fillers often agglomerate in the insulating matrix, resulting in abnormal B value. Overall, the Zener current rises with filler content due to an increase of the internal tunneling sources (Figure 11.15A). These results are in agreement with those reported by Van Beek and Van Pul [76,77], although different expressions are employed to analyze the electrical responses.

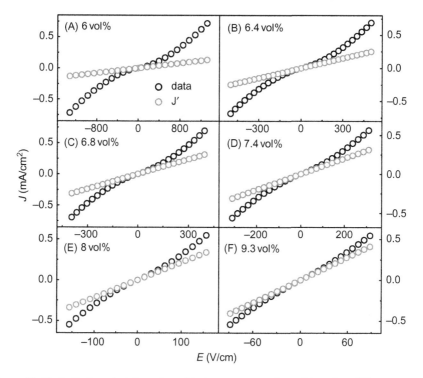

Figure 11.14 *J−E* characteristics of graphite/epoxy composites with various filler contents. The black open squares are experimental data; red open squares are linear current density *J'* deduced from $\sigma(p)E$. (For interpretation of the references to color in this figure legend, the reader is referred to the web version of this book.)

Based on the study of GNP/epoxy nanocomposites, He and Tjong [79] proposed that the width of the forbidden band and the degree of the band tilting are two main factors governing the generation of mobile charge carriers. A narrower forbidden band reduces the tunneling distance, facilitating the interband tunneling, as illustrated in Figure 11.16. Similarly, a large electric field decreases the tunneling distance by tilting the band more seriously. For the carbon/polymer nanocomposites, the width of the forbidden band is influenced by the nature of polymer matrix, while the internal field strength is determined by the dispersion and geometry of conducting fillers. In order to suppress the Zener effect, the polymer matrix with a wide forbidden band is preferred (Figure 11.16B). Also, a poor dispersion of conductive fillers within the insulating matrix can decrease the internal field strength, therefore inhibit the Zener effect.

11.3.5   Zener Tunneling in Graphene-Based Polymer Nanocomposites

The nonlinear conduction behavior of graphene/polymer nanocomposites is very intense, as shown in Figure 11.17. This strong nonlinearity is caused by the intense

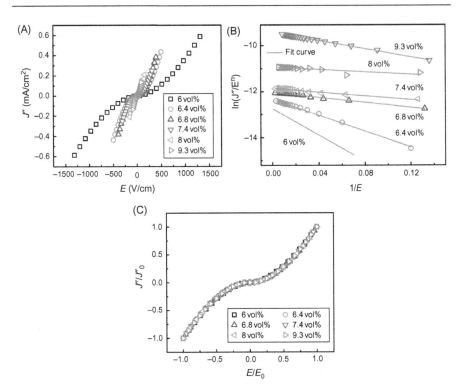

Figure 11.15 (A) Plot of J'' versus E for GNP/epoxy composite samples with various filler contents. (B) Relationship between $\ln(J''/E^n)$ and $1/E$, where the fitting results are shown in red solid lines. n is evaluated from the data in (A). (C) Normalized relationship of J'' versus E. E_0 and J''_0 are the maximum field strength and corresponding Zener current density.

Table 11.4 Parameters Characterizing Zener Current of The GNP/epoxy Composites

p (vol%)	n	$\ln A$	B (V/cm)
6.0	1.75 ± 0.02	-12.74 ± 0.32	28.69 ± 4.21
6.4	1.86 ± 0.03	-12.34 ± 0.30	17.23 ± 3.15
6.8	1.85 ± 0.03	-12.03 ± 0.27	5.52 ± 1.32
7.4	1.82 ± 0.03	-11.86 ± 0.26	4.08 ± 1.02
8.0	1.84 ± 0.03	-10.93 ± 0.21	2.64 ± 0.34
9.3	1.73 ± 0.02	-9.43 ± 0.15	8.89 ± 1.73

Zener tunneling current [81], which is caused by the intrinsic monolayer nature of graphene and its good dispersion in the polymer matrix, as indicated by the SEM micrographs (Figure 11.18) and the low percolation threshold of the composites (Figure 11.19).

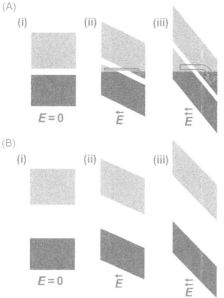

Figure 11.16 Transition of electrons from valence band to conduction band for insulating matrix with a (A) narrow and (B) wide forbidden band. (i)−(iii) illustrate the cases in which the insulating matrix subjected to no electric field, intermediate electric field, and large electric field, respectively.

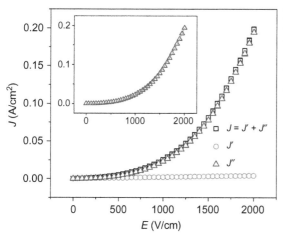

Figure 11.17 $J-E$ characteristic of graphene/PVDF composite with $p = 1.4$ vol%. The inset shows the agreement between J'' with Zener tunneling density $AE^n \exp(-B/E)$.

11.4 Concluding Remarks

Based on the discussion above and the study of the electrical conduction in other carbon filler/polymer systems (Figure 11.20), it can be concluded that the conduction behavior of carbon filler/polymer composites can be well described by a parallel two-resistor (PTR) model, as illustrated in Figure 11.21. One is a static ohmic resistor (R_s), which is predicted by the percolation theory and given by $R_s = l/\sigma(p)$ S, where l and S are the sample thickness and electrode area, respectively. Another is a dynamic nonohmic resistor (R_d), which is caused by Zener tunneling and given

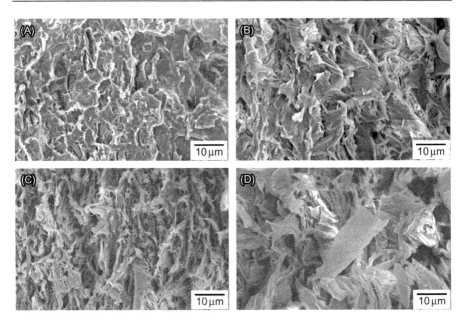

Figure 11.18 SEM micrographs of PVDF loaded with (A) 0.4 vol%, (B) 0.5 vol%, (C) 0.8 vol%, and (D) 1.4 vol% graphene sheets.

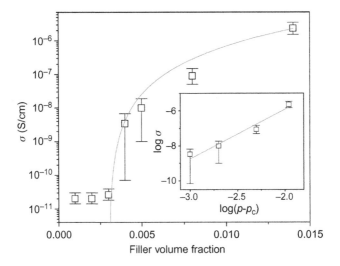

Figure 11.19 Static conductivity of the graphene/PVDF composites showing percolative behavior. The red solid curve is the fitting based on $\sigma = \sigma_0(p-p_c)^t$. Inset is the plot of log σ versus $\log(p-p_c)$.

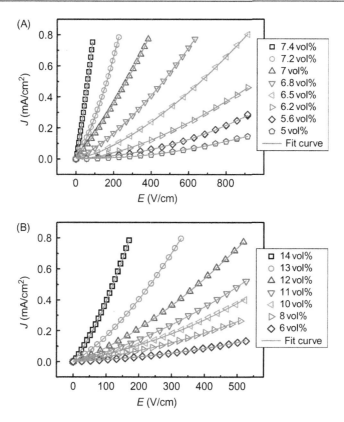

Figure 11.20 Good agreement between the model and experimental $J–E$ characteristics of (A) GNP/PMMA and (B) CB/epoxy composite systems. The open symbols are data points; the solid lines are fitting curves.

Figure 11.21 PTR model of CNF/HDPE composite responding to external electric field. R_s is the ohmic resistor relating to linear conductivity σ_0 and R_d is nonohmic resistor corresponding to Zener tunneling.

by $R_d = l \cdot \exp(B/E)/SAE^{n-1}$, in which the resistance is modulated by external electric field.

The larger the resistance R_s compared to R_d, the more current will flow through R_d, hence the nonlinearity in the $J–E$ response is more pronounced. In another aspect, when the field strength is weak, R_d is very large and the conduction of charge carriers is mainly through R_s. Therefore, the electrical conduction is linear approximately. By increasing the field strength to a crossover electric field E_c, in which R_d is comparable to R_s, the appearance of nonlinearity can be defined. So the crossover current density is $J_c \sim \sigma(p)^{n/(n-1)} \cdot A^{1/(1-n)}$ for $E_c \gg B$. Thus J_c is not only related to the linear conductivity (conductance) but also affected by the tunneling frequency of charge carriers. The result indicates that it is insufficient to judge the applicability of NLRRN or DRRN model by only evaluating the critical exponent of linear conductivity (conductance).

Corresponding to the two resistors, the electrical conductivity of the carbon/polymer composites can be divided into two parts: ohmic (σ_O) and nonohmic (σ_{NO}), both of which can be readily deduced from Eq. (11.10),

$$\sigma_O = \sigma(p) \tag{11.11}$$

$$\sigma_{NO} = A(p)E^{n(p)-1}\exp(-B(p)/E) \tag{11.12}$$

In the case of $p \ll p_c$, the gap width between the isolated clusters is large, thus the effect of field emission is negligible. So σ_{NO} is near to zero and $\sigma \approx \sigma_O$. At $p \gg p_c$, there are only small number of isolated clusters present inside the system. In another words, most conducting clusters are connected to form a conducting network, so $\sigma_O \gg \sigma_{NO}$ and $\sigma \approx \sigma_O$. That is why the nonlinear conduction behavior is particularly pronounced near p_c.

The physical properties of polymer nanocomposites depend greatly on the dispersion of carbonaceous fillers within the polymer matrix. Microscopic techniques, such as SEM and TEM, can only detect the filler dispersion in a small-scale dimension. Practically, it is useful to have a reliable method to evaluate the degree of carbonaceous filler dispersion within the whole polymer matrix. He and Tjong [82] showed that the Zener tunneling parameters can serve as the indicators for the degree of filler dispersion in the polymer matrix. They prepared MWNT/epoxy nanocomposites with different degrees of filler dispersion (Figure 11.22) and tested the electrical responses of these composite samples (Figure 11.23). The results indicated that the dispersion of nanotubes affects the nonlinear conductivity of these composites markedly. When the MWNTs are poorly distributed in the epoxy matrix in the form of bundles, as demonstrated in Figure 11.24A, the average gap between these bundles is large, so a conducting network throughout the system cannot be formed easily, yielding low static conductivity. Furthermore, such wide insulating gaps lead to a high potential barrier B, which suppresses the tunneling of charge carriers from one bundle to another. Thus the number of successful tunneling per second is low, leading to small A values (Tables 11.5 and 11.6). For the case where CNTs are disentangled and dispersed uniformly (Figure 11.24C), they connect one another to form a conducting network readily. This accounts for the high conductivity of the nanocomposites concerned. In addition, the average insulating gap between the MWNTs becomes narrower, resulting in a much lower potential barrier B and facilitating charge carriers to tunnel. This increases the tunneling frequency and gives rise to large A values. For the samples with intermediate level of MWNT dispersion (Figure 11.24B), the conductivity and the Zener tunneling parameters

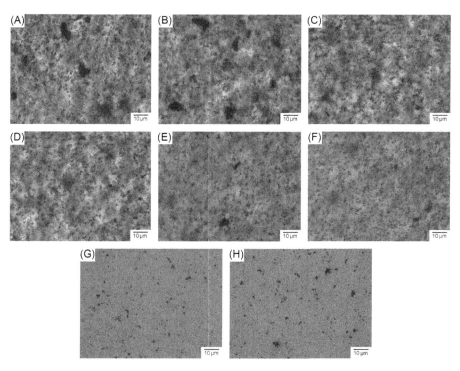

Figure 11.22 Optical micrographs of 2 vol% (A, C, E, and G) and 3 vol% (B, D, F, and H) MWNT-loaded epoxy resin. The composite samples were fabricated by different routes (A, B: I; C, D: II; E, F: III; G, H: IV), leading to different degree of MWNT dispersion.

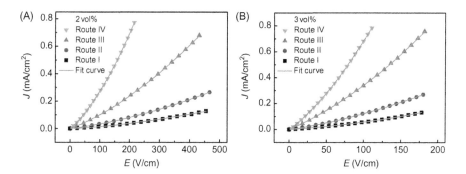

Figure 11.23 $J-E$ characteristics of MWNT/epoxy composites with (A) 2 vol% and (B) 3 vol% MWNT fabricated by different routes. The red solid lines are the curve fitting to Eq. (11.10).

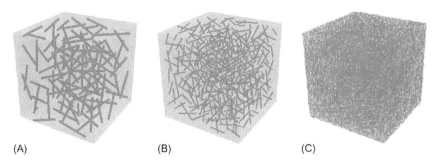

(A)   (B)   (C)

Figure 11.24 Schematics for physical configurations of (A) poor, (B) intermediate, and (C) homogeneous dispersion of MWNTs in the epoxy matrix. The green sticks represent MWNTs, and the gray region represents the epoxy matrix. (For interpretation of the references to color in this figure legend, the reader is referred to the web version of this book.)

Table 11.5 Parameters Characterizing $J-E$ Response of The MWNT/Epoxy Composite ($p = 2$ vol%) Specimens Fabricated by Different Routes

Route	σ (mS/cm)	A	B (V/cm)	n
I	1.15×10^{-4}	7.40×10^{-7}	43.24	1.90
II	2.59×10^{-4}	1.60×10^{-6}	16.74	1.86
III	7.71×10^{-4}	4.92×10^{-6}	6.43	1.84
IV	1.72×10^{-3}	2.66×10^{-5}	3.20	1.79

Table 11.6 Parameters Characterizing the $J-E$ Response of The MWNT/Epoxy Composite ($p = 3$ vol%) Specimens Fabricated by Different Routes

Route	σ (mS/cm)	A	B (V/cm)	n
I	4.34×10^{-4}	3.19×10^{-6}	66.15	1.95
II	9.31×10^{-4}	8.97×10^{-6}	46.60	1.85
III	2.33×10^{-3}	2.64×10^{-5}	5.86	1.82
IV	3.89×10^{-3}	8.21×10^{-5}	1.05	1.78

take intermediate values. This situation agrees well with the results reported by van Beek and van Pul [76,77] (Table 11.7), where the n value decreases monotonously as the level of carbon black dispersion is improved. Thus n serves as an important indicator for the state of nanotube dispersion in the polymer matrix.

Finally, it is concluded that the composite samples with a homogeneous nanotube dispersion display strong Zener effect. Therefore, the nanotube dispersion can be described in terms of Zener tunneling parameters, i.e., A, B, and n. These parameters, together with the static electrical conductivity, provide important information about the filler dispersion and hence the percolating nature of the carbon nanostructures within the composites.

Table 11.7 Internal Field Emission Parameters A, B, and n, and E/J Values for Different Degrees of Carbon Black Dispersion in Rubber

Sample	Dispersion	ln A	B (V/cm)	n	E/J (Ωcm)
1A	Good	-12	0.35	1.25	6.9×10^4
1B	Moderate	-17	1.00	1.80	1.4×10^6
1C	Poor	-32	50	2.00	2.2×10^{11}

Source: Reproduced with permission from Ref. [77].

References

[1] Han H, Kim CG. Low earth orbit space environment simulation and its effects on graphite/epoxy composites. Compos Struct 2006;72(2):218−26.

[2] Park SY, Choi HS, Choi WJ, Kwon H. Effect of vacuum thermal cyclic exposures on unidirectional carbon fiber/epoxy composites for low earth orbit space applications. Compos Part B 2012;43(2):726−8.

[3] Chung DDL. Electromagnetic interference shielding effectiveness of carbon materials. Carbon 2001;39(2):279−85.

[4] Wu J, Chung DDL. Increasing the electromagnetic interference shielding effectiveness of carbon fiber polymer−matrix composite by using activated carbon fibers. Carbon 2002;40(3):445−7.

[5] Soldano C, Mahmood A, Dujardin E. Production, properties and potential of graphene. Carbon 2010;48(8):2127−50.

[6] Sengupta R, Bhattacharya M, Bandyopadhyay S, Bhowmick AK. A review on the mechanical and electrical properties of graphite and modified graphite reinforced polymer composites. Prog Polym Sci 2011;36(5):638−70.

[7] Krupka J, Strupinski W. Measurements of the sheet resistance and conductivity of thin epitaxial graphene. Appl Phys Lett 2010;96:082101.

[8] Yang YL, Gupta MC, Dudley KL, Lawrence RW. A comparative study of EMI shielding properties of carbon nanofiber and multi-walled carbon nanotube filled polymer composites. J Nanosci Nanotechnol 2005;5(6):927−31.

[9] Al-Saleh MH, Sundararaj U. Electromagnetic shielding mechanism of CNT/polymer composites. Carbon 2009;47(7):1738−46.

[10] Liang J, Wang Y, Huang Y, Ma Y, Liu Z, Cai J, et al. Electromagnetic interference shielding of graphene/epoxy composites. Carbon 2009;47(3):922−5.

[11] Zhang B, Dong XM, Fu RW, Zhao B, Zhang MQ. The sensibility of the composites fabricated from polystyrene filling multi-walled carbon nanotubes for mixed vapors. Compos Sci Technol 2008;68(6):1357−62.

[12] Wu Z, Chen X, Zhu S, Zhou Z, Yao Y, Qu W, et al. Enhanced sensitivity of ammonia sensor using graphene/polyaniline nanocomposite. Sens Act B Chem 2013;178 (1):485−93.

[13] Novoselov KS, Fal'ko VI, Colombo L, Gellert PR, Schwab MG, Kim K. A roadmap for graphene. Nature 2012;490(7419):192−200.

[14] Zhou SY, Gweo GH, Fedorov AV, First PN, Heer WAD, Lee DH, et al. Substrate-induced band gap opening in epitaxial graphene. Nature Mater 2007;6(10):770−5.

[15] Berger C, Song Z, Li X, Wu X, Brown N, Naud C, et al. Electronic confinement and coherence in patterned epitaxial graphene. Science 2006;312(5777):1191−6.

[16] Hernandez Y, Nicolosi V, Lotya M, Blighe FM, Sun Z, De S, et al. High-yield production of graphene by liquid-phase exfoliation of graphite. Nature Nanotechnol 2008;3 (9):563−8.

[17] Li B, Zhong WH. Review on polymer/graphite nanoplatelet nanocomposites. J Mater Sci 2011;46(17):5595−614.

[18] Nieto A, Lahiri D, Agarwal A. Synthesis and properties of bulk graphene nanoplatelets consolidated by spark plasma sintering. Carbon 2012;50(11):4068−77.

[19] Hammel E, Tang H, Trampert M, Schmitt T, Mauthner K, Eder A, et al. Carbon nanofibers for composite applications. Carbon 2004;42(5−6):1153−8.

[20] Mukhopadhyay K, Porwal D, Lai D, Ram K, Mathur GN. Synthesis of coiled/straight carbon nanofibers by catalytic chemical vapor deposition. Carbon 2004;42 (15):3254−6.

[21] Melechko AV, Merkulov TE, McKnight TE, Guillorn MA, Klein KL, Lowbdes DH, et al. Vertically aligned carbon nanofibers and related structures: controlled synthesis and directed assembly. J Appl Phys 2005;97(4):41301.

[22] Endo M, Kim YA, Hayashi T, Fukai Y, Oshida K, Terrones M, et al. Structural characterization of cup-stacked-type nanofibers with an entirely hollow core. Appl Phys Lett 2002;80(7):1267.

[23] He LX, Tjong SC. Effect of temperature on electrical conduction behavior of polyvinylidene fluoride nanocomposites with carbon nanotubes and nanofibers. Curr Nanosci 2010;6(5):520−4.

[24] Li Q, Xue Q, Hao L, Gao X, Zheng Q. Large dielectric constant of the chemically functionalized carbon nanotube/polymer composites. Compos Sci Technol 2008;68 (10−11):2290−6.

[25] Li YL, Kuan CF, Chen CH, Kuan HC, Yip MC, Chiu SL, et al. PMMA/functionalized graphene oxide nanosheets composites. Mater Chem Phys 2012;134(2−3):677−85.

[26] Geng Y, Liu MY, Li J, Shi XM, Kim JK. Effects of surfactant treatment on mechanical and electrical properties of CNT/epoxy nanocomposites. Compos Part A 2008;39 (12):1833−76.

[27] Kirkpatrick S. Percolation and conduction. Rev Mod Phys 1973;45(4):574−88.

[28] Bergman DJ, Imry Y. Critical behavior of the complex dielectric constant near the percolation threshold of a heterogeneous material. Phys Rev Lett 1977;39(19):1222−5.

[29] Webman I, Jortner J, Cohen MH. Thermoelectric power in inhomogeneous materials. Phys Rev B 1977;16(6):2959−64.

[30] Stephen MJ. Mean-field theory and critical exponents for a random resistor network. Phys Rev B 1978;17(11):4444−53.

[31] Nan CW. Physics of inhomogeneous inorganic materials. Prog Mater Sci 1993;37 (1):1−116.

[32] Reghu M, Yoon CO, Yang CY, Moses D, Smith P, Heeger AJ, et al. Transport in polyaniline networks near the percolation threshold. Phys Rev B 1994;50 (19):13931−41.

[33] Wu JJ, McLachlan DS. Scaling behavior of the complex conductivity of graphite-boron nitride percolation systems. Phys Rev B 1998;58(22):14880−7.

[34] Fraysse J, Planès J. Interplay of hopping and percolation in organic conducting blends. Phys Stat Sol B 2000;218(1):273−7.

[35] Scher H, Zallen R. Critical density in percolation processes. J Chem Phys 1970;53 (1):3759−61.

[36] Liu CH, Fan SS. Nonlinear electrical conducting behavior of carbon nanotube networks in silicone elastomer. Appl Phys Lett 2007;90(4):041905.

[37] Bardhan KK. Nonlinear conduction in composites above percolation threshold— beyond the backbone. Physica A 1997;241(1−2):267−77.

[38] Bardhan KK, Chakrabarty RK. Identical scaling behavior of DC and AC response near the percolation threshold in conductor−insulator mixtures. Phys Rev Lett 1994;72 (7):1068−71.

[39] Lin H, Lu W, Chen G. Nonlinear DC conduction behavior in epoxy resin/graphite nanosheets composites. Physica B 2007;400(1−2):229−36.

[40] Celzard A, Furdin G, Mareche JF, McRae E. Non-linear current−voltage characteristics in anisotropic epoxy resin−graphite flake composites. J Mater Sci 1997;32 (7):1849−53.

[41] Lu W, Wu D, Wu C, Chen G. Nonlinear DC response in high-density polyethylene/ graphite nanosheets composites. J Mater Sci 2006;41(6):1785−90.

[42] Chen G, Weng W, Wu D, Wu C. Nonlinear conduction in nylon-6/foliated graphite nanocomposites above the percolation threshold. J Polym Sci B 2004;42(1):155−67.

[43] Stroud D, Hui PM. Nonlinear susceptibilities of granular matter. Phys Rev B 1988;37 (15):8719−24.

[44] Rammal R, Tremblay AMS. Resistance noise in nonlinear resistor networks. Phys Rev Lett 1987;58(4):415−8.

[45] Gefen Y, Shih WH, Laibowitz RB, Viggiano JM. Nonlinear behavior near the percolation metal−insulator transition. Phys Rev Lett 1986;57(24):3097−100.

[46] Bardhan KK, Chakrabartiy RK. Identical scaling behavior of DC and AC response near the percolation threshold in conductor−insulator mixtures. Phys Rev Lett 1994;72 (7):1068−71.

[47] Kenkel SW, Straley JP. Percolation theory of nonlinear circuit elements. Phys Rev Lett 1982;49(11):767−70.

[48] Zheng Q, Song Y, Wu G, Yi X. Reversible nonlinear conduction behavior for high-density polyethylene/graphite powder composites near the percolation threshold. J Polym Sci B Polym Phys 2001;39(22):2833−42.

[49] Gefen Y, Shih WH, Laibowitz RB, Viggiano JM, Gefen, et al. Respond. Phys Rev Lett 1987;58(25):2727.

[50] Chakrabarty RK, Bardhan KK, Basu A. Nonlinear $I-V$ characteristics near the percolation threshold. Phys Rev B 1991;44(13):6773−9.

[51] Aharony A. Crossover from linear to nonlinear resistance near percolation. Phys Rev Lett 1987;58(25):2726.

[52] Celzard A, McRae E, Furdin G, Marêché JF. Conduction mechanisms in some graphite−polymer composites: the effect of a direct-current electric field. J Phys Condens Matter 1997;9(10):2225−37.

[53] Song Y, Zheng Q. Effect of voltage on the conduction of a high-density polyethylene/ carbon black composite at the NTC region. Compos Sci Technol 2006;66(1):907−11.

[54] Gupta AK, Sen AK. Nonlinear DC response in composites: a percolative study. Phys Rev B 1998;57(6):3375−88.

[55] Pike OH, Seager CH. Electrical properties and conduction mechanisms of Ru-based thick-film (cermet) resistors. J Appl Phys 1977;48(12):5151−69.

[56] Hill RM. Hopping conduction in amorphous solids. Phil Mag 1971;24(192):1307−25.

[57] Pollak M, Riess I. A percolation treatment of high-field hopping transport. J Phys C Solid State Phys 1976;9(12):2339−52.

[58] Shklovskii BI. Non-ohmic hopping conduction. Sov Phys Semicond 1976;10:855.

[59] Talamantes J, Pollak M, Baron R. Moderate-field variable-range hopping transport. J Noncryst Solids 1987;97−98(Part 1):555−8.

[60] Apsley N, Hughes HP. Temperature- and field-dependence of hopping conduction in disordered systems, II. Phil Mag 1975;31(6):1327−39.

[61] Lampert MA, Mark P. Current injection in solids. New York: Academic; 1970.

[62] Radhakrishnan S, Chakne S, Shelke PN. High piezoresistivity in conducting polymer composites. Mater Lett 1994;18(5−6):358−62.

[63] Jonscher AK. Electronic conduction in dielectric films. J Electrochem Soc 1969;116 (6):217C−26C.

[64] Sheng P, Sichel EK, Gittleman JL. Fluctuation-induced tunneling conduction in carbon−polyvinylchloride composites. Phys Rev Lett 1978;40(18):1197−200.

[65] Kim GT, Jhang SH, Park JG, Park YW, Roth S. Non-ohmic current−voltage characteristics in single-wall carbon nanotube network. Synth Met 2001;117(1−3):123−6.

[66] Zener C. A theory of the electrical breakdown of solid dielectrics. Proc Roy Soc Lond A 1934;145(855):523−9.

[67] Kane EO. Zener tunneling in semiconductors. J Phys Chem Solids 1959;12(2):181−8.

[68] Duke CB. Tunneling in solids. New York: Academic; 1969.

[69] McAfee KB, Ryder EJ, Shockley W, Sparks M. Observations of Zener current in germanium $p-n$ junctions. Phys Rev 1951;83(3):650−1.

[70] Chynoweth AG, McKay KG. Internal field emission in silicon $p-n$ junctions. Phys Rev 1957;106(3):418−26.

[71] Chynoweth AG, Feldmann WL, Lee CA, Logan RA, Pearson GL. Internal field emission at narrow silicon and germanium $p-n$ junctions. Phys Rev 1960;118(2):425−34.

[72] Chynoweth AG, Logan RA. Internal field emission at narrow $p-n$ junctions in indium antimonide. Phys Rev 1960;118(6):1470−3.

[73] Chynoweth AG. Prog Ser 1960;4:95.

[74] Franz W. Handbuch der physik, 17. Berlin: Springer-Verlag; 1956. p. 155

[75] Balberg I. A comprehensive picture of the electrical phenomena in carbon black-polymer composites. Carbon 2002;40(2):139−43.

[76] Van Beek LKH, Van Pul BICF. Internal field emission in carbon black-loaded natural rubber vulcanizates. J Appl Polym Sci 1962;6(24):651−5.

[77] Van Beek LKH, Van Pul BICF. Non-ohmic behavior of carbon black-loaded rubbers. Carbon 1964;2(126):121.

[78] He LX, Tjong SC. Nonlinear electrical conduction in percolating systems induced by internal field emission. Synth Met 2011;161(5−6):540−3.

[79] He LX, Tjong SC. Universality of Zener tunneling in carbon/polymer composites. Synth Met 2012;161(23−24):2647−50.

[80] He LX, Tjong SC. Zener tunneling in conductive graphite/epoxy composites: dielectric breakdown aspects. Express Polym Lett 2013;7(4):375−82.

[81] He LX, Tjong SC. Low percolation threshold of graphene/polymer composites prepared by solvothermal reduction of graphene oxide in the polymer solution. Nanoscale Res Lett 2013;8(1):132.

[82] He LX, Tjong SC. Carbon nanotube/epoxy resin composite: correlation between state of nanotube dispersion and Zener tunneling parameters. Synth Met 2012;162 (24):2277−81.

Printed in the United States
By Bookmasters